WITHDRAWN

FRANCE AND THE EXPLOITATION OF CHINA
1885–1901
A STUDY IN ECONOMIC IMPERALISM

FRANCE
AND THE
EXPLOITATION
OF
CHINA
1885-1901

A Study in Economic Imperialism

ROBERT LEE

HONG KONG
OXFORD UNIVERSITY PRESS
OXFORD   NEW YORK
1989

Oxford University Press

Oxford New York Toronto
Petaling Jaya Singapore Hong Kong Tokyo
Delhi Bombay Calcutta Madras Karachi
Nairobi Dar es Salaam Cape Town
Melbourne Auckland

and associated companies in
Berlin Ibadan

First published 1989
Published in the United States
by Oxford University Press, Inc., New York

British Library Cataloguing in Publication Data

Lee, Robert, 1952–
France and the exploitation of China, 1885–1901: a study in
economic imperialism.
1. China, 1644–1912
I. Title
951'.03
ISBN 0-19-582708-2

Library of Congress Cataloging-in-Publication Data

Lee, Robert S. (Robert Stuart), 1952–
France and the exploitation of China: a study in economic
imperialism, 1885–1901 / Robert S. Lee.
p.   cm. — (East Asian historical monographs)
Includes bibliographical references (p.   ).
ISBN 0-19-582708-2
1. Investments, French — China — History —
19th century. 2. China —
Economic conditions, 1644–1912. 3. France — Colonies —
Asia.
I. Title. II. Series.
HG5472.L44 1989
332.6'7344051 — dc20
89-23196
CIP

Printed in Hong Kong by Liang Yu Printing Factory Ltd.
Published by Oxford University Press, Warwick House, Hong Kong

To Ceridwen

'l'heure était venue de substituer aux généraux et aux amiraux, l'ingénieur et le banquier qui sont aujourd'hui les vrais conquérants de la Chine.'

Auguste Gérard, 1897

Preface

IMPERIALISM remains a lively historiographical issue a generation after the dissolution of the European territorial empires in Africa and Asia. Most of this English-language historiography draws its empirical evidence from the British imperial experience and emphasizes formal territorial empires, that is the vast number of relatively undeveloped states which were either conquered or obliged to acknowledge European suzerainty, mostly during the nineteenth century.

Yet France was, after Britain, the world's second greatest imperial power, and, together with her ally Russia was Britain's only serious rival in formal empire-building on the Asian mainland. Moreover imperialism was not confined to formal empire; and Latin America, the Ottoman empire, and China, all of which preserved their political integrity and sovereignty during the period of the territorially acquisitive 'new imperialism', were deeply affected by the aggressive aspirations of the industrialized Western world.

This book, then, is an attempt to correct two imbalances in the English-language historiography of imperialism, and so concentrates on France, the second imperial nation, and China, then as now the world's most populous state, then as now perceived in the West as a potentially vast market, and then perceived by some in the West as a potential field for further territorial conquest. France, whose domestic economy was very different from that of Britain, naturally had different imperial aspirations and different means of achieving them. Hence the interaction between France and China differed greatly from that of Britain and China, and this book seeks to illuminate some of those differences.

I wrote the book using mostly French archival sources, although I also consulted British archives. A study of imperialism is almost by definition Eurocentric, since it was the European powers which constantly took the initiative in attempting to resolve their economic and social tensions through the domination of weaker societies. Nevertheless I have sought to give some consideration to Chinese responses and so owe a great debt to a large historiography on Chinese affairs, much of which has been produced by the Harvard school.

The sources consulted have used a wide variety of systems of transliteration of Chinese names. Most English-language sources have used a combination of Wade-Giles and Post Office forms. Late nineteenth-century French documents sometimes used these forms, sometimes standard French orthography, or a variant thereon, often a combination of all three. In this book *pinyin* forms have been used throughout, including quotations. All French quotations have been translated into English and their Chinese names rendered in *pinyin*. However in titles, both in the notes and bibliography, the original forms have been maintained. A list of Chinese names, listing *pinyin*, Wade-Giles or Post Office, and French forms is included in a glossary and should resolve any difficulties.

A less troublesome problem has been the changes in some geographical names over the last century. The rule I have followed is that purely geographical names have been rendered in their modern forms, whereas names with a political content have retained their contemporary forms. Thus the metropolitan province of China appears as Zhili, its name prior to 1912, not Hebei, its present name, while the capital of Yunnan is always Kunming, its modern name, not Yunnanfu. There are a few exceptions to this rule: Annam is offensive to modern Vietnamese, so Vietnam is substituted; the Chang Jiang is here described as the Yangzi Jiang in the interests of easy recognition; and for the same reason the Namti River retains its French and Vietnamese rather than any Chinese name. The Hong Ha always appears with its Vietnamese name, even though in China it is called the Yuan Jiang and it is more familiar to English-language readers as the Red River. 'Our Hong Ha' unfortunately lacks the imperial resonance of 'notre fleuve Rouge', but is more appropriate in a modern translation.

The book had its origins as a doctoral thesis written at the University of Sydney. I was then the recipient of an Australian Commonwealth Government Postgraduate Award. I am indebted both to the Commonwealth Department of Education and to the University, in particular the History Department's Postgraduate Studies Committee, for their support in undertaking the research, especially in enabling me to spend a year in Paris. The transformation of the thesis into a book was mostly done while I was on study leave from my teaching at what was then the Macarthur

Institute of Higher Education and I am grateful to my colleagues at the University of Western Sydney, Macarthur for their support and encouragement with the project. While on study leave in Paris I was able to stay at the premises of the Institut francophone de Paris, and I wish to express my gratitude to M. Michel Fleury, Mme Françoise Auffray, and the staff of the Institut for their warm welcome, support, and assistance. M. Michel Bruguière, whose pioneering research on the Chemin de fer du Yunnan deserves wider recognition in the anglophone world, was most generous with his time and was a source of much information on the topic. The assistance and unfailing good humour of the archival staff in the French Ministries of Finance, Foreign Affairs, and Overseas France, as well as in the Archives Nationales are also gratefully acknowledged.

Special thanks are due to a few people whose involvement with the book has been more intimate. Dr Bruce Fulton of the University of Sydney supervised the original thesis on which it is based. My mother has been a marvellous typist and both my father and my wife, Ceridwen, have provided helpful and learned critical comments. This book would not have been written without these people's support, but its failings are certainly my own responsibility.

ROBERT LEE
Sydney
December 1988

Contents

Maps

Tables

Abbreviations

AN	Archives nationales
Ch. de D.	Chambre des Députés
C.P.	Correspondance politique
Dép.	Dépêche
Dép. com.	Dépêche commerciale
Dép. pol.	Dépêche politique
DDF	France, Ministère des Affaires Etrangères, Commission de publication des documents relatifs aux origines de la guerre de 1914, *Documents diplomatiques français (1871–1914)*
FO	United Kingdom, Public Record Office, Foreign Office Archives
HMSO	Her Majesty's Stationery Office
IC	Indo-Chine
IN	Imprimérie nationale
J.O.	France, Assemblée nationale, *Journal officiel. Débats parlementaires*
MAE	Archives of the Ministère des Affaires Etrangères
MC	Archives of the Ministère de Colonies (now Département de la France d'Outre-Mer)
MF	Archives of the Ministère des Finances
n.d.	no date
NF	nouveaux fonds
n.p.	no page
NS	nouvelle série
s.d.	sans date
s.n.	sans numéro
Tél.	Télégramme
t.	tome
v.	volume

France and China
in the Nineteenth Century

THE period from 1885 until the end of the century marks both the final phase of French territorial expansion on China's frontiers and the beginnings of serious French exploitation of China herself. It begins with the acquisition of Tonkin and ends with the movement towards Anglo-French co-operation in the exploitation of the south-western Chinese province of Yunnan. Despite the drama of these years and despite the changes the period brought in France's role in East Asia, French imperialism in China at this time has not been the subject of any recent substantial published study.[1]

The French conquest of Tonkin (1883–5) was something of a watershed. It was undertaken not so much for the value of Tonkin itself but rather to establish a safe, convenient, and profitable trading route into south-western China. Although some French citizens regarded Tonkin as opening the way for further conquests at China's expense, in a very real sense its conquest had been undertaken not in a spirit of territorial acquisitiveness, but with an eye to the commercial and other economic prospects it would open in China.

As it turned out, France's possession of Tonkin, in spite of considerable and sustained efforts by the French government, did little to improve her poor commercial position in China. After 1885, however, other French economic interests in China, especially investments, expanded considerably, and accelerated rapidly after China's defeat by Japan in 1895. This process was started, to some extent, as a result of the terms of the treaty which, from the point of view of international law, ended the war of conquest. But the growing importance of investment was the result partly of the inability of French industry to produce cheaply the European goods which were in demand in China; partly of the need for French capital to find outlets abroad; partly of the implementation of official policy which encouraged such investment; and partly of China's political and economic problems which, especially after 1895, obliged her to look to Western powers for capital.

This growth in investment and the interaction between private interests and government policy which helped to produce it is one of the major themes of this book. A second is the continuing neo-mercantilist attempt to use Tonkin to create a favourable, indeed quasi-monopolistic trading position in southern and south-western China. A third concerns the subordination of territorial ambitions to broader economic considerations.

Until the end of the century there was a strong and widely held belief that France itself should annex part of the Chinese empire. Many of those who hoped to achieve this aim were resident in Indo-China. Paul Doumer, who was Governor-General in Hanoi at the crucial period, is the most notable example. Contrary to such ambitions was the growing conviction that French interests would best be served by seeking to exploit, in competition and if necessary in association with other powers, the entire Chinese empire rather than by attempting to conquer, govern, and exploit a part of that empire. This policy of maintaining Chinese independence found powerful support in the diplomatic and business milieux of Paris. In the end it was the outlook of the metropole which triumphed.

The final decision was not taken until late in 1899. The spectacular events in Beijing during the following year, despite their enormous impact on popular European consciousness of China, in no way altered that decision. For this reason, the Boxer Rebellion does not play a large role in this account. Many local factors beyond French control, notably the attitudes of other imperialist powers and the resistance of the Chinese people and government, helped to decide French policy, but, ultimately, the French government gave priority to economic considerations. As opposition to annexation on the part of these economically minded decision-makers increased, there occurred what was both a reversion to earlier mercantilist attitudes to Asia and an anticipation of the modern type of exploitation of weak nations which Dr Sukarno would later call neo-imperialism.

The emphasis placed on the economic impulses for French policy in China is not intended to deny that other factors such as national pride or religious zeal were at work. They were, but the evidence suggests that they were not decisive. Both these factors, for example, were at play in the maintenance of the so-called

French religious Protectorate in China. Although an imperial edict according Chinese Christians certain rights had been issued under French pressure in 1846, the Protectorate only dated from the Tianjin and Beijing treaties of 27 June 1858 and 25 October 1860. The pretext for the French invasion which led to those treaties was the murder of a priest, père Chapdelaine in Guangxi. His residence there was quite illegal, but that technicality did not worry Napoleon III, whose aim was, as an anti-clerical republican succinctly put it, 'to use the religious protectorate as a counter-weight to the influence England exercised through her trade'.[2] The relevant clauses obliged China to protect Catholic missionaries and not to indulge in anti-Christian propaganda. In reality it was a Chinese Protectorate of missions, but France supervised that Protectorate and issued Catholic missionaries' passports. The papacy never formally asked for nor granted France the Pro-tectorate of Catholics in China, but on the whole accepted its reality with gratitude.

The Protectorate did have a certain logic: about three-quarters of the Catholic missionaries in China were French and 19 of the 40 missions in the country were under French control.[3] Except for a Portuguese bishop in Macao, the church in China had no terri-torial organization. Each mission was led by a monsignor, chief of whom was Alphonse Favier in Beijing. He once assured a French diplomat that 'the Government can count on me and my coadjutor who are, by the grace of God, good Frenchmen who in our church business will never forget the primordial interests of France in China'.[4] He was as good as his word, and when in 1886, 1893, and again in 1900, Leo XIII considered sending an apostolic legate to Beijing, Favier, successfully opposed the moves.[5] Like normally anti-clerical French politicians, he realized that such a move would much weaken the force of the French Protectorate. The 1886 pro-posal was the most serious and provoked the strongest French reaction. The initiative for direct diplomatic relations between the Vatican and China came from the Chinese side. Li Hongzhang, who had negotiated the Treaty of Tianjin which ended the Sino-French War in 1885, floated the idea with the clear aim of weakening French influence. The Vatican was receptive to the proposal and Leo even nominated an Italian, Monsignor Agliardi, to the Beijing post. The French reaction, however, was swift and savage. Despite his belief that relations between the Church and the secular republic should be improved, Foreign Minister Charles de Freycinet threatened to withdraw the ambassador to the Holy See the moment Agliardi left for Beijing. Leo, who shared de Freycinet's concern for improved relations, backed down. In defending his actions, de Freycinet provided the classical republican defence of the Protectorate:

In acting thus, the government is not pursuing an aim of religious propaganda, which is as opposed to its principles as to the rules which govern politics at the moment; we are only proposing to use the relations and incessant progress accomplished by the missionaries among the Chinese people for the benefit of the name of France. Our trade in China has not grown sufficiently for us to create these predominant interests, and we must recognize that if we renounce the Protectorate of Catholic missions, our role in the Celestial Empire will in fact be considerably lessened.[6]

Nevertheless, the religious Protectorate continued to sit uneasily with the secular and frequently anti-clerical quality of the Third Republic. Jules Ferry once described it as 'only a burden with no compensations', and in 1893 Foreign Minister Casimir-Perier supported the papal proposal for an apostolic legation in Beijing. His successor, Gabriel Hanotaux, however, rallied to the Protectorate and the bureaucrats in the Foreign Ministry consistently supported it, in the opinion of one observer, because they feared any loss of influence, in the opinion of another, because of the malign influence of 'the irresponsible clerical *camarilla* which runs the Quai d'Orsay'.[7] Camarilla or not, the Ministry's views were clearly expressed in advice it gave to Delcassé when he was asked in the Chamber to justify subsidies to religious orders in China. It argued two points, which were both doubtful. Firstly, 'it is manifest that the requests to our representatives for support from all sides ... can only fortify our prestige in the eyes of the Chinese and consequently contribute to support our views so far as general French policies in the Far East are concerned'. Secondly, the missions taught French in their schools and 'French language education is one of the most powerful means for facilitating our penetration into the regions neighbouring on our Indo-Chinese possessions'.[8]

This 'prestige' could be a dangerous thing, for while the missions had a high public profile in China, they were never especially popular and generally provoked resentment which often flared into violence. The destruction of the cathedral of Notre Dame in Tianjin and massacre of Chinese Christians and some foreigners including the French Consul on 21 June 1870 were early and notorious examples of such hostility. Auguste Gérard, Minister in Beijing from 1894 to 1897, was determined to use the reconstruction of the church in 1897 as a symbol not only of Chinese expiation but also of the French prestige to which his colleagues in Paris were prone to allude. The church rejoiced in the name of Notre Dame des Victoires and the tombs of the 'martyrs' of 1870 were placed in the lateral naves. Gérard insisted on a spectacular inauguration on the anniversary of the riots despite popular rumours that the foundations were laid on the bodies of stolen children. Similar rumours had prompted the riot a quarter of a century earlier. The Chinese authorities, terrified of the threat to order that Gérard's stipulations would pose, beheaded a man accused of stealing children, issued notices exonerating the priests, and moved so many troops into Tianjin that it looked like 'a city under siege'. Marines from a French gunboat provided an honour guard, but a British offer to provide a further honour guard was declined with thanks, as Gérard believed that 'the ceremony had to conserve a clearly French character and it was inappropriate to show the English colours beside ours'.[9]

Gérard felt that the whole event was a great demonstration of national prestige in the face of popular hostility, but many other Frenchmen would agree with J.L. de Lanessan, a former Resident-General in Tonkin, in questioning the efficacy of such a performance. 'There are other', and he implied, better 'means than missions for establishing in China the prestige of France'. He argued that the Protectorate brought France much trouble and no prestige, if anything, quite the reverse. For when British, American, or German diplomats talked to the Chinese, it was normally about commercial or industrial matters of some genuine interest to all but the most reactionary Chinese politicians. But all too often French diplomats merely talked about missionaries, whom the Chinese despised. For although there were about 700,000 Chinese Catholics, there was amongst them 'not one scholar, not a sole man enjoying the least influence in his village'.[10] In fact, this was an exaggeration and influential Chinese Catholics did exist, including Jing Zhang, Minister in Paris during the late 1890s, but they were few in number and most educated Chinese found Christianity distasteful. The anti-clerical journalist, Paul Boell, spoke for many in both France and China when he asserted that the missionaries were singularly poor agents of French or

Western civilization because they taught obscurantist nonsense.[11]

Auguste François, a former Prefect and Consul in southern China, was one of those most critical of the Protectorate. While travelling to his post at Longzhou in Guangxi at the end of 1896, he met a young missionary fresh from Paris, dressed in Chinese clothes but speaking not a word of the language, accompanied only by a Chinese guide who was not a Catholic and who spoke neither French nor Latin. The ill-prepared priest, père Mazel, was headed into wild and inhospitable country. François was not suprised to hear that within a few months Mazel had been murdered. When he reached Longzhou he found the vicar apostolic of the province, Monsignor Chouzy, living at the consulate because amongst the 'completely hostile population' no one would sell or even lease him a house. François believed that missionary activity, because of the overt hostility it aroused, was contrary to wider French interests in China. 'French religious and commercial interests are in a sense opposed to each other', he suggested, also because missionaries 'consider the establishment of our nationals in China as harmful to their own influence'.[12]

Other Consuls may not have been as harsh or direct in their judgements, but were also constantly aware of the difficulties of supervising the Protectorate. The Berthémy Convention of 20 February 1865 permitted missionaries to buy buildings, but this right was often contested by local populations, sometimes with litigation, sometimes more crudely by arson. The arrogance of missionaries often exacerbated social tensions, so Chinese officials both despised and feared them. It was the unenviable and unceasing task of French Consuls, who were often ideologically hostile to both parties, to mediate between the two. In cases of murders of missionaries or serious injury to them or their property, French demands typically were threefold: exemplary punishment of the culprits either in the cangue or by beheading; demotion of 'negligent' local officials; and payment of an indemnity, which, until 1897, was usually fairly modest. At times these demands would be reinforced by the visit of a French gunboat to the scene of the incident.

The activities of enthusiastic but ignorant missionaries and arrogant converts who sought to use their adopted religion to flout Chinese law and social norms meant that conflict was frequent. In 1896, a Consul in Hankou reported that for the first time since 1891 he had no missionary claims outstanding in the area for which the consulate was responsible. Georges Dubail, having coped with problems of missionaries' 'doubtful titles' and their 'perhaps excessive zeal', thought all issues resolved when he left his post at Shanghai in August 1896, but his successor Georges Servan de Bezaure, found that they all resurfaced to plague him. The work also had its hazards: in 1897 Frédéric Haas, Consul at Chongqing, was ordered to the Tibetan frontier to resolve a dispute involving missionaries. He succeeded, but in the process contracted fever and a liver abcess which forced his repatriation.[13]

The difficulties the Protectorate created for French diplomacy prompted many to question its value. Stephen Pichon, who followed Gérard as Minister in Beijing, was not one of its critics. In fact he ardently defended it. But even he was forced to admit that: 'The religious Protectorate absorbs the bulk of our diplomatic activity in China and ... it is, as a result, the principal source of our expenses here.' While he took pride in the fact that France was the only power willing to help Chinese Catholics, the resources this absorbed worried him. He therefore suggested that the Legation take two per cent of any indemnity paid to missions to defray expenses. Paris told him that the idea was illegal and could not be considered. Nevertheless its very conception is evidence of the frustration the Protectorate could create even in the minds of its supporters.[14]

Although the Protectorate brought only the vaguest and most dubious benefits to French diplomacy in return for the labour, expense, and aggravation its supervision entailed, it did provoke jealousy on the part of other powers and also resentment from the Chinese. In November 1889 both Germany and Italy asked France if they could issue passports to their own nationals who went to China as missionaries. Foreign Minister Goblet replied that France had no wish to impose her protection on anybody, and thereafter the Protectorate was to some extent shared.[15] Germany was most solicitous of the German Catholic missionaries in southern Shandong, although their head, Father Anzer, found his government less concerned for his Chinese converts than the French had been.[16] Ironically imperial Germany, the only country seriously to challenge the French Protectorate, was even less Catholic than the Third Republic. The curious dual protection of Anzer's mission by a Protestant monarchy and a secular republic ended dramatically.

In November 1897 two of Anzer's German colleagues were

murdered. The event was used by the German government as a pretext to claim massive compensation. It demanded, and received, an indemnity of 222,000 taels (compared with the 15,000 taels paid in the Mazel case), the lease of south Shandong's best port, Qingdao, and two railway concessions.[17] The incident marked not only the beginning of the struggle for concessions, but also the beginning of the end of the French Protectorate. Early in 1898 France agreed no longer to intervene in support of Catholics in southern Shandong, but at the same time the French felt they had to emulate the extravagance of German compensation claims. Thus when père Berthelot and two Chinese followers were killed in Guangxi, the French demanded a railway concession and construction of an expiatory chapel as well as the usual three forms of amend. The murders thus became bargaining points in the struggle for French penetration into the provinces bordering on Tonkin.[18]

The relative moderation of French demands following Berthelot's murder compared with German reaction to the Shandong murders was largely the result of very real French fear at the manic pace of the 'scramble for concessions' which the German moves had precipitated. But if the 'scramble' worried the French, it was devastating for the Chinese government, whose policy of playing the imperialist powers off against each other had gone disastrously wrong. In an attempt to recover the political initiative and to ensure that missionary difficulties would never again provide the pretext for such wild demands, an Imperial Decree of 15 March 1899 recognized Catholicism and allowed clergy to make direct contact with local officials. This would reduce the level of involvement of French Consuls and also provide local Chinese officials with an incentive to settle minor disputes involving missions as quickly and as fairly as possible.[19] Its general effect was to augment the legal status of Catholicism in China while weakening the French Protectorate.

The Quai d'Orsay, however, still held that the Protectorate was of value. As one bureaucrat wrote with exquisite hypocrisy as late as April 1900:

While other powers only seek material advantages in China, we are undertaking there a civilizing task which, far from prejudicing our economic interests, can only assist in their development, for it furnishes effective opportunities to demand not only fair compensation but at the same time the settlement of industrial or commercial matters.[20]

The folly of that argument was about to be proved, for a savage Chinese popular reaction to the symbiotic cultural and economic aggression of the West was fermenting. More than any Imperial Decree, the Boxers, ignorant and obscurantist as they were, amply demonstrated the hazards of close identification with missionary endeavours.

Together with the religious Protectorate, the treaties of 1844, 1858, and 1860 conferred on France the privileges of the most-favoured-nation clause, under the terms of which France shared in the commercial and other advantages secured from China by third powers, in effect by Great Britain. This assured France equality of access to the treaty 'ports' of China, where French merchants could conduct trade on the same terms as those from other countries. These privileges remained the legal basis of French activity in China until the almost simultaneous collapse both of the French empire in East Asia and of the Third Republic itself.

The argument that economic factors bore most heavily on decision-making processes implies something of a Eurocentric approach to imperialism. The relative importance of events on the periphery of European expansion, and of events and pressures operating in the metropolitan power, have been the subject of a considerably recent historiography.[21] As an Australian, that is a citizen of a former British colony which, in the late nineteenth and early twentieth centuries greatly influenced British policy in the Pacific, the present writer is unlikely to underestimate the significance of the periphery. Nevertheless, it must be emphasized that the French population of Indo-China, consisting mainly of officials and soldiers, was not nearly as powerful a force as, for example, the British populations of Australasia or South Africa.[22] Although there was a relatively strong and talented corps of military officers in Tonkin after 1885, these men were never able to become a peripheral force which actually determined policy, as were their colleagues in the Western Sudan.[23] Some certainly tried but the verdant plateaux and malarial valleys of Yunnan were destined not to experience the fate of the sands of the Western Sudan.

The importance of peripheral factors is rightly emphasized in the case of the Garnier affair by Ella Laffey.[24] But the Garnier affair in a very real sense ultimately came to nothing. Whatever its value to propagandists and romantics, it did not lead to any extension of French territorial control, nor even to any real improvement in conditions for French commerce. The decision to conquer Tonkin was taken not in Saigon but, as Kim Munholland argues, in Paris, extraordinarily enough by the Minister responsible, Admiral Jauréguiberry.[25] Furthermore, this work will argue that the decision not to conquer Yunnan was similarly taken in Paris, similarly by the Minister responsible, and in spite of strong peripheral pressure to do so.[26] Despite its great causal importance in Africa and the Pacific, the periphery was less significant in the formulation of French policy in East Asia.

Even historians who have taken a Eurocentric approach to French imperialism in East Asia, however, have tended to emphasize factors other than the economic impulses which to this writer seem dominant. Thus, in an important analysis Henri Brunschwig argues that imperial expansion was essentially part of the French search for a prestige which had been lost in 1870.[27] In support of this contention he produces statistics which indicate the very small percentage of French trade with and investment in her colonies.[28] He acknowledges the existence of economic pressure groups and the use of economic arguments in support of the establishment of empire, but suggests that these date from Jules Ferry's defence of the acquisition of Tonkin in July 1885, in which Ferry appealed to Tonkin's, and its Chinese hinterland's, value as potential markets for French goods.[29] In this argument Brunschwig follows the lead made by John F. Cady who, after examining religious and economic factors, suggests that 'the taproot of French imperialism in the Far East from first to last was national pride — pride of culture, reputation, prestige and influence'.[30] Certainly it is not difficult to find evidence in support of this conclusion in the form of contemporary imperialist rhetoric which spoke continually of national glory. Bismarck was well aware of this aspect of French imperialism, and so encouraged it, as was the radical nationalist Déroulède, who objected to the attempt to compensate for the loss of Alsace and Lorraine by acquiring colonies in the most bitter terms: 'I have lost two children and you offer me twenty servants.'[31]

The emphasis which Cady and Brunschwig placed on prestige has been challenged effectively by a number of writers. Prestige may have contributed to the decision to conquer Tonkin, but it was hardly significant in determining French policy in China itself. And even though the possession of Indo-China did bring the Third Republic some prestige, Dieter Brötel rightly suggests that French motivation was complex and that economic factors were far more important.[32] C. A. Julien made the same point as long ago as 1946.[33] The complex role of the Geographical Societies, where passions as diverse as national prestige, scientific research, exotic exploration, and economic gain were fused into an intoxicating mixture, have been examined.[34] The work of John Laffey on the interest of the Lyon Chamber of Commerce in China and Indo-China has led him seriously to challenge Cady's approach and to conclude that:

contrary to the views of Cady and his followers, economic motivation had a role of prime importance in the expansion of France in the Far East.[35]

W. C. Hartel has attempted to identify the businessmen and corporations which supported imperialist endeavour through organizations such as Geographical Societies, Chambers of Commerce, and colonial committees and to estimate their influence on governments.[36] Even Fieldhouse admits that in the early 1880s, 'French interest in Tongking was undoubtedly economic', although he rightly suggests that at this stage it was still largely a commercial interest, — the hope that 'the Red River [Hồng Ha] would give France a shorter, cheaper and above all monopolistic route to markets and sources of raw materials in southern China'.[37]

An account of the background to the Sino-French peace treaty of 1885 is essential to an understanding of French objectives in China during the years 1885–1901. To a large extent the story begins with the annexation of part of Cochin-China in 1862 which seemed to offer an alternative means of access to open competition with Britain in the treaty ports.

Just four years after this annexation, an official expedition left Saigon under the command of Doudart de Lagrée. Its aim was to ascertain whether the Mekong River was a navigable stream which could be used to transform the newly founded colony into the outlet for the trade of south-western China. The expedition found that it was not navigable but, after de Lagrée's death, his replacement, Francis Garnier, discovered that the Hong Ha flowed from the Yunnan plateau to the Gulf of Tonkin. Another potential route had been found. Garnier's report of the expedition was a significant document in the development in France of the myth of the putative wealth of Yunnan and south-western China generally.[38] This myth was to be a potent force in determining both

French and British policy in the region for the next 30 years.[39] In 1873 Garnier died in a melodramatic attempt to open the Hong Ha to French trade. In the wake of the incident, under the terms of the Treaty of Saigon of 1874, Vietnam agreed to open the entire length of the Hong Ha, including the ports of Hanoi and Haiphong, to international trade.[40] However, along the upper reaches of the Hong Ha power lay less in the hands of the Hue government than in those of the Black Flag armies (independent remnants of Taiping forces operating in northern Tonkin under the leadership of Liu Yongfu). To preserve their own independence, they ensured that trade on the river was very insecure, indeed impossible for Europeans.

The revival of French interest in Tonkin which was ultimately to lead to its conquest began with the accession to power of the moderate republicans in 1879. In July 1881, during the first ministry of Jules Ferry, funds were voted for the dispatch of an expedition to reclaim the right of navigation conferred by the treaty of 1874. Initially a small detachment of seven hundred men was sent from Saigon to Hanoi. The war escalated and hostilities ultimately reached as far north as Fuzhou.

Chinese support for the Black Flags and Chinese claims to suzerainty over Vietnam led to a needlessly prolonged undeclared war with China.[41] It was not until June 1885 that Patenôtre, the French Minister in Beijing, and Li Hongzhang, signed at Tianjin a convention by which China relinquished her claims to suzerainty over Vietnam and allowed the French to trade along the Tonkin border at customs rates lower than those in the treaty ports.[42]

Although interest in the acquisition of Tonkin was displayed by the two de Freycinet governments, especially by their Minister for the Marine, Admiral Jauréguiberry, and by the 'grand ministère' of Gambetta, it was during the second ministry of Jules Ferry (1883–5) that the conquest of Tonkin was completed. In doing so Ferry was pursuing objectives which were firmly established republican policy: nevertheless, his contemporary notoriety as the 'tonkinois' and his vigorous and articulate arguments in support of its acquisition make him a highly significant figure in an analysis of the aims of the French in both Tonkin and China itself.

T. F. Power has argued that Ferry's main interest was in the prestige which would accrue to France as a result of an extension of her authority in Indo-China.[43] Certainly Ferry argued that French control of Tonkin was necessary for the security of Cochin-

China in similar terms to those he used to describe the situation of Tunisia *vis-à-vis* Algeria during his first ministry.[44] This suggests that he assumed that colonies were intrinsically a good thing, but for Ferry the economic possibilities of China itself were probably an even more important consideration.

To Ferry, China's chief attraction for France was the allegedly large and prosperous population which was believed to inhabit the south-western provinces adjacent to Tonkin. This population he saw as constituting a potentially captive market for French industry. His vision was mostly, although not entirely, mercantilist, which was not really surprising at a time when, as Ferry himself was acutely aware, tariff barriers were being erected with increasing enthusiasm on both sides of the Atlantic.[45] With France in control of Tonkin and hence the easiest means of access to south-western China, the use of discriminatory tariffs on goods in transit through Tonkin rather than actual French administration of the area could lead to the establishment of a French commercial monopoly over a considerable portion of the Chinese empire. Ferry, like so many other Europeans and Americans, became captivated by the chimerical vision of four hundred million potential consumers of Western goods. As Ferry himself stated in his defence following the fall of his ministry:

the creation of the outlet with which we are now concerned will give us, in a certain future, free and even privileged exchange with China and, with that, markets of four hundred million inhabitants . . . [earlier described by Ferry as] not poor blacks like the inhabitants of equatorial Africa, but . . . one of the most advanced and wealthy people in the world.[46]

There was also a more modern and prescient element to Ferry's imperial ambitions, which in 1885 was in fact incorporated into the peace treaty. He cited the investment opportunities which would become available for French surplus capital as a factor in support of his Tonkin policy:

I say that France, which has always abounded in capital and has exported considerable quantities abroad — in fact you can count in billions the capital exports of this great country — which is so rich; I say that it is in France's interests to consider this side of the colonial question.[47]

It has been argued, notably by Henri Brunschwig and T. F. Power,[48] that Ferry's use of economic arguments was merely part of an attempt to defend the aggressive policies which he had followed and which had led to his fall. His concern for the possibility of the creation of new opportunities for investment or markets for French products was neither genuine nor consistent, but part of his rhetoric only after the fall of his ministry. Both writers see these arguments first emerging with clarity in the defeated Ferry's speech to the Chamber on 28 July 1885.

Yet, while Ferry's most complete statements of an economic theory of imperialism are to be found in that speech and in *Le Tonkin et la mère-patrie* written in 1890, Ferry, as a Lorrainer, always had been concerned for industrial expansion and sought to bolster the industrial interests of the north-east.[49] As early as November 1883 Ferry had justified the continuation of hostilities in Tonkin on the ground that its acquisition would be a step towards the opening up of Chinese markets to French industry:

Cast your eyes on the map of the world and see with what vigilance, with what eagerness the great nations who are your friends or your rivals are reserving outlets for themselves ... See with what eagerness each of these industrial races ... strain to take their share in the still unexplored world, in that Africa, in that Asia that holds so much wealth and particularly in that vast Chinese empire. It is not a question of wanting to conquer that great Chinese empire ... But we must be at the gateway of that rich region ... And it is for that that we admire and thank the vigilance, the wisdom or deep instinct that pushed our predecessors toward the mouth of the Hong Ha and which established for them as a goal the possession of Tonkin.[50]

Moreover, at the time of Ferry's first ministry the aim of using Tonkin as a means of commercial entry into China had been expressed. In July 1881 Antonin Proust, the *rapporteur* of a bill to provide reinforcements in Indo-China, followed a reference to the need to defend Cochin-China with these comments:

We must, besides, neglect nothing to obtain an uncontested influence in the Indochinese peninsula, which must ultimately open up commerce between Yunnan and the South China Sea. In fact the European powers have been constantly preoccupied for a long time with the establishment of new commercial routes into central China.[51]

Thus the motives behind Ferry's aggressive Tonkin policy were threefold. Firstly, he believed in the necessity to sustain French prestige in the East, which was not an unreasonable concern when France had so little else and was still suffering from the memory of French impotence at the time of the Tianjin massacre of 1870.

Secondly, and largely as a result of the first concern, he wished more firmly to establish the security of existing French possessions in Indo-China. Thirdly, and most importantly, he expected the new Protectorate to provide access to new markets for French goods and new avenues for the investment of surplus French capital, not just in Tonkin, but more significantly within the borders of the Chinese empire itself.[52] Moreover Ferry was not alone. Admiral Jauréguiberry had acted with very similar motives when he was responsible for colonial matters in the de Freycinet governments.[53]

In fact France failed to develop a large export trade in China. This was a great disappointment to the ardent imperialists who followed Ferry and a source of constant concern to French foreign ministers and their agents in Asia. As late as 1900 there were only twenty-four French businesses established in Shanghai, the centre of European economic life in China. Valued at 64,196,067 francs, their activities included the provision of services such as insurance, banking, hotels, and restaurants; international trade in commodities like silk and wine; and retailing bicycles and food products.[54] Virtually all these businesses, apart from the silk exporters, provided luxury goods and services to the European and comprador populations of the city. There were not four hundred million potential consumers of Bordeaux and the age of the bicycle as mass transportation in China was far in the future.

The poor performance of French goods in the Chinese market was the result of France's unique pattern of industrialization in the late nineteenth century. For while Britain, and later Germany and the United States of America, tended to specialize in cheap mass-produced consumer goods, made in factories by relatively un-skilled labour, French industry concentrated on higher value-added goods of superior quality and price, often produced in smaller workshops. This specialization was particularly marked in export goods. The nature of commodities exchanged with France's largest trading partner, Great Britain, in 1888 reveals the extent of this specialization and the differences between the two economies.

The figures in Table 1.1 suggest that while French industry was not underdeveloped compared with that of Britain, it was very different and tended to complement rather than rival its neighbour.[55] Certainly France did have large factories and did produce both cheap consumer goods, typically hardware and textiles, and heavy industrial goods such as ships, machinery, and

Table 1.1 Percentages of Commodities Traded between Britain and
France, 1888

Category	GB	France
1. Raw materials, for example, coal, wool, hides	15	7
2. Intermediate goods, for example, iron, copper, cotton and other thread, and cloth	55	8
3. Finished industrial products, for example, silk, leather, machinery, hardware, clothing	20	43
4. Processed food, for example, wine and spirits, flour, sugar	2	17
5. Unprocessed food, for example, butter, eggs, grain, and vegetables	2	16
6. Unclassified items	6	9

Source: Patrick O'Brien and Caglar Keyder, *Economic Growth in Britain and
France 1780–1914: Two Paths to the Twentieth Century* (London, George
Allen and Unwin, 1978).

railway equipment. Such products were not, however, areas of
special French expertise and efficiency, and did not enjoy great
success as exports. The proportion of French exports as a percentage of those of eleven industrial nations in 1899 clearly shows both
French failure in these fields and French success in more valuable
luxury consumer goods.[56] (See Table 1.2.)

Since luxury goods, almost by definition, carry cultural as well as
material value, the products at which French industry excelled
were, on the whole, exported more to wealthy than to poor
countries, and were very much attuned to European taste. This
meant that French exports were consumed more in Europe than
elsewhere, and, ironically, in view of the common French percep-
tion that her great rival in Asian markets was British industry,
were consumed more in England than anywhere else. Moreover,
while other industrializing countries like Germany, Belgium, and
the United States of America followed the British lead in increasing the proportion of their exports to places outside Europe,
France did not. Between 1880 and 1910 only Africa, where France
had the bulk of her colonies, emerged as a significant market for
French products.[57] (See Table 1.3.)

In the early nineteenth century British trade in China was
dominated by exports of Indian opium, but later in the century
trade changed from opium to cotton cloth, the cheap manchester

Table 1.2   Percentage of French Exports of Various Products Produced
in Eleven Industrial Nations, 1899

Products	Per Cent	Products	Per Cent
Tobacco and alcohol	45.0	Chemical products	13.4
Clothing	33.5	Non-ferrous metals	11.0
Books, films, etc.	26.4	Metal goods	
Bicycles, motor cars,		(unspecified)	10.4
etc.	21.0	Capital goods	7.4
Skin, wood, paper	20.5	Ferrous metals	5.6
Minerals	16.7	Railway equipment and	
Textiles	15.8	ships	4.7
		Electrical goods	4.6
		Agricultural machinery	3.1

Note: The eleven countries are Belgium, Canada, France, Germany, Italy, Japan,
Luxembourg, Sweden, Switzerland, UK, and USA.
Source: H. Tyszynski, 'World Trade in Manufactured Commodities 1899–1950',
The Manchester School of Economic and Social Studies, No. 3, September
1951, in Paul Blairoch, 'La Place de la France sur les marchés inter-
nationaux', in Paul Levy-Leboyer (ed.), *La Position internationale de la
France: aspects économiques et financiers, XIX–XX siècles* (Paris, Ecole
des Hautes Etudes en Sciences Sociales, 1977), p. 49.

with which the first industrial society had been created. In the five
years from 1894 to 1898 an average of £3,606,673 worth of cotton
cloth was exported from Britain into China. Total British exports
to China over that period averaged £5,322,081 per annum, so
cotton cloth constituted no less than 67.8 per cent of all British
exports to that market. Against English cotton French industry
could not compete effectively. Nevertheless it was the one
European product for which there was a large market in China and
it also dominated French exports. During the same period cotton
products exported from France into China averaged 13,730,733
francs per annum out of an average total of 21,624,882 francs, or
63.5 per cent. Thus the domination of cotton was almost as great in
French as in British exports, albeit on a much smaller base. British
cotton exports to China were worth 6.57 times more than French
exports at this time.[58]

French officials and politicians were well aware of the problem.
Despite their high quality, French products were over-priced and
ill suited to Chinese tastes, lamented the Consul in Guangzhou in
1889, while he urged French industry to adapt to the needs of

Table 1.3 Destination of Exports of Various European Countries by Percentage

From	To Europe	North America	South America	Asia	Africa	Oceania
1880						
France	71.7	9.2	10.2	2.2	6.7	0.2
Germany	91.3	6.6	0.9	0.9	0.2	0.1
Belgium	92.9	2.8	3.0	1.1	0.2	0.0
UK	35.5	15.8	10.2	25.3	4.3	8.4
1910						
France	69.8	7.4	6.9	3.5	12.3	0.1
Germany	73.9	8.9	7.8	5.8	2.3	1.0
Belgium	81.9	4.2	5.9	4.8	2.3	0.6
UK	35.0	11.6	12.5	24.4	7.4	8.6

Source: Paul Bairoch, 'La Place de La France sur les marchés internationaux', in Paul Levy-Leboyer (ed.), *La Position internationale de la France: aspects économiques et financiers, XIX–XX siècles* (Paris, Ecole des Hautes Etudes en Sciences Sociales, 1977), p. 42.

Chinese consumers. He concluded that 'our commerce ... must decide to manufacture for this market where it hopes to sell its products'.[59] The message reached the highest levels: 'Our imports [into Guangzhou] in 1888 have mainly consisted of fine wines and champagne, cognac, perfume, and fashion articles.... Our cottons are much too expensive', Foreign Minister Eugène Spuller bluntly told Premier Tirard.[60]

Some Consuls actively sought to promote French trade. The irrepressible and enthusiastic Frédéric Haas, whose impetuosity in Mandalay in 1885 had helped provoke the British conquest of Burma, even organized an exhibition of French goods in his lonely post of Chongqing in 1896. He had no success, so sent back to Paris drawings of Chinese ricebowls, teapots, and soup-spoons in the vain expectation that some manufacturer might attempt to make them in enamel and export them to China.[61] A very persuasive man, he cajoled two French companies into establishing shops in Sichuan, one in Chongqing, the other near the Tibetan frontier. The latter met with such popular and official hostility that it was forced to close and move its operations to the provincial capital of Chengdu. He boasted that 'I have succeeded in founding two French houses in Sichuan', but was forced to admit that 'the level of business will be very modest'. Both shops sold watches, fashion accessories, and luxury groceries, French specialities to be sure, but hardly the sort of goods to find a huge market in the Sichuan of 1896, much less on the Tibetan frontier.[62] Both shops caused their share of political difficulties and were the subject of popular protest, as had been the French businesses Haas had encouraged in Burma more than a decade earlier, but the consequences of his quixotic plans were less dramatic in Sichuan.[63]

While Haas was certainly the most energetic consular promoter of small French trading enterprises in China, he was not the only one. In 1896 a Chinese official in Fuzhou, hoping to indulge in some enterpreneurial activity to supplement his income, asked Claudel, manager of the French Vice-consulate, if he could arrange to provide some samples of cotton sheeting. Six months later MM. Thézard et fils agreed to forward the samples, the request having reached them via the Foreign Ministry, the Ministry of Commerce, and the Elbeuf Chamber of Commerce. Some months later the sample reached Fuzhou, only to be refused by the official on the not unreasonable ground of excessive delay. Thézard insisted that Claudel pay the bill of 230.65 francs himself and declined to express any regret at the Vice-consul's personal loss. Hanotaux was genuinely shocked at the manufacturer's dismissive attitude towards both the China market and the Vice-consul, while Claudel confined himself to the acid but accurate observation:

Consuls are sometimes accused of not being sufficiently concerned with the interests of our commerce. Your Excellency will kindly recognize that the example furnished by this little affair will hardly encourage them.[64]

A more serious form of official encouragement for the export of French goods into China came with the application of the French customs system to Indo-China in 1888. Although transit duties were only a quarter of import duties, they were intended to ensure that French goods predominated the import trade into Yunnan along the Hong Ha. The transit duties proved to be ineffective. They were often imposed with excessive zeal and were high enough to produce considerable resentment on the part of Chinese, English, and even some French merchants, but were too low to offer real protection to French exports. Despite those failings, Governor-General Paul Doumer defended the duties vigorously in 1899:

Tonkin must not become a public route open to all merchants; it must be both a market and a means of penetration into China for French products, both raw materials and manufactures, for which the metropole seeks the regular sale. Only the application of a customs tariff protecting French interests will allow the realization of these *desiderata*... It is annoying that our industrialists and merchants have not yet made any serious effort to capture the Yunnan market; but, as steps are going to be taken shortly, it would be silly to abolish the transit duties now, and with them, the protection due to our national interests.[65]

Although the French export trade to China never amounted to much and some attempts to develop it bordered on farce, the same was not true of trade in the other direction. For China was the major source of a crucial commodity for the French luxury goods industry. That commodity was raw silk. The French silk industry had been centred on the city of Lyon since before the Revolution. Throughout the nineteenth century it was the largest silk industry in Europe. Its production from 1885 to 1900 inclusive averaged 5,759,750 kilograms per annum, compared with an average 5,739,188 kilograms of its only serious rival, Milan.[66] Until the

Table 1.4 Percentages of Silk Sold by the Largest Suppliers to the Lyon Industry, 1865–1899.

Period	France	China	Japan	Italy
1865–9	27.80	16.61	17.26	21.59
1870–4	30.59	24.11	12.91	21.55
1875–9	18.59	35.71	16.36	19.17
1880–4	15.84	31.20	17.76	26.02
1885–9	13.24	33.53	17.77	22.75
1890–4	11.28	34.75	21.04	14.87
1895–9	11.01	38.11	18.17	11.33

Source: Calculated from a table in John F. Laffey, 'French Imperialism and the Lyon Mission to China (unpublished Ph.D. thesis, Cornell University, 1966), p. 114.

mid-nineteenth century the bulk of this industry's raw materials came from southern France. However French sericulture was devastated from 1854 by *pébrine*, a disease endemic to silkworms. In 1855, French cocoon crops were 19.8 million kilograms, but were only 7.5 million in 1856, and 5.5 million in 1865. Although French sericulture never recovered, the Lyon silk weaving industry remained buoyant. During the Second Empire Paris was the world centre of fashion, and the Cobden–Chevalier Treaty of 1860 removed the British 15 per cent tariff on French silks, thereby dooming the British industry. French success in this luxury industry meant that by 1874 it was consuming about a third of the world's raw silk, but had to import three-quarters of its needs.[67]

In its search for alternative supplies, the Lyon industry turned first to Italy and then to the eastern Mediterranean. The spread of *pébrine* throughout Europe meant that East Asia was an increasingly attractive source of raw silk. In 1852 Lyon bought 52 bales (each bale weighed about 60 kilograms) from China, the first-ever imported from East Asia: by 1869 China was the industry's most important supplier, the quality and competitive price of its product more than compensating for its remoteness. China's domination of Lyon's silk supplies remained unchallenged throughout most of the second half of the nineteenth century. (See Table 1.4.)

It was this raw silk which constituted the bulk of Sino-French trade. In 1899 no less than 207,820,910 francs of the 242,497,000 francs worth of Chinese exports of France, or 85.7 per cent, was silk. This high level of dependence on Chinese raw silk meant that

Table 1.5   French Trade with China, 1890–1899

Year	Imports from China (francs)	Exports to China (francs)
1890	130,798,377	12,400,384
1891	135,859,134	10,344,940
1892	157,377,141	7,244,486
1893	156,243,220	5,696,600
1894	120,984,089	19,971,563
1895	158,492,673	15,279,976
1896	106,020,679	23,335,679
1897	168,118,517	29,484,450
1898	149,620,000	20,053,000
1899	242,497,000	25,071,000

Source: Commerce générale de la France avec la Chine, Note, juin 1900, MAE NS 563, 63–4.

Table 1.6   Chinese Trade with Britain, Germany, and France in Francs, 1895

	Imports from	Exports to
France	15,280,000	158,493,000
Britain	137,161,000	85,512,000
Germany	44,288,000	24,514,000

Source: Calculated from O. P. Austin, *Commercial China in 1899: area, population, production, telegraphs, transportation routes, foreign commerce, and commerce of the United States with China* (Washington, Government Printing Office, 1989), pp. 1748–51.

France had a consistently unfavourable balance of trade with China. (See Table 1.5.)

Trade between French Indo-China and China, once it began to develop, did nothing to relieve this unfavourable balance. In 1900 China exported goods to the value of 90,787,000 francs, while imports from Indo-China were worth only 74,700,000 francs. No single commodity dominated Chinese exports — gold worth 10,787,000 francs was the most valuable — but Indo-Chinese rice, worth 54,000,000 francs in 1900, clearly dominated trade in the other direction.[68] The French need for Chinese raw silk to sustain one of the most important of her luxury industries meant that in the mid-1890s France alone among the great imperialist powers had an unfavourable trade balance with China. (See Table 1.6.)

Conversely, for China France was an important export market and silk an important export commodity. In 1887 silk (37 per cent of the total, most of it raw) displaced tea (35 per cent) as China's leading export and remained so through to 1911.[69] The significance of silk in sustaining China's generally dismal trade position during the late nineteenth century was clear. (See Table 1.7.)

Although these silk exports meant that Sino-French trade was conducted on terms which appeared highly unfavourable to France, in reality France benefited greatly from them. China was a cheap and reliable supplier of an important raw material, even when the two nations were at war. Once in France, the raw silk was transformed into expensive luxury goods many of which were exported. This high-added value contributed to economic growth and employment in an area of traditional French expertise. French silk exports averaged 252,551,000 francs per annum between 1894 and 1898 inclusive. Easily the largest customer was Great Britain, a successful exporter to China, but incapable of sustaining a silk finishing industry. Britain bought an average 124,057,600 francs worth of silk each year at this time, or 49.12 per cent of total French silk exports.[70] These exports to Britain alone were worth a little more than France's trade deficit with China, which averaged 119,002,250 francs over the same period. French industry may not have been able to acquire a large market in China, but it was certainly adept at catering to the desires of the wealthier elements of English society. Imports from China were essential to this lucrative business. French politicians and officials need not have been as worried as they were about the apparently poor results of trade with China. It was money well spent.

Trade with China brought France other direct economic benefits as well, notably the earnings of her shipping companies and merchants. China did not share in these invisible earnings to any great extent at all. By the mid-1890s Haiphong-based French shipping companies dominated the trade on the Guangdong coast between Hong Kong and Haiphong, including the island of Hainan. Elsewhere on the China coast, however, the French role was more modest. Messageries Maritimes maintained a regular mail liner service, but it was subsidized throughout this period. In view of France's role as China's most important European customer, French shipping performed poorly. Germany, Japan, Scandanavian countries, and the United States of America were all able to increase their shipping tonnages at Chinese ports in

Table 1.7 China's Foreign Trade: Imports, Exports, and Silk Exports, 1885–1900

Year	Net Imports	Exports	Trade Defecit	Raw Silk Exports	Raw Silk as Percentage of Total Exports
1885	88,200	65,006	23,194	15,526	23.4
1886	87,479	77,207	10,273	21,852	28.3
1887	102,264	85,860	16,403	24,607	28.7
1888	124,783	92,401	32,382	23,765	25.7
1889	110,884	96,948	13,937	28,642	29.6
1890	127,093	87,144	39,949	24,491	28.1
1891	134,004	100,948	33,056	29,884	29.6
1892	135,101	102,584	32,518	30,341	29.6
1893	151,363	116,632	34,731	29,325	25.1
1894	162,103	128,105	33,998	33,604	26.2
1895	171,697	143,293	28,404	38,724	27.0
1896	202,590	131,081	71,509	31,672	24.2
1897	202,829	163,501	39,327	44,461	27.2
1898	209,579	159,037	50,542	45,413	28.6
1899	264,748	195,785	68,964	71,583	36.6
1900	211,070	158,997	52,074	39,732	25.0

Note: The *Haiguan* tael, the silver-based currency used by the Chinese Imperial Maritime Customs, declined against the gold-based franc from a value of 6.64 francs in 1885 to 3.90 in 1900.

Source: Calculated from tables in Hsiao Liang-Lin, *China's Foreign Trade Statistics, 1864–1949* (Cambridge, Mass., Harvard University Press, 1974) pp. 22–3, 109, 190–1.

the late nineteenth century, while French shipping remained stagnant.[71]

This was largely due to the tendency of French silk buyers to rely on foreign firms to handle their raw material. During the 1889–90 silk season at Guangzhou, for instance, some 15,680 bales of raw silk were exported to Europe, virtually all of it to France. One French company, Cozon et Giraud, handled 1,542 bales, just 9.83 per cent of the total, while four English and four German firms handled the balance.[72] A second French firm was established there in 1892, but little changed.[73] At Shanghai the situation was similar. In the 1897–8 season some 71,000 bales were exported from Shanghai to France, but the French companies active in the trade there, Ulysse Pila and Lieber Chauvin Chevalier shipped only 13,000 bales or about 18 per cent.[74]

While the low level of French exports to China was the result of the specialized nature of the French economy, the failure of French service industries to exploit the potential of the silk trade was perhaps more the product of cultural factors. The revolutionary land settlement and Napoleonic family law continued to ensure that there was relatively little internal or external migration and very modest population growth in France compared with Britain or Germany in the nineteenth century. In France there was not the same large, restless entrepreneurial class anxious to find wealth and caring little for where it was found as there was in Britain, Germany, and the United States of America. Commerce in remote parts of the world did not exercise great appeal to ambitious young middle-class Frenchmen. Secure investments and government service were more attractive options than emigration, especially as in Asia in general and in China in particular, English was the medium of most international trade. A French Consul in Hankou asked in 1892: 'Why don't our merchants who really want to have commercial relations in this part of China, send out some young men who know English and business practices, and as a result can represent them in these distant countries and make French products known there?'[75] French capitalists on the whole were not greatly interested, and young men like those needed would find work in Dover or London which would be more congenial and more lucrative than in Shanghai or Tianjin, let alone Hankou. Consequently, in spite of silk, the French presence in trade and services in late nineteenth century China remained modest.

Thus the hope that French goods would find favour with the Chinese market was unrealistic and trade never developed in the way Ferry had expected. However, French capital could and subsequently did welcome the investment opportunities which were to emerge after 1885. In this way the economic aspects of what could be described as Ferry's 'war aims' were realized. This development and the rejection of the territorial ambitions which some considered Ferry's aims implied are the major themes of this book.

Although Ferry had emphasized the economic benefits France was expected to gain in southern China, it was in the north that her capitalists first began to take advantage of the peace. The first part of this work examines the development of French economic interests in central and northern China after 1885. The second part concentrates on those interests located near the Tonkin frontier and on the territorial aspirations with which they were entwined. Ultimately the policy of seeking investments and influence throughout China proved to be incompatible with the trend towards territorial domination of Tonkin's hinterland. This contradiction in French imperialism became especially acute during the last years of the nineteenth century when it appeared as though the Chinese empire might be dismembered and divided amongst the imperialist powers. However economic and political imperatives drove French decision-makers to resolve this contradiction in such a way as to anticipate what has become the norm in economic relations between rich and poor countries in the twentieth century.

Part I

The Diffusion of French Investment in Northern and Central China

2

The Quest for Industrial Hegemony
1885–1896

IF it had been in a spirit of neo-mercantilism that Jules Ferry provoked war with China, it was in a similar spirit that the terms of the peace treaty were decided. As early as the signing of the abortive Li–Fournier convention in May 1884, the French negotiators had decided that commercial and industrial advantages were of more worth than the indemnity which would have been exacted by more traditional diplomacy. By Articles 3 and 4 of that convention France renounced an indemnity in return for open trade between Vietnam and China subject to tariffs to be later decided on a basis favourable to French commerce. Ferry defended such terms in the Chamber in highly mercantilist language:

Was the renunciation of a monetary indemnity too high a price to pay for these considerable advantages? Would a money payment have the same value in the nation's eyes as a treaty promising good neighbourly relations and a commercial and political alliance, which would bring neither humiliation nor bitterness but would open ... outlets for our production? We didn't think so ...[1]

The Li–Fournier convention was never ratified, but the Treaty of Tianjin of 9 June 1885 by which the war was terminated included industrial as well as commercial provisions. This ensemble was designed to ensure that France would acquire from her victory in the war new markets for her industry, which were seen to be of ultimately greater value than a cash payment. This industrial provision was Article 7, by which the Chinese government agreed to approach French industrialists when it decided to construct railways.[2] The article was vaguely worded and never really achieved its intended purpose so far as the French were concerned. Its significance lies in its nature as an attempt to create through diplomacy a mercantilist framework in which China's modernization would be executed for the benefit of French industry. The use of a diplomatic obligation to achieve this end was in itself an admission of the uncompetitive state of French industry in the late nineteenth century. The executive of the organization representing French heavy industry, the powerful Comité des Forges admitted as much in a submission to the Foreign Office occasioned by the negotiation in China of the commercial convention which was to supplement the treaty. The Comité argued, in effect, that the inferiority of France's industrial and commercial position could only be remedied by the creation of quasi-mercantilist privileges under the government's aegis:

The greatest industrial nations of the old and new worlds are contending for the new outlets which the conquest of Tonkin has just created. English products are favoured in these countries because of Britain's exceptional commercial and maritime situation; German products are benefiting there, compared to ours, from the advantages of lower prices of labour and raw materials, as well as from the favourable transport arrangements which the German government, whose objective is to secure overseas markets, does not fail to obtain for them.

To struggle successfully against these powerful competitors, French industry and commerce therefore need to find in their government the most effective support at all levels ... We especially ask that the final treaty interprets the vague wording of Article 7 of the Treaty of Tianjin in French interests.[3]

The French negotiator in Tianjin, Cogordan, attempted to meet the Comité's wish in this respect. Specifically, French drafts provided for the right of French citizens or protegés to exploit mines and establish industries in the three southern provinces of Yunnan, Guangxi, and Guangdong, and defined the privilege granted to France in Article 7 of the 1885 treaty as the reservation of a thousand kilometres of the future Chinese railway network for construction by French industry.[4] Li Hongzhang, however, refused to entertain these propositions of Cogordan who had to content himself with the vague promise that France would not be forgotten when China needed foreign help in constructing her railways.[5] Certainly no such clauses appeared in the convention Li and Cogordan signed on 26 April 1886. Nor did Ernest Constans, the commercially adept envoy sent to Tianjin to renegotiate the convention in order to permit trade in opium and salt across the Sino-Vietnamese frontier, pursue the question of industrial privileges any further.[6]

It was thus on the rather flimsy basis of the text of Article 7 as it stood in the 1885 treaty that French industry claimed for itself a privileged position in Chinese railway construction, a position based on treaty obligations to which China was bound to conform by international law. Although the benefits of this article proved to be chimerical, the expectations it raised were enormous. In the mid-1880s France, alone amongst the Western powers, had an unfavourable balance of trade with China. For although she imported large quantities of Chinese raw silk, French exports had not been able to win any considerable share of the Chinese markets. In 1884 a French Consul in Tianjin, referring to France's 'notorious inferiority compared with other European and American powers', suggested that this could best be overcome by the establishment of a syndicate of industrialists and merchants which would then appoint agents to represent it permanently in China.[7]

At that time Tianjin had become the main centre of political and industrial activity in northern China. It was there that Li Hongzhang, Viceroy of Zhili (now Hebei) and Commissioner of the Northern Ports, had his residence, and directed his attempt for the modernization of China's military and industrial capacity through the use of Western technology. Until 1886 most of the profits to be obtained from Li's activities had gone to British, German, and, to a lesser extent, American companies. However, essentially in response to the opportunities which the treaty of 1885 seemed to offer French industry, the second half of the decade saw one lavishly financed syndicate and other lesser groups representing the products of France in Tianjin.

The large Parisian discount banking house, the Comptoir d'escompte de Paris operated a branch office in Shanghai and was the only French bank in China at that time. It was under the aegis of this bank that a powerful industrial syndicate was formed in Paris to exploit the privileges which diplomacy was believed to have acquired.[8] However, this Syndicat de la Mission de l'Industrie française en Chine was established as much on official as on entrepreneurial initiative. Paul Ristelhueber, the new Consul at Tianjin, returned to Paris to interest the financial and industrial community in implementing Article 7, and it was this model of official initiative and encouragement that was to be repeated frequently by French diplomats during the following 15 years. It was thus not only with the force of Article 7 behind it, but also with the very real interest of the French diplomatic community in China, that the syndicate's mission, headed by the civil engineer, Jean-Marie Thévenet, arrived in Tianjin late in March

1886. Initially the syndicate seemed to enjoy enormous success. It had absorbed the modest group headed by the Paris art dealer, Siegfried Bing at the end of 1885, and the efforts of the Decauville group to interest the Chinese in a portable light railway which had been brought to Tianjin in May 1886 came to nothing. Thévenet had considerable prestige, as well as an enormous salary of 80,000 francs per annum, plus allowances and commissions, and, with diplomatic support assured, the syndicate seemed to be destined to success.[9]

Although the initial plan was to devote most of its energy to railway construction, Thévenet decided against such a policy, probably as he wished to respect Li Hongzhang's declaration that the first main line to be constructed from Tianjin to Kaiping was only to be a trial, to see how the innovation operated in China and whether it would be accepted by the people. The line was therefore constructed under the direction of the engineer in charge of the railway built to transport coal from Kaiping, Charles Kinder, whose preference was for British material. At least one French diplomat considered that many subsequent difficulties would have been avoided had Li been required to honour the terms of Article 7 from the beginning.[10]

However, if Li quickly and skilfully deflected Thévenet's interest away from railways, he quickly gave him other projects. Soon after Ristelhueber had introduced the two men, Li asked Thévenet to supply two iron bridges to replace the boat bridges across the Beihe in front of his residence. This early success was followed in October 1886 by the signing of a contract for the construction of a naval dockyard at Port Arthur (now Lushun) at a cost of 1,150,000 taels, or about six and a half million francs. The anglophone press in China reacted bitterly to this contract and claimed that China had handed over the construction of her only naval station in northern China to her enemies, and that the French engineers to be employed were in fact spies. Some Western diplomats in Beijing even raised the matter with the *Zongli Yamen*, the board in Beijing responsible for dealings with foreigners. The fracas did not appear to deter Li, for a month later he ordered two military balloons from Thévenet's syndicate.[11] China's need for the base was uncontested. For although the Chinese navy was larger than the Japanese, it was not as effective a fighting force, not simply because it was not as well commanded and as well manned, but because it lacked the ancillary services

Table 2.1   Western Industrial Equipment Ordered in Tianjin, 1886–1890

Country	Total Orders in Francs	Notes
Germany	9,921,900	Mostly Krupp canons, also some rails and cement.
USA	644,000	One locomotive, sleepers, and oil for the Kaiping railway.
UK	1,713,040	Rails, two locomotives, and other items for Kaiping railway.
France	15,344,186	6,349,560 francs for the Port Arthur base; two loans given by the Comptoir d'escompte, one of 520,000 taels (2,912,000 francs) to the Shandong government, one of 500,000 taels (2,800,000 francs) to the Imperial House; a lighthouse at Weihaiwei; two dredges for the Huang He; purchase of the Decauville light railway; and various copper, iron, cartridge, and arms orders.

Source: Lefèvre (Tianjin) à Lemaire, 20 mai 1890, Archives of the Ministère des Affaires Etrangères, Chine, Affaires diverses commerciales Boîte 328 (1890).

necessary to keep such a fleet in efficient operation. In 1888 Japan possessed four dry docks whereas the only one in China was that under construction by the French at Port Arthur. As Chinese navy ships all had unsheathed iron bottoms, which required docking three times a year for the ship to maintain her maximum speed, the need for a facility such as Port Arthur was urgent.[12]

As for the choice of the French syndicate as the contractors, Li was probably wishing to appease the French government by appearing at least to honour Article 7. France's status as the protecting power of Catholic missionaries in China gave her the capacity to create considerable diplomatic difficulties for China, something Li, a member of the *Zongli Yamen* as well as Viceroy of Zhili, was anxious to avoid. He therefore favoured French industry in the years immediately following the conclusion of the war between the two powers. Thus from the beginning of 1886 until May 1890, Li placed more orders with French industry than with those of the other Western powers combined. (See Table 2.1.)

Moreover, Thévenet's apparent desire to obtain orders even at a loss in the hope that more profitable ones would follow, and Li's desires both to satisfy the French government and to obtain the equipment he needed at the lowest price, were quite compatible. The *Chinese Times* subsequently alleged that an engineer in Li's service had drafted plans and made estimates for the construction of the Port Arthur base, estimates that an independent engineer had later verified. These specifications were then given to Thévenet by Li's assistant, Zhou Fu.[13] Thévenet then offered to do the work for ten per cent less without making his own survey.[14] The other two engineers concerned were almost certainly British, as there is an element of national sentiment in the allegation, but the massive loss of 1,600,000 francs which the syndicate was to make on the Port Arthur contract suggests that Thévenet's estimates were unrealistically discounted to eliminate competition.[15]

It was such practices which enabled the syndicate to make such an impressive, if ephemeral, impact on the industrial scene in Tianjin in the mid to late 1880s. In 1886 and 1887 its position appeared to be unassailable. Ristelhueber was largely successful in ensuring that no other French concerns interfered with its operations. Decauville was induced to sell its experimental light railway, albeit at a loss. Schneider et cie., the large metallurgical company of le Creusot, had sent a naval engineer, Fliche, to Tianjin at the same time with a similar brief to that of Thévenet. Ristelhueber constantly urged Fliche to associate le Creusot's interests with those of the syndicate instead of competing with it, and, following Fliche's return to France, such an arrangement was reached early in 1887.[16] To strengthen the syndicate's position, Thévenet suggested to his employers, the Comptoir d'escompte, that they open a branch in Tianjin which would be able to lend money to Chinese officials who would buy industrial material from the syndicate with the proceeds of these loans. His advice was taken and the agency opened in October 1886. Before the end of the year it had lent 520,000 taels to the Shandong government for the improvement of the Huang He and nearly two million francs to other high functionaries.[17]

Although there was not a lot of business done at Tianjin in 1887, French industry was patronized to the extent of about two and a half million francs or more than a third of the total. This included two dredges for the Huang He ordered from Fives-Lille and four large calibre canon for Weihaiwei from le Creusot. Thévenet entertained hopes that a plan he had designed for the improvement of the Huang He at the enormous cost of 100 million francs might be implemented by the Shandong government following disastrous floods that year; and there was also the expectation that the Chinese navy would soon require more ships. Ristelhueber therefore optimistically reported to Paris that 'the present situation allows us to conceive rather fine hopes for the future'.[18] This impression had been confirmed by Li's declaration to the Commander of the British naval division in the China seas that he wished to please everyone, and would address himself to Germany for the requirements of the army, to England for the navy, and to France for the supply of industrial equipment.[19]

Ristelhueber, however, had every reason to endeavour to present the syndicate's affairs in the best possible light, as his personal interest in them was considerable. This interest was revealed by Paul Boell, who was the correspondent in China of *Le Temps* from 1889 to 1892. Boell had had access to the books and correspondence of the syndicate and concluded that from 1886 to 1889 Ristelhueber received payments from Thévenet totalling 83,000 francs.[20] Additionally Thévenet had asked Denfert-Rochereau, the chairman of the Comptoir d'escompte, to allow Ristelhueber the enormous retainer of three thousand francs per month.[21] If, however, Denfert-Rochereau did not accede to this request, he was certainly generous with his influence in other ways, for early in 1889 Ristelhueber was appointed a Consul-General in the abnormally short time of three years after his promotion to the grade of first class Consul. He wrote to Thévenet: 'I am very grateful to M. Denfert for what he has done for me and to you who certainly caused his proposal.'[22]

When Thévenet returned to Paris in 1889 he was promoted to the grade of first class in the Corps de Ponts et Chaussées and decorated for services rendered to French industry.[23] However, the orders worth 8,720,000 francs which he obtained were ultimately to result in a loss for the syndicate of about four million francs, or 45 per cent.[24] A later Consul at Tianjin bitterly judged Thévenet's behaviour:

When one examines the contracts M. Thévenet signed with the Chinese, one wonders how an engineer in chief of the Corps de Ponts et Chaussées could subscribe to such engagements and one is led fatally to the painful conclusion that M. Thévenet's only motive was to assure himself the high commission which each order he obtained would bring, without regard in any way for the ultimate profits or losses to the builders, the only true interested parties.[25]

The precarious position the syndicate had been placed in began to become evident in the spring of 1889. In March the Comptoir d'escompte de Paris crashed, and, although this need not necessarily have prejudiced the fortunes of the syndicate, it was a harbinger of what was to follow and had a disastrous effect on French prestige in China. The bank's agency in Tianjin was closed suddenly on one day's notice, a singularly inept act politically as Li and many of his colleagues kept large deposits with the bank. So great was the emotion in high political circles in Tianjin when the bank refused to give depositors access to their funds that the legation had to intervene to ensure that the bank continued operation until the depositors were paid the 400,000 taels (about two and a half million francs) owed them. Following the closure of the branch, French banking remained unrepresented in Tianjin, and indeed throughout China except for Shanghai. The Comptoir national d'escompte, which was essentially the old Comptoir d'escompte de Paris reconstituted with government assistance, decided not to re-open its predecessor's agency there.[26] The damage to prestige was compounded in 1890 by the purchase of the Comptoir's building for use as an agency by the Deutsche-Asiatische Bank.[27]

By April Li was complaining to the French Minister in Beijing, Gabriel Lemaire, about the non-delivery of some canon he had ordered from le Creusot, which should, by contract, have arrived at the same time as others ordered from Krupp's, which were at that time ready for shipment.[28] In May Thévenet returned to Paris, leaving his deputy, Réné Griffon, to continue work at Port Arthur as best he could. This he did, but only by reducing the work force drastically, hence slowing construction, and by using the funds which had been set aside as a guarantee of ten per cent of the value of the contract for maintenance of the facility for one year after commissioning and repair of any major breakdowns arising out of faults in construction for ten years.[29] The spending of this guarantee was later to lead to considerable resentment against French industry on Li's part. Eventually the syndicate itself went into dissolution in December 1889. In less than four years it had spent about two million francs in general costs in Tianjin alone, a vast sum out of all proportion with the business actually conducted. Georges de Bezaure, for many years French Consul, first in Tianjin then in Shanghai, later suggested that this ostentatious expenditure had been made to impress the mandarins with the syndicate's wealth and power, unaware that the Chinese were themselves quite frugal in the administration of businesses, and that successful European firms operating in China such as Jardine Matheson and Company or Butterfield and Swire followed the Chinese example in this.[30]

The resentment against the French which the Comptoir and the syndicate had caused was further exacerbated when the director of armaments at the Port Arthur arsenal was accused by the Imperial Censors in Beijing of purchasing Krupp canon after accepting bribes from the German company. The accused official was Li's nephew, and Li himself was incensed to discover that it had been an agent of le Creusot who had made the denunciation to the censors.[31]

In an attempt to re-establish an effective industrial presence in Zhili after these setbacks, one of the companies of the old syndicate, Fives-Lille, together with le Creusot, formed the Association industrielle française at the beginning of 1890. Thévenet was placed at the head of this new organization, but was to remain in Paris and representation in Tianjin was shared by Griffon who, as a civil engineer, would manage the interests of Fives-Lille such as bridges, railways, and ports, and a naval engineer, Taton, who would represent the naval construction and artillery interests of le Creusot.[32] This structure looked logical but was in fact based on exigencies other than those of sound management. Griffon was appointed because he was already on the site and, if the judgement of one French Consul was right, because Thévenet hoped that he would thereby take the blame for the latter's mistakes.[33]

The circumstances behind the appointment of Taton were more complicated. Following an inspection of the work at Port Arthur in 1888, Li had approached the legation about the possibility of a first-class marine engineer and three petty officers being seconded to the project from the French navy.[34] The Minister, Lemaire, conveyed Li's request to Paris and enthusiastically supported it, largely on the grounds that it would maintain France's industrial advantage at Port Arthur, of which he wrote: 'There is a mine there which we are already exploiting to the exclusion of all com-

petitors, and its riches will not be exhausted for a long time.'
Lemaire further suggested that the engineer's salary be supplemented by the syndicate to ensure that Li's parsimony would not
deter a good potential applicant.[35] Thévenet himself prepared
a paper supporting such an appointment and described it as 'a
precious opportunity to repeat in northern China, the great work
of Griquel at Fuzhou, to the great honour of our country [and] to
the great profit of our national industry.'[36] The precious opportunity, however, was badly bungled by both the French bureaucracy and the new Association industrielle. It was not until May
1890 that the Ministry of the Marine decided on the secondment of
Taton, who was a second-class marine engineer, to China for two
and a half years.[37] In the meantime Thévenet informed Li that the
engineer to be appointed would be paid half by the Zhili government and half by the Association industrielle.[38] This proposition,
based on a suggestion by Lemaire, who should have known better,
was unacceptable to Li, who wanted to ensure the loyalty of his
European employees by being their sole benefactor. Already unhappy with Thévenet for other reasons, Li never answered his
letter, and when Ristelhueber introduced Taton to the Viceroy as
the engineer especially chosen by the French Ministry of the
Marine to serve him, Li replied that his requirements had not been
met, and that, in any case, as it had taken so long for the engineer
to have arrived he had changed his intentions and decided to place
Port Arthur under the management of Chinese engineers, trained,
ironically, in Fuzhou.[39] Even Ristelhueber was shocked by the
inability of the metropolitan bureaucracy and industry to implement a plan so favourable to French interests as that which Li had
proposed.[40] However, although the delays were largely the fault of
the bureaucracy, it was Lemaire and Thévenet, both men of
experience in China, who had introduced the unacceptable concept of divided responsibility into the appointment.

Meanwhile Griffon's work at Port Arthur was gradually being
completed and was due to be handed over to the Chinese in
November 1890, a year later than stipulated in the contract of
1886. At Li's request, Ristelhueber arranged for the Commander
of the French fleet in East Asia, Admiral Besnard, to inspect the
works. Captain de Cornulier-Lucinière, an officer on Besnard's
staff appointed to report on the installation, found that it was on
the whole well constructed and ready to begin operations. Some
works had not yet been completed, and, more seriously, there was

a leak in one of the caissons and a floor flooded during high tides. Additionally a 45-tonne crane had not yet arrived due to an accident in France. He concluded that these delays and faults were almost all due to instances of *force majeure* and that the Chinese would have no cause to regret having confined its construction to French industry.[41] Whatever the quality of Griffon's work, there remained the problem that he had only completed it by spending, with Li's permission, the guarantee of some 32,000 taels which the syndicate had deposited in the former Tianjin branch of the Comtoir d'escompte. The problem of the replacement of these funds was to lead to much bitterness both between the former colleagues, Griffon and Thévenet, and towards French industry generally on Li's part. Required to report to the throne about the Port Arthur project, Li Hongzhang asked Ristelhueber, then *Chargé d'affaires* in Beijing, who was to provide the guarantee now that the Comptoir was in liquidation.[42] In Paris, where the syndicate as well as the Comptoir was in liquidation, the latter argued that it had merely held the sum on behalf of the syndicate, whereas the former claimed that it was the Comptoir's responsibility.[43] The only certainty was that Griffon had spent the money and, after considerable acrimonious correspondence with his employers, he resigned as a representative of the Association industrielle française as from 21 June 1891.[44] The matter was never resolved to Li Hongzhang's satisfaction. As late as June 1893 the *Zongli Yamen* unsuccessfully approached the French legation in Beijing with a claim of some 19,000 taels for repairs to a bad crack in the dry dock which was allegedly the result of faulty workmanship.[45]

However, the problem of the Port Arthur guarantee was only one of a number of causes for Li's dissatisfaction with the work of the French industrialists whom he had once favoured. Following a tour of inspection he made of the European-style defensive installations he had been responsible for constructing around the Bohai Sea, Li Hongzhang expressed satisfaction with the work of the French at Port Arthur, but was far more impressed with that of the German officer, von Hanneken at Weihaiwei and Dalian.[46] Apart from the question of the guarantee, Li was disillusioned with the personnel the Association industrielle française had provided for the operation of the workshops at Port Arthur. A French foreman named François had arrived on the site in April 1891 and had been dismissed the following month by the arsenal's director, Gong Yuheng, on the grounds of his incompetence. Following François' departure, there were no longer any French citizens working at Port Arthur, and de Bezaure accurately observed that as a result immediate orders worth some 120,000 taels which should have gone to French industry would be placed elsewhere.[47]

A further source of Chinese dissatisfaction was the inadequacy of two dredges which Sheng Xuanhuai had ordered for the Shandong government in 1887. Sheng (1844–1916) was at that time *daotai*, the official responsible for trade and relations with foreigners in general, at Yantai (then Chefoo), as well as manager of the China Merchants Steam Navigation Company, a protégé of Li Hongzhang, and probably the greatest industrialist of the late Qing period. Thévenet promised in the contract that the manufacturing company, Fives-Lille, would deliver them in 14 months at a cost of 167,500 taels. It was not until June 1890, 31 months later, that the dredges arrived at the mouth of the Huang He, along which it was intended that they would operate. At the mouth, however, they stayed, unable to cross the bar and quite useless for the purpose for which they were constructed. Fives-Lille had constructed suction dredges which could not move the heavy silt of Chinese rivers. A Chinese report succinctly observed that they pumped 'too much water but not enough sand'.[48] There was no remedy for this failing, and the fault lay with Thévenet, who had not considered the possibility that an alteration of conditions at the mouth of the river could make the operation of the dredges he had specified impossible, and who had permitted the contract to contain a clause stating that payment would be due when the dredges were delivered to the site where they were to work. The failure of the contract to contain any clause protecting the company from *force majeure* made the task of Bernard Lavergne, the syndicate's liquidators' agent in China, a difficult one. In attempting to extract the 1,050,000 francs owing for the dredges from the Chinese, Lavergne had the strong support of the Minister, Lemaire, but was unable to obtain any payment whatever.[49]

Lavergne's and Lemaire's task was made more difficult by the behaviour of Griffon, who initially refused to involve himself in the affair and after his departure from the Association industrielle française encouraged the Chinese in their resolve not to accept delivery of the dredges.[50] The deteriorating relations between Griffon and his former employers had further disastrous effects on French political and industrial prestige in 1891 and 1892. Since 1886 a retired artillery officer, Croizade, had represented a group of French industrialists at Tianjin. He had had little success, however, largely due to the official patronage which was accorded at that time to the Thévenet syndicate. In June 1890 the group expanded to include some companies which had belonged to the syndicate then in liquidation and became known as the Groupe française en Chine. As such it continued ineffectively until the unemployed Griffon joined forces with Croizade in October 1891. Very quickly it received orders for bridge girders worth nearly a million francs on the Shanhaiguan railway.

The terms of the contracts were very favourable to the Groupe, which did not have to submit tenders but merely to allocate construction to the member company best equipped to undertake it. That company would then charge what it considered to be a fair price. This uniquely good arrangement was concluded for political reasons: the Groupe expressed its gratitude to the Consul, the Vicomte de Bezaure, who was alleged to have considerable influence over Li Hongzhang, while de Bezaure reported that it had been on Li's specific instructions that Griffon and Croizade had received the orders.[51] In fact it was Li Hongzhang who was most cunningly playing the game of politically motivated industrial patronage in Tianjin. For while, as Lemaire put it, the French were thrown a few bones, the vast majority of the orders for the railway went to Britain, as its engineer-in-chief, Kinder, wished. Moreover, such orders as were given to Griffon merely served to increase the jealousy and resentment of the Association industrielle's representatives, Taton and Lavergne.[52]

This resentment was exacerbated by the civil suit Griffon brought against his erstwhile employers in an attempt to extract from them 35,097 francs which he claimed was owing to him as unpaid salary. The case was heard in Tianjin by the Consul, de Bezaure, under the rules of extraterritoriality which Europeans then enjoyed in China, and in February 1892 he ruled in favour of Griffon's claim. Lavergne refused to pay, so de Bezaure ordered the seizure and sale of the Association industrielle's premises in Tianjin and the handing over of the proceeds, some 23,000 francs to Griffon.[53] The bitter debate between the rivals partly conducted in the local anglophone press was only terminated by the decision of the Association industrielle to withdraw from China altogether, and late in June 1892 Lavergne and Taton received telegraphic orders to dispose of the company's assets and return to Paris as soon as the question of the dredges was settled.[54] They had not succeeded in acquiring a single order for the companies they represented.

Le Creusot and Fives-Lille did not, however, withdraw from Tianjin before pressing their claims at the highest level. The metallurgist Henri Schneider and Thévenet put pressure on Ribot to recall de Bezaure following his judicial decision which was so unfavourable to their interests.[55] Additionally Lemaire, who along with Ristelhueber had been responsible for the execution of the policy of complete support of the Thévenet syndicate to the exclusion of all other French interests in the late 1880s, suspected that de Bezaure was prejudiced in favour of Griffon. Ribot therefore ordered de Bezaure's recall to Paris in the hope that his removal would lessen passions in Tianjin. He emphasized, however, that the recall was not a means of censure against de Bezaure and instructed his successor, Jules Raffray, to be rigorously neutral.[56] This solution, however, failed to take any account of the attitude of the Chinese to such a change. For, although de Bezaure may not have had any influence on Li Hongzhang, relations between the two men at least were cordial. Moreover, Li was favourably disposed towards Griffon because of the latter's willingness to stay in China and take personal responsibility for his work at Port Arthur, conduct which contrasted favourably with Thévenet's abrupt departure.[57] Angered by rumours that de Bezaure's removal had been engineered by Thévenet in Paris,[58] Li Hongzhang took the unprecedented step of wiring the Chinese *Chargé d'affaires* there, Jing Zhang, to request him to protest to Foreign Minister Alexandre Ribot about the change in personnel. Additionally he persuaded his colleagues of the *Zongli Yamen* to approach Lemaire about the matter.[59]

Confronted with the difficulties at Tianjin, the problems with Fives-Lille's dredges on the Huang He, and the continuing inability of French industry to gain any benefit from Article 7 of the 1885 treaty, Lemaire decided to accompany Raffray to his new post at Tianjin and explain the French official position on these issues to Li Hongzhang himself. This *démarche* was ill conceived and badly executed by Lemaire, who was later described by Ribot as 'a very mediocre agent'.[60] Li initially refused to see Raffray, although he subsequently relented to the extent of allowing the introduction to be made and then informing the new Consul that all affairs were to be discussed with the *daotai*, the ubiquitous Sheng Xuanhuai, and that he would grant no further audiences.[61] The antagonism between Li and Lemaire was such that the latter refused to fly the consulate's flag at half mast when Li's wife died, while Li loudly proclaimed him to be an imbecile during an interview on 8 May 1892. Lemaire should have known better: he had joined the department as a student interpreter in Shanghai on his sixteenth birthday in 1855, and had spent practically all his career in China.[62]

As for the other issues with which Lemaire was concerned, he felt unable to make serious claims for the implementation of Article 7 as the only French engineer involved in railway equipment, Griffon, now preferred to work without diplomatic support.[63] Negotiations with Sheng Xuanhuai about the two dredges which occupied most of Lemaire's unhappy fortnight at Tianjin were no more conclusive. The Chinese refused to pay for them on the grounds that Fives-Lille had modified the specifications in the contract in a way which, while it increased their power, also increased their draft. The French insisted that delivery had only been prevented by the change in conditions at the mouth of the river, a case of *force majeure*. Sheng offered to buy the dredges for 80,000 taels and use their engines in new ships for his China Merchants Steam Navigation Company. Lavergne refused this as the contract price was 123,000 taels and Fives-Lille had additionally been spending 2,000 taels each month for maintenance of the dredges for the previous 26 months.[64] Lemaire's visit to Tianjin thus achieved nothing and brought French prestige to its nadir. Two articles in the generally francophobe *Hong Kong Daily Press* fully exploited Lemaire's failure and sought to characterize the visit as a disaster for French aspirations.[65]

The campaign to force the Chinese to pay Fives-Lille for its inefficient dredges was to continue to have a debilitating effect on French diplomacy in China for more than a year after the confrontation at Tianjin. Negotiations on the French side were placed, nominally at least, in Raffray's hands, although it was on Lemaire's orders that he issued an ultimatum to Sheng that unless they were accepted and the contract price paid by 30 December 1892 they would be abandoned and all maintenance would cease. Sheng, who had already increased his offer to 83,000 taels, remained unmoved, and the appearance of this threat was carried out, although Lavergne retained one employee to keep deterioration to a minimum. In April Lemaire approached the *Zongli Yamen*, which acknowledged the inadequacy of Sheng's offer. Eventually, after considerable haggling, Lavergne reluctantly accepted Sheng's proposal of 100,000 Shanghai taels or 95,000 Tianjin taels in June 1893.[66]

In view of the amount of ill will they created, these efforts, ultimately resulting in the gain of only 15,000 taels, seem to have been badly misplaced energy. Certainly the behaviour of Lavergne and Lemaire compared very poorly with that of a German company which had constructed a dredge in Shanghai also for use on the Huang He. This dredge proved to be no more effective than the French ones, and the Chinese refused to accept it. The Germans opened negotiations, and quickly agreed to reduce the price by almost half without invoking the assistance of their Consul or Minister. As Sheng told Lavergne, this was a far better way of conducting business in China,[67] and certainly no orders were addressed to French industry by Chinese officials from the time of Lemaire's visit to Tianjin until after his departure from Beijing. Indeed, a deputation of members of the Groupe française en Chine received in Paris by Foreign Minister Jules Develle in May 1893 pointed out this lack of orders and argued that this damaging situation for French industry had been created by de Bezaure's recall from Tianjin and Lemaire's insistence on full payment for the dredges.[68] Paul Boell characteristically indulged in colourful hyperbole in what was nevertheless a substantially correct, albeit exaggerated, description of French diplomacy in China in 1892 and early 1893:

Would you believe that *all* our affairs in China are presently subordinated to a miserable warranty settlement on two dredges delivered by Fives-Lille, which they want to force the Chinese to accept contrary to the contract terms. It is this *questionable* affair, to say the least, which has occupied alone, for nearly two years, all M. Lemaire's attention ... Our influence in China may perish, M. Lemaire seems to say, but Fives-Lille's shareholders must get rid of their *lemons!*[69]

The settlement of Fives-Lille's claim and the departure of Lavergne in June 1893 brought to an end the activities of the Thévenet syndicate and its immediate descendant, the Association industrielle. Originally established to exploit Article 7 of the 1885 treaty, they had had curiously little involvement in railway building, despite repeated attempts by French diplomats to cajole the

Chinese into acknowledging that the article represented a legal obligation towards France. For although Li Hongzhang was willing to place orders with French industry when he could ensure that the prices and conditions were advantageous to him, neither he nor any other Chinese would admit to the principle of a privilege. Thus when Lemaire approached Prince Qing, who was then the head of the *Zongli Yamen*, to remind him of the terms of Article 7 in May 1890, the latter amiably stated that he would of course consult French industrialists and use their equipment provided their prices were the lowest. Lemaire refused to accept this argument, knowing that on this basis the supplies for the two lines then under consideration (Hankou to Beijing and Tianjin to the Russian frontier) would come from countries other than France. He therefore forced Qing to read the original text of the treaty and reminded him of the history of Article 7 and France's generosity in renouncing an indemnity in favour of such a provision. Qing indicated that France could have every confidence in China's word, but must for her part make honourable and advantageous conditions for China. Additionally, France could not hope for a monopoly of railway construction. Lemaire replied that no such claim was being made, but that 'we would count only on having the first and the largest part'. In response to this modest claim, Prince Qing moved on to a discussion of the costs and difficulties of the project and concluded by stating that it had not yet been decided when work would begin.[70]

Such tactics on the part of French diplomats were far from efficacious and did little more than remind the Chinese of the fact that France had been able to impose a victor's peace on them in 1885. Neither of the lines which had prompted Lemaire's intervention in 1890 was proceeded with at that time, but an Imperial Decree of 22 April 1891 authorized the extension from Linxi to Shanhaiguan and beyond of the modest rail network Li Hongzhang had had constructed to serve the Kaiping coal mines. De Bezaure quickly saw Li to insist that at least part of the line's construction be reserved for France. The Viceroy told de Bezaure the unlikely story that Giers, the Russian Foreign Minister, had advised the Chinese Minister in St Petersburg not to allow foreigners, and the French in particular, to construct railways following difficulties the Russians had themselves experienced in purchasing railways the French had constructed on their territory. Li added that a Chinese company would build the new line and concluded by reminding de Bezaure of his own difficulties with French industrialists.[71]

This hostility towards French industry, arising both from a resentment of Article 7 and from dissatisfaction with the performance of the Thévenet syndicate, was evident in the specifications of the tenders for equipment for the line. Tenders for its one large bridge were to be lodged in London with Sir Benjamin Baker, the engineer of the Forth bridge; the contract signed with Whittall and Company, the London agents of Jardine Matheson and Company; and it was indicated that preference would be given to a design manufactured in the USA and Britain but not in France. Similarly the profile of the new line's rails was specified as Sandberg, a type unknown in France.[72] Confronted by de Bezaure with the accusation that these specifications were designed to exclude French industrialists, Li Hongzhang stated that Kinder had been responsible for them, and that Kinder's success in building the line from Kaiping to Linxi inspired confidence. Li then compared Kinder's reliability with that of Thévenet, who, Li said, had left his post when difficulties arose, and concluded by claiming that he was free to adjudicate the tenders as he wished. In these circumstances de Bezaure glumly concluded that the privilege conferred by Article 7 was of practically no value:

It is now certain that the Chinese government seeks, by all means, to elude the obligations of the treaty of 1885. From its point of view it would be an act of humiliation and weakness to give Frenchmen some of its railway construction. I believe that only an energetic diplomatic action could bring to an end the painful situation in which the Chinese have placed our industrialists who have come here to seek, trusting the treaties, orders for their special expertise.[73]

A similar request was made in Paris by the directors of the Etablissements Eiffel, a member of the Groupe which employed Croizade and Griffon in Tianjin.[74] Ristelhueber, instructed in accordance with Eiffel's wishes, reminded the members of the *Zongli Yamen* of the terms of Article 7. Prince Qing, however, informed the *Chargé d'affaires* that the article did not constitute a privilege for France, but merely signified that China would use French products provided their quality was equal to and price cheaper than those of other manufacturers. Ristelhueber protested that such an interpretation was useless, but was unable to extract any concessions from the *Zongli Yamen*.[75]

Li Hongzhang's interpretation of Article 7 was no more acceptable to Ristelhueber than that given in Beijing. Li stated that if
China wanted the co-operation of French industry, her officials
would approach the representatives of French companies, and were
under no obligation to operate through the intermediary of the
French government. Moreover the French could not object if the
Chinese chose to appeal to other countries for assistance. More
specifically, Li concluded that as the line under consideration was
an extension of the existing Kaiping to Linxi line, it was only
logical to maintain continuity of the English technqiues and material which had been used up to that point.[76]

In practice, however, Li Hongzhang still sought if not to please
at least to mollify all, and very soon after writing to Ristelhueber
in these discouraging terms, placed the order discussed above for
bridge girders with Griffon and Croizade. Such a concession,
granted by Li very much with political considerations in mind, fell
far short of what the Quai d'Orsay defined at this time as the import of Article 7:

China has undertaken a formal engagement with us: if this engagement
does not constitute an exclusive privilege in our favour, that is only in the
sense that if the Imperial government is dissatisfied with our nationals'
offers or supplies, it may turn to their competitors. But it must start by
consulting French industrialists.[77]

Even this derisorily trivial success for the principle of privilege
led to complaints to the Foreign Office from the British manufacturers whose orders were cancelled as a result. The British
Minister in Beijing, Sir John Walsham, was instructed to discover
the exact nature of the French monopolistic claims. Lemaire sent
both him and the *Zongli Yamen* a copy of the definition above.
Neither the British nor the Chinese were willing to accept such an
interpretation, and the latter denied that they were in any way
obliged to buy French material. Lemaire was forced to admit to
Ribot that nothing could be done to ensure the exercise of France's
putative rights.[78]

The inability of Article 7 to secure for French industry the expected advantages was emphasized by Griffon's continued, indeed
increased, success following his rupture with French diplomats in
May 1892. Raffray recognized that British success in equipping Li's
railway owed much to the influence of Kinder, and hoped that
Griffon could achieve something similar for France:

M. Griffon, who is *persona grata* with the Viceroy, seems to prefer to act alone and without the immediate support of the legation and consulate, and perhaps he is right: the official execution of Article 7 of the 1885 treaty oppresses the Chinese like a war indemnity and offends their self-respect. It is possible that M. Griffon will ultimately obtain its real execution unofficially, more easily than if he had claimed it officially.[79]

Certainly for a short time Griffon appeared to maintain some kind of effective French industrial presence in Tianjin, and was even appointed a consulting engineer for the Shanhaiguan railway in July 1892.[80]

However, even this very tenuous success could not be exploited due to the poor quality of the girders ordered from Griffon in 1891. The order was placed with the Etablissements Eiffel and the shipment arrived in Tianjin late in 1892. The girders, however, were found to be chipped as the steel was of an inferior grade, apparently of Belgian origin. Li did not wish to be too strict in his judgement on equipment ordered through Griffon, as its purchase had been decided upon largely for political reasons.[81] However, a subsequent shipment, which arrived in September 1893, was of such bad quality that Kinder refused to use it and threatened to resign if required to do so. Li deferred to his engineer-in-chief's wish and replacement girders were ordered from England. Raffray described what had happened as 'a disastrous affair for French industry'.[82]

If the year 1894 brought considerable changes in Chinese attitudes to Western technology and in China's political and military prestige, it also saw a transformation of French official policy at least in the area. A new Foreign Minister, Gabriel Hanotaux, and a new Minister in Beijing, Auguste Gérard, both men committed to an imperialistic policy, determined to take advantage of China's weakness and sought to extend the field of French investment, notably in southern China, where Ferry had anticipated securing so many highly profitable commercial and industrial outlets ten years before. However Gérard, who was nothing if not inventive in his often successful attempts to obtain new concessions for potential French investors, shared with his predecessors a belief in the importance and opportunities of Article 7. Additionally, the continued expansion of the railway network instigated by Li Hongzhang demanded further equipment from Europe and the presence of Griffon in Tianjin until his death in October 1896 ensured an element of continuity with the policies of the period from

1885 to early 1894. Thus when news arrived in Paris of the Imperial Decree authorizing construction of a railway from Tianjin to Lugouqiao (near Beijing), Marcelin Berthelot, Foreign Minister for five months in late 1895 and early 1896, wired Gérard in fairly pressing terms asking him to ensure that the equipment required for the new line was ordered from France.[83] Gérard's action was consistent with that of his predecessors: he wrote to the *Zongli Yamen* reminding the ministers of the terms of Article 7 of the 1885 treaty, and asked Griffon to come to Beijing to prepare the French case. As the new line was only an extension of the system already existing in Zhili, Griffon considered that it would be best to solicit the undertaking as a whole, including both construction and supply of equipment. To complement this approach Griffon prepared plans for a different route between the two cities from that proposed by Kinder. Gérard sent Griffon's proposal to the *Zongli Yamen*, with the request that it be forwarded to Hu Yufen, the former Guangxi provincial judge whom the Imperial Decree had appointed Director-General of the new line. Griffon himself sent a copy to Li Hongzhang, and Jean du Chalyard, then French Consul in Tianjin, was instructed to support Griffon's proposal.[84]

Gérard himself was sceptical as to the prospects for success of all this activity,[85] and indeed the day after he had sent Griffon's plans to the ministers in Beijing, tenders were called in Tianjin for equipment for the line. Despite a promise to the contrary which du Chalyard had extracted from Hu Yufen, the specifications were seen by the French, not unreasonably, as favouring British industry. The insistence on Sandberg profile rails; the provision that tenders for the bridges required be lodged with Sir Benjamin Baker in London; and the early closing date for tenders (March 1896) were all considered to be discriminatory against French industrialists.[86] The calling of tenders signalled the beginning of a French diplomatic campaign, once again conducted in both Tianjin and Beijing, aimed specifically at having the specifications altered and more generally at coercing the Chinese into acknowledging the force of Article 7. Gérard began by complaining to the *Zongli Yamen*, but the ministers were both unable and unwilling to discuss the relative merits of Sandberg profile and flat-bottomed rails or any such issue and informed him that the railway's construction was the responsibility of Hu Yufen and the Military Affairs Bureau. The campaign only had very modest results, for the Chinese found the threats to their sovereignty which the French demands implied more offensive than the more subtle and effective economic penetration brought by British engineers. Gérard commented emotively, but nevertheless with some accuracy:

In these industrial matters of which it is ignorant, the Chinese government is at the mercy of its subaltern staff, all more or less associated with foreign industrialists or financiers. Against all these obscure influences, neither reason, right nor treaties has any hold. If in spite of all this a part of the job is given us, it will be because Hu Yufen and his advisers, confronted with our efforts in Tianjin as well as in Beijing, will feel it prudent to grant us a little, to throw some ballast overboard.[87]

Gérard's activity led to some minor orders being placed with Griffon,[88] a large order for sleepers of Russian and Japanese origin going to Philippot, a Tianijin-based French banker, and the postponement of the closing date for tenders by nearly two months. However, further setbacks continued: the line's locomotive superintendent, G. D. Churchward, specified that the new motive power required should conform to British designs with a few concessions to American practice.[89] Additionally Hu Yufen took the opportunity to claim that Article 7 applied only to railways to be constructed in the provinces adjoining Tonkin. This in turn resulted in Gérard making further unproductive protests to the *Zongli Yamen*, which consistently maintained that the line's construction was not its responsibility, but that of the Directors of Military Affairs and the Imperial Railways. This was an interpretation which Gérard refused to accept, pointing out that it was the faithful execution of a provision contained in a treaty which was at issue, something that was very much the responsibility of the *Zongli Yamen* and not of the officials constructing the line.[90]

Until 1896, French industry had not submitted any tenders in response to the open invitations of the Chinese Imperial Railways, but preferred to rely on the principle of privilege. However, as Gérard had managed to persuade the Chinese to postpone the closing date for the submission of tenders for, amongst other items, 9,700 tons of rails, for the first time the French, in the form of Griffon and Paul Dujardin-Beaumetz, a representative of the Comité des Forges then in China, agreed to participate. Together they submitted a tender price of 37.75 taels per ton. However, four companies, German and English, made lower offers and the contract was awarded to Arnhold Karberg, whose price of 35 taels was the lowest. Immediately it was informed of its failure, the Comité des Forges wired Dujardin-Beaumetz and instructed him to protest on the grounds that the delivery time (by the end of December the same year) was impossibly short and that the type of rail required was purely English. However, neither he nor Antoine Grille, an engineer sent to China by Fives-Lille to work on another project, nor Gérard agreed with the terms of this protest. The two engineers observed that the modification of French rolling mills required to manufacture the rails was a simple and inexpensive operation and Gérard, by his refusal to forward the Comité des Forges' protest to the Chinese, accepted the engineers' criticism of their employers' lack of initiative.[91]

While welcoming the fact that French industry was willing to compete in submitting tenders, Gérard pointed out the inadequacy of its performance in China and suggested that it adopt a similar *modus operandi* to those of successful British, German, and American companies. To extract any benefit from Article 7, Gérard considered that a three-point programme should be instituted:

1. that our industry be ready to submit tenders, making the relatively cheap modifications to its machinery which acceptance of the specifications determined by the Chinese railway management implies.
2. that our industry has, like English, German, and American industry, permanently settled and accredited representatives in Shanghai, Tianjin and Hankou.
3. that the representative of our industry be properly accredited with the Russo-Chinese Bank, which has been founded specifically for the development of French as much as Russian interests in the Far East.[92]

In response to Gérard's comments, Hanotaux wrote to Baron Reille, the chairman of the Comité des Forges, indicating his agreement with Gérard's approval of the Comité's decision to submit a tender, his criticism of the subsequent protest and his plan of action for the future, omitting however the suggestion of accreditation to the Russo-Chinese Bank.[93]

Although Gérard encouraged a far more innovative industrial policy in 1896 than had Lemaire, he nevertheless retained a belief in the efficacy of Article 7 despite the massive evidence, including his own experience, to the contrary. Thus when in November 1896 the *Beijing Gazette* published a report to the throne from Li's successor in Zhili, Wang Wenshao, recommending the creation of a school to teach railway technology to Chinese students at Shanhaiguan, Gérard again claimed privileges for France on the basis of Article 7. He informed the *Zongli Yamen* that at least one of the three European teachers at the school should be a French citizen who would teach in the French language. The Chinese ministers evasively replied that the organization of the school was Wang's responsibility. Resourceful as Gérard was in obtaining outlets for French industry, he still considered Article 7 as important a weapon as had Lemaire, an opinion which complemented his belief in pursuing a 'strong' policy in Beijing. He thus informed Hanotaux that:

> the Chinese government is bound to elude and reduce as much as possible the scope of this article. ... I do not believe it any less necessary to insist at each opportunity, on the value of this same Article 7 and on the advantages it confers on us.

Despite this assertion, the policy was in fact slowly being abandoned, largely under the impact of Gérard's own successes achieved by pursuing other means. If it was not until Stephen Pichon's arrival in Beijing as the new French Minister in 1898 that the failure of Article 7 was officially acknowledged, it was nevertheless under Gérard that it ceased to be regarded as the foundation of French industrial policy in China. Griffon's death in Yantai in October 1896 removed an element of continuity with the policy of the 1880s. The eulogy Gérard wrote is notable more for the list of impressive plans the engineer had prepared than for his achievements, of which the Port Arthur naval base was the most substantial.[95] Both Gérard and the Consul at Tianjin, du Chalyard, expressed regret that, following Griffon's death, French industry would have no permanent representative in north China apart from Philippot, a banker employed by the Comptoir national d'escompte, who was hardly competent in technical matters.

After ten years of reliance on Article 7, France's industrial and commercial situation in China had declined *vis-à-vis* that of the other Western powers. The small French colony in Tianjin had remained stationary, while those of other Europeans were growing; there were few French companies trading in the area and no French banks; prices for French products were higher than those of competing nations, even if their quality was superior; and, above all, the memory of the failure of the Thévenet syndicate continued to work against the French and was exploited by competitors. Moreover the Chinese seemed to du Chalyard to feel more bitterness against the French for their defeat in 1884–5 than gratitude for the triple intervention in which France participated to relieve China from the Japanese demands of 1895.[96] Although he did not admit as much, it is probable that the emphasis French diplomacy had placed on Article 7 of the treaty which concluded the war of 1884–5 was largely responsible for this continued Chinese bitterness. Du Chalyard did, however, proceed to criticize indirectly the usefulness in China of a privilege such as that contained in Article 7. Apart from during the last couple of years of Lemaire's ministry in Beijing, Griffon worked closely with French diplomats, and following the example of his erstwhile employer, Thévenet, sought to obtain orders through political influence. This approach was very much in the administrative traditions of the Corps des Ponts et Chaussées of which he was a member. It was, however, singularly inappropriate for Chinese conditions. Du Chalyard pointed out that a Chinese official such as a *daotai* or viceroy would naturally dislike discussing business matters with the representatives of a foreign power which could place diplomatic or even military pressure on him. Sheng Xuanhuai had recently told him that it was for this reason that he preferred to do business with Swiss, Danish, or Belgian businessmen. Generally mandarins preferred to make arrangements with European businessmen who came to them without official support to request concessions in a deferential way rather than to demand them as a legal right. Furthermore, it was easier for the Chinese official to negotiate a bribe with a private businessman than with a foreign diplomat. Du Chalyard concluded his indictment of Griffon's methods and, by implication, of the basis of French industrial diplomacy in China for the preceding ten years, by recommending a much more restrained policy, ironically at a time when the powers which had hitherto followed such policies were themselves about to adopt much more aggressive stances. He described the correct attitude of an industrial representative:

In a word, he must do his business himself, and save the intervention of representatives of his country for cases where he needs support for just claims or to ensure respect for his unrecognized rights. It is in this way that, before our very eyes, English and German Ministers and Consuls have proceeded; and their nationals have had no cause as yet to complain about this attitude, judging by the commercial results they are obtaining.[97]

In view of the price disadvantage which French products suffered in comparison with English, German, and American ones, a problem which du Chalyard himself acknowledged, it is doubtful as to whether such a policy would have achieved much success. His recommendation strangely echoes the comments which the anglophone press, confident that liberal policies of free competition would ensure British dominance, had made several years earlier. Thus the Tianjin *Chinese Times* had welcomed French concern in providing outlets for its industry as a more secure basis for France's diplomatic position in China than the 'politico-ecclesiastic pretensions' of the religious Protectorate. However, it objected to the official patronage which the Thévenet syndicate enjoyed on the classic liberal grounds that 'the function of the Statesman is to remove obstacles to free intercourse. . . . It is the business of the trader himself to push his own wares.'[98] After suggesting that French industry would have been more successful in China without the enormous official support Ristelhueber in particular had given to the Thévenet syndicate, the paper vitriolically condemned the principle of privilege:

The original sin . . . lies in the introduction of the vicious article . . . in the Tianjin Treaty of 1885. Either the clause is the mere outcome of one of those characteristic struggles for a shadow, to which the French and Chinese are both so prone, and means simply nothing (as the Chinese consider), or else it means that the Chinese government binds itself to buy dear and bad articles from France and refuse cheap and good ones from elsewhere — there is no middle way: and the nation which tries to build up its commerce on such a rotten foundation as that need not be surprised if its success is as the crackling of thorns under a pot.[99]

In the late nineteenth century British businessmen could still afford self-righteously to espouse liberal trading policies and at the same time be assured of profitable sales and investments in countries open to imperialist exploitation. French industry, which was protected domestically after 1892, could have no such confidence. However, given the wide currency of the neo-mercantilist beliefs which had led Ferry and his contemporaries into inserting a clause like Article 7 into the settlement of the war of conquest of Vietnam, French industrialists and statesmen felt impelled to find export markets wherever possible.[100] In the Darwinist language fashionable at the time, another Anglo-Chinese paper observed that: 'In the modern struggle for existence the French have become so imbued with the commercial spirit that their national policy has been to a considerable extent subordinated to the promotion of trade.' The *Hong Kong Daily Press* went on to criticize the semi-official nature of the Thévenet syndicate, its ostentatious extravagance and its subsidized unprofitable operations. It further pointed out that the credibility of French diplomacy in China had been staked on the success of the syndicate, a situation which led to frequent invocations of Article 7 and threats of difficulties with missionary claims on the part of French officials.[101]

If these British criticisms of Article 7 of the 1885 treaty were based essentially on self-interest, those of du Chalyard were not, and there is little doubt that it was at best an ineffectual measure and probably a counter-productive one. Certainly the only major concession or order to emerge from it was that of the Port Arthur naval base and it was completed only at a considerable loss for the syndicate concerned. The French successes in obtaining the first concession for a foreign-owned railway in China (Dong Dang to Longzhou, in Guangxi), and then a majority share in the first main line from Beijing to Hankou, both had little to do with the claims of Article 7 but much with the changed circumstances after the Treaty of Shimonoseki and the triple intervention. The attempt to obtain such an extensive privilege by a Western power was a unique one in China. Perhaps the only comparable examples of domination of sections of the Chinese modern infrastructure by one foreign power were the customs service and the telegraph system. The former, although international in the composition of its personnel, was effectively under British control, while the latter was Danish-owned and operated at the request of the Chinese government. In neither case was a monopoly or privilege imposed on the Chinese by treaty, yet in both cases a far more real privilege existed than the French were able to obtain by virtue of Article 7. The offensive and humiliating nature of the Article in Chinese eyes combined with the price disadvantage of French goods suggested that the expectations French industry held for its future in China after 1885 would probably never be realized. The mistakes of Thévenet, of the members of the various syndicates in France, and of French diplomats in north China, however, all ensured that this bizarre attempt at latter day mercantilism would fail.

3

The Struggle for the Beijing–Hankou Railway Concession, 1896–1898

ALTHOUGH serious Chinese interest in railway construction had emerged at about the same time as France obtained the very vague railway privileges in the treaty of 1885, there was little connection between the two. Indeed, those Chinese who hoped to construct railways did so not out of any desire to comply with Article 7 of the Sino-French Treaty, or even from a desire to appear to comply with it, but rather to maintain and extend their own regional power bases. The two Chinese officials most inclined to use the railway to this end were Li Hongzhang, then Governor-General of Zhili, and Zhang Zhidong, Governor-General at Guangzhou. Zhang's energetic prosecution of the war over Vietnam during 1884 and 1885 indicates that he was far from being a friend of France. Ironically, however, the rivalry of these two officials was to lead to extensive French involvement in what was to be the largest Chinese railway project of the Qing dynasty, the Beijing–Hankou railway.

Apart from a temporary and illegal British line from Shanghai to Wusong, the first railway in China was built under the sponsorship of Li Hongzhang at the Kaiping coal mines at Tangshan in Zhili. By 1888 this mineral line had been extended to Tianjin, about a hundred and fifty kilometres away, and was operating as a full-scale railway known as the Chinese Railway Company. The company, although under official supervision, was owned by Chinese capitalists, rather like the China Merchants Steam Navigation Company, and employed a young English engineer, Charles Kinder, to supervise technical matters.[1] Li proposed to extend the line from Tianjin to Tongzhou near Beijing, a route which would obviously be highly profitable. This proposal, however, aroused passionate opposition from many quarters and for diverse reasons. Boatmen on the Beihe feared for their future employment; conservative officials and the censors opposed railway construction because of their dislike of foreign innovations in general; while Li's political enemies opposed the line because of the military and financial benefits he would reap.[2]

It was in this context that Zhang Zhidong wrote a memorial to the throne which used the very arguments of the opponents of Li's proposal to support the construction of railways in the interior. Clearly, such railways would not make Beijing more vulnerable to Western attack, nor put boatmen out of work as the Tianjin–Tongzhou line would. Zhang argued that a great interior railway should be built from Lugouqiao near Beijing to Hankou on the Yangzi. This line, he suggested, would be of great economic and strategic benefit to China. Zhang's memorial was approved and in August 1889 he was appointed Governor-General at Wuchang, Hankou's twin city and the administrative capital of Hunan and Hubei. From here he was to supervise the construction of this, China's first main-line railway. Two million taels were allocated towards its construction, which was estimated to cost 30 million taels in all.[3]

The French Minister in Beijing, Lemaire, was quickly informed of the proposal to construct the railway and was led to believe that the Chinese were interested in borrowing the enormous sum of 200 million francs from French bankers in order to pay for it. Additionally, Lemaire believed that France would be asked to supply the necessary equipment.[4] If true, such reports would mean that Article 7 of the 1885 treaty was going to be honoured in a spectacularly profitable way. Foreign Minister Eugène Spuller contacted a number of French banks, including the Comptoir d'escompte, and informed Lemaire in a telegram of extravagant length that the bankers were prepared to lend the sum required, provided that serious guarantees, including the customs revenue, were offered to secure the loan.[5] The apparent enthusiasm for the project from both the French and Chinese is surprising in view of the bad reputation French industry had come to acquire in China during the late 1880s. Lemaire was not reluctant to use diplomatic pressure to maintain this momentum, and ensured that both Zhang Zhidong and Li Hongzhang were reminded of the terms of Article 7 of the 1885 treaty.[6]

By the end of the year, however, the proposed railway had been shelved. While an anglophone Beijing newspaper blamed its postponement on the irresolution of the Imperial government, it had been military considerations which prompted the Chinese to devote their railway building energies elsewhere.[7] Concern that Japan might wage war to increase her influence in Korea led to the reallocation of the two million taels from the Hankou railway to the extension towards the Great Wall and Korea of the existing Tangshan–Tianjin line.[8] The decision to concentrate railway construction in the north was amply justified, both by the Japanese threat and by the emergence in Zhili of an industrial infrastructure under the patronage of Li Hongzhang. The fundamental ineffectiveness of Li's innovations, however, was demonstrated in the Sino-Japanese War of 1894–5, when Li's railways, coal mines, and Western style army and navy proved unable to resist Japanese attack. Li's failure in that war led to his dismissal from the governor-generalship of Zhili and consequently to the shift of the centre of industrial patronage from Tianjin to Wuchang, where Zhang Zhidong continued to pursue his version of self-strengthening.

Zhang's most important interest was the Hanyang Iron Works, which he had established as a government-owned enterprise in 1890. Four years later an iron ore mine was opened at Daye to supply the foundry.[9] Zhang had intended to build the Beijing–Hankou railway largely with materials from his Hanyang Iron Works, thus providing a market for the latter and ensuring the maximum possible benefit to Chinese industry from the former.[10] The collapse of Li Hongzhang's industrial and military base in the north in 1895 prompted Zhang to revive his proposal for the Beijing–Hankou line. The central government, well aware that one of the reasons for China's defeat had been her inability to transport southern forces to the theatre of war in the north, was sympathetic to Zhang's proposal which he made in a telegram on 29 August 1895. Within two months construction of the railway was authorized a second time by imperial decree. At that time Zhang's chief Western adviser was a German engineer named Hildebrandt who, naturally, wasted no opportunities to remind Zhang of the efficiency of Germany's railways and industry. Zhang therefore suggested that such foreign assistance as would be needed could come from Germany.[11]

A copy of Zhang's telegram was leaked to Gérard who, predictably enough, reminded the *Zongli Yamen* of the terms of Article 7 of the 1885 treaty. Additionally, Gérard urged the few representatives of French industry in China, notably Griffon, to approach Zhang with offers of assistance.[12] The prospects for

French involvement in the project were not good at this time. Gérard believed that neither Griffon's personality nor his industrial backers were equal to the task and lamented: 'It is not enough for the Government of the Republic to practise protection in favour of our industry: our industry must defend itself even more.'[13] Moreover the imperial decree of 1895 which authorized construction of the railway had specified that a Chinese company raise ten million taels to commence construction. The intention was to keep foreign influence on the railway to a minimum and, as far as possible, Chinese capital and equipment were to be used. The rails were to be made at the Hanyang Iron Works and only the rolling stock and steel bridges were to be ordered from foreign companies.[14]

These intentions were not realized, mainly due to the reluctance of Chinese investors to place their capital in the care of officials.[15] Additionally, despite the clarity of Chinese intentions, foreigners in China were constantly putting forward their own proposals to build the railway. Albert W. Bash arrived from New York representing the syndicate which later became the American China Development Company, and was particularly active. At Gérard's insistence, Réné Griffon also attempted to interest Chinese officials in French technical expertise and material.[16] Additionally, another French syndicate, represented at Shanghai by Emile de Marteau, had proposed to Zhang Zhidong that they jointly operate the Hanyang Iron Works. Nothing, however, came of this proposal, and in May 1896 Zhang sold the works to Sheng Xuanhuai.[17]

Sheng was then a director of both the Chinese telegraphs and the China Merchants Steam Navigation Company. He had been closely associated with Li Hongzhang's industrial enterprises in Zhili and had been Li's customs *daotai* in Tianjin. His purchase of the Hanyang Iron Works marked the beginning of his association with Zhang. In return for taking over the unprofitable iron works Zhang arranged for Sheng to be appointed Director-General of the new Chinese Imperial Railway Administration on 20 October 1896. In taking over responsibility for railway construction, Sheng enabled Zhang to avoid the embarrassment of having to appeal for foreign capital as by October it was obvious that Chinese investors were insufficiently interested in the enterprise. Sheng's plan for raising the necessary capital was contained in a memorial dated 12 October. He proposed to establish a Chinese company which would ultimately have a capital of 40 million taels (about 160 million francs). He proposed to borrow seven million from Chinese capitalists and 13 million from the Imperial government. These 20 million would enable the first half of the line to be constructed. This railway would itself be the guarantee for a loan of 20 million taels obtained abroad to complete the line. This foreign loan would be amortized in 25 years, after which time the railway would be entirely in Chinese hands. Sheng intended to obtain both the foreign loan and the material for the line from a power without political ambitions in China. Both the United States and Belgium were industrial nations and Sheng was interested in either of them, especially the latter, as 'Belgium was but an iron and steel manufactory, and acknowledged to be a small country, without any wish for aggrandisement, and ... borrowing money from them would be most advantageous and attended with but little risk'.[18]

Thus, in spite of Article 7 and Chinese determination to build the railway, the prospects of French industry were not good. Additionally Réné Griffon had just died, so the major French engineering companies were unrepresented.[19] This was especially unfortunate, as on 27 October ministers of the *Zongli Yamen* informed Gérard that foreign capital was not barred from the venture and that henceforth it was Sheng's sole responsibility to raise the necessary funds. Thus, despite the preference for small powers, France was not yet excluded. The only good news Gérard could report was that Bash clearly lacked the authority and backing to conclude a deal for the Americans.[20]

As Sheng's attitude was crucial, the French Consul in Tianjin, du Chalyard, approached him on 19 November about the possibility of French involvement. Sheng stated that he wanted to discuss matters not with consuls and ministers, but with the representatives of foreign bankers and industrialists and that capital would be obtained from those financiers offering the best deal. Similarly, any material required for the construction of the line which could not be made in China would be bought from those foreign manufacturers whose prices were most competitive. Sheng concluded by advising du Chalyard to persuade his compatriots to hurry, as he was already negotiating with British, German, and American companies. However, Sheng was being less than frank about his putatively commercial *modus operandi*. For, on a recent visit to Beijing, he had approached the Belgian Minister, Baron de Vinck, about the possibility of Belgian involvement. De Vinck himself told Gérard about Sheng's advances in some detail: Sheng professed to be impressed with the work of Belgian engineers seconded from the Belgian engineering company, John Cockerill, who were employed at the Hanyang Iron Works. As Cockerill had already lent funds to the Hanyang Works at a rate of eight per cent, Sheng asked if de Vinck could wire Brussels to discover if Cockerill would be willing to lend the newly created Imperial Railway Administration one hundred million francs. In return Sheng was willing to give preference to Belgian equipment and personnel, and offered by way of guarantee the railway's dividends and plant. De Vinck, although unimpressed with the guarantees, wired Sheng's request to Brussels on 9 November; Sheng promised not to make any other arrangements until a reply was received. Despite de Vinck's reserves on the question of adequate guarantees, there could be little doubt that the proposal would be well received in Brussels: Leopold II believed that Belgian capital should be invested in China and had recently obtained the appointment of his friend and former agent in the Congo, Emile Francqui, as Consul in Hankou, 'especially ordered to pursue in China the execution of an industrial and economic programme conceived by the King himself'.[21]

De Vinck, sceptical as to the willingness of Belgian financiers to take on alone so large a project in a remote foreign country, and Gérard, sensing an opportunity for French finance and industry to become involved under the cloak of Belgian neutrality, together conceived a policy of Franco-Belgian co-operation. Despite some reservations, Gérard considered that such a policy could ensure that China would, albeit indirectly, honour Article 7 of the 1885 treaty:

I am tempted to wonder if, in the circumstances, given the hesitation which China feels in approaching us, a great power, there would not be an advantage for our industralists and financiers in possibly taking into partnership Belgian companies. China's distrust of the great powers, in industrial and financial matters especially, the desire of China to attenuate and elude Article 7 of our 1885 treaty is so marked, that this partnership with a neutral power's companies seems to me to be a way of pushing aside the prejudices and resistance which threatens to be such an obstacle. I only make this suggestion to Your Excellency with all kinds of reservations, and with the very distinct feeling that what it offers may be insufficient or even unwanted. I would not have dared to make it if I did not feel how much the very weakness and pusillanimity of China, in making her distrust the strong and pushing her towards the neutral, obliges in some ways a great power, to preserve her industrial and economic interests, to put on, temporarily at least, this mask of neutrality.[22]

A flurry of triangular telegraphic correspondence was initiated by de Vinck and Gérard between Beijing, Brussels, and Paris. Both the Belgian and French governments were receptive to the association envisaged by their ministers in Beijing, so the centre of negotiations had shifted by the end of the year from *yamens* and legations in Beijing to government and private offices in Brussels and Paris.[23] On 30 December 1896 the Belgian Minister in Paris, Baron d'Anethon, told Hanotaux that the Belgian Société générale and the Cockerill company of Séraing-Liège had formed a syndicate which intended to lend a hundred million francs to Sheng's administration. D'Anethon further asked Hanotaux for the assistance of French finance in the project. Hanotaux replied that the government was sympathetic to such co-operation between the two nations' financiers and industrialists. He suggested that it would be necessary first to find a French bank to join the syndicate, and, once French capital had participated in the loan, any orders for railway equipment would be divided between French and Belgian manufacturers in the same proportions as the two countries had furnished capital.[24]

Hanotaux took an active role in the search for a suitable French partner. At first he approached the powerful and French-financed Russo-Chinese Bank. However, its commitments to the Chinese Eastern Railway precluded it from financing a second great Chinese railway.[25] Thereby assured of the benevolent neutrality of the powerful Sinorusse, Hanotaux contacted the Comité des Forges. The Comité's secretary, Pinget, informed the Foreign Minister that Jules Gouin, president of the Banque de Paris et des Pays-Bas was interested in the project. Discussions between Gouin and Baron Hély d'Oissel of the French Société générale began in Paris on 10 January 1897.[26] The Banque de Paris had been closely involved in the establishment of the Russo-Chinese Bank, and the Gouin family's directorships, notably of the Société de Construction des Batignolles, meant that Gouin would have easy access to both French finance and heavy engineering.[27]

From this point negotiations proceeded rapidly, both in China and Europe. On 16 January 1897 de Vinck and Gérard received telegrams informing them of the acceptance in principle of the loan by the French and Belgian financiers. As the Belgian diplomatic network in China was fairly thin compared with the French, this information was passed on to the Belgian Consul in Shanghai, Frère, through the French Consul-General, de Bezaure, and using French telegraphic codes. To Frère's astonishment, Sheng wanted to commence negotiations about the loan immediately, without waiting for the arrival of engineers from Europe.[28] After consulting the Belgian bankers, Frère was authorized to negotiate, and early in February he and Sheng signed what de Bezaure described as a 'provisional convention' for a loan to be issued at a rate of five or five and a half per cent. The details of the agreement were to be finalized after the arrival of the delegates from Europe. The main points to be decided then would be the interest rate and the sources of equipment, as the Belgians wanted its supply confined to Cockerill, whereas Sheng wished to see the Hanyang Iron Works supply at least some of it.[29]

In spite of this tentative agreement, Sheng continued to negotiate with other aspirants. A former Minnesota senator, William D. Washburn had arrived early in January, representing a genuinely powerful American syndicate. The British and Germans had men already working on the railway: Charles Kinder had constructed the northern railways under Li Hongzhang's aegis while Hildebrandt was doing surveys around Hankou on behalf of Zhang Zhidong. The German offer, however, although financially lucrative, was conditional upon Germany being permitted to establish a coaling base on the China coast. This was exactly the kind of thing that Sheng was trying to avoid. Although his duplicity in negotiation was already becoming legendary, he was being honest when he told de Bezaure that 'we want to stay masters in our own house'. He went on to explain:

I wish to build the railway's infrastructure myself and make the rails and most of the equipment in the Hanyang factories ... In approaching Belgium, China is finding a way of eluding the exactions and competition of the great powers. She has found it equally dangerous to displease or to satisfy them ... Your country's capitalists may form partnerships with the Belgians and help them. In that case it would be very important to act secretly to avoid awaking the susceptibilities of the Imperial government. I am strongly in favour of this partnership, but openly I can and I will deal only with the Belgians.[30]

These sentiments did not, however, prevent Sheng from keeping his options open and approaching another French syndicate even after coming to the provisional arrangement with Frère. Emile de Marteau was an Austrian subject but represented the not very powerful but impressively named Syndicat des études industrielles et des travaux publics en Chine. In March 1897 Sheng indicated to de Marteau that he wished to receive from him a definite proposition for financing and constructing the Beijing–Hankou railway. In Shanghai de Bezaure confessed his embarrassment at his inability to support de Marteau, while in Paris de Marteau's employers' attempts to obtain support from the Quai d'Orsay were unsuccessful.[31] Hanotaux considered that France's best opportunity lay in association with the Belgians, and decided that no French company would threaten that arrangement.[32]

While Sheng was continuing with such ruses in Shanghai, serious negotiations were under way in Paris and Brussels. It had been decided that a Franco-Belgian company would be created to build the railway. However, even before the Société d'études des chemins de fer en Chine was constituted, difficulties involving French industry began to emerge. Following the initial discussions in January, the Banque de Paris et des Pays-Bas had attempted to bring French industrial concerns into the syndicate. Despite Gouin's connections, however, by early March, when it had been hoped to create the new company, only Edmond Duval of Fives-Lille had shown any serious interest. Fives-Lille was interested in supplying bridges and locomotives, but no French concern was interested in supplying rails at anything like a competitive price. The protection which French industry enjoyed meant that greater profits were to be made supplying the domestic market, and so French industrialists were not, on the whole, anxious to compete abroad.[33] This attitude, and the economic reality behind it, was to be the source of much of the subsequent friction between the two partners in the railway. Moreover, it was the essential uncompetitiveness of French industry, an uncompetitiveness fostered by increasing tariffs during the 1890s, which, more than anything else, was responsible for the disappointing French performance in securing markets in East Asia generally. The Société d'études des chemins de fer en Chine was eventually formed at the end of March. Capitalized at 250,000 francs, its French members were the Banque de Paris et Pays-Bas, the Société générale, the Comptoir national d'escompte, the Société de Crédit industriel et commercial, the Banque Parisienne, the Banque Internationale, and MM. Höttinguer et cie. The main Belgian companies involved were the Société John Cockerill and the Belgian Société générale.[34] Even before the company's formation, two Belgian engineers, Masy and Rizzardi, had left for China where they were authorized to sign an agreement with Sheng. They were to be followed in April by two colleagues, Walin and Dufourny, the latter being the head of the mission.[35] At Hanotaux's request, the directors of the Société d'études agreed to include a French engineer in the second part of the mission, and on 17 April 1897 a French citizen, d'Orival, sailed for Shanghai with his Belgian colleagues.[36] By that time Masy and Rizzardi were already negotiating with Sheng: they had been authorized to sign an agreement with him even though they were engineers and not financiers.[37] These negotiations took place in Hankou and by 30 April Sheng and the Belgians had agreed on the terms of a provisional contract. However, it was not signed for another month, partly because Sheng constantly sought to achieve better terms by delaying, and partly because of objections from the British, American, and German legations in Beijing, which protested that the proposed contract violated the most-favoured-nation clause. Despite these protests, Gérard was unimpressed with the contract and commented at length on its inadequacies, notably on the low interest rate of four per cent, although this was compensated to a certain extent by the rate of emission being reduced to ninety. Gérard was also dissatisfied with the provision that material be obtained in Belgium and suggested that it be reworded to refer to the Belgian company instead, thereby permitting French industry to share in the potential bonanza.[38] The provisional contract was eventually signed in Hankou on 27 May 1897. The French Vice-consul in Hankou, Claudel, emphasized the provisional nature of the document and urged that, in view of the competition from other powers, it be given a definitive form as soon as possible.[39]

It was certainly not a contract which inspired great enthusiasm amongst European capitalists. As well as lending money to China at the low rate of four per cent, the contract provided as a guarantee for repayment of the loan and interest only the existing Chinese railways. China was not regarded as a secure field for investment in the 1890s and it was difficult to attract capital unless the government in Beijing offered absolutely watertight guarantees. Additionally, the contract left control of the railway in

Chinese hands. Dufourny, Walin, and d'Orival arrived in Shanghai within a few days of the contract's signature and made no secret of their disappointment with the document and even announced their intention to denounce it completely. They were dissuaded from this drastic course by de Bezaure in Shanghai and later by Gérard and de Vinck in Beijing. The diplomats agreed that it was a poor contract from a commercial point of view, but pointed out that it had eliminated British, American, and German competition, at least temporarily, and urged the industrialists to. improve their position by negotiating supplementary conventions.[40] This they decided to do, especially in view of the clause in the provisional contract which provided for revision within two months of its signature.

The commercially unattractive nature of the contract led to other powers being more hostile and suspicious of the deal than was normal even in the overheated atmosphere of China in 1897. Thus the observation of the Beijing correspondent of *The Times* that 'the contract appears to be unworkable, because, complete control being vested in Chinese hands, no confidence can be felt in the security to be offered by the syndicate', increased British concern as to its political implications. Sir Claude Macdonald, the British Minister in Beijing, was well aware of French and Russian support for the Société d'études. Salisbury wanted the concession blocked, although Macdonald believed this was impossible. None the less, Salisbury began threatening the Chinese, and ordered Macdonald to tell the *Zongli Yamen*:

that it would be impossible for Her Majesty's Government to continue friendly relations with them in regard to their fleet and other matters if, while conceding preferential advantages to Russia in Manchuria and to Germany in Shandong, they offered those or other foreign Powers special privileges or openings in the Yangzi region. If the Beijing–Hankou line were to be financed by the Russo-Chinese Bank, as seemed probable, it ceased to be a commercial or industrial enterprise, and became a political movement against British interests in the Yangzi region.[41]

The British protest was the product not so much of an investor who has lost a good opportunity, but of a power whose increasing territorial interests were threatened. For *The Times* as well, one of the most worrying aspects of the contract was the stipulation that the Russo-Chinese Bank was to be the railway's banker, thereby bringing French and Russian interests into the heart of the Yangzi valley. The threat posed was political:

> With these two powers actively interesting themselves in the Belgian syndicate it would be extremely rash to assume that the railway will not be begun, notwithstanding its lack of attraction either for European or for Chinese capitalists. Whether it is finished or not within any reasonable period might be a matter of secondary importance if partial construction afforded occasion and excuse for diplomatic interference of an exclusive kind.[42]

Such British fears were unfounded. Central China was not Manchuria, where the Russians were building railways with so little concern for profit that they did not even bother to charge fares.[43] Moreover the French, and even less the Belgian governments, had none of the territorial ambitions along the route of the Beijing–Hankou line that the Russians did along the Chinese Eastern Railway. The French government certainly wished to extend French influence in China, but mostly in economic ways. The King of the Belgians, too, was in it for the money, but was a good deal more enthusiastic than French or Belgian financiers.[44] He spoke to de Montholon, the French Minister in Brussels, of the enterprise as 'a source of immense profits for those who have the perspicacity to pursue its completion'. At the same time, he threatened to look for another partner if French capitalists would not agree within a week to invest the required funds.[45] The following morning Leopold told de Baüer, the manager of the Brussels branch of the Banque de Paris et des Pays-Bas that he wanted him to use all his influence to ensure that French capital committed itself quickly. De Baüer replied that he was unimpressed with the contract, notably with its low interest rate and lack of guarantees, and that his bank would not be prepared to be involved on this basis.[46] Even the Belgian Foreign Minister, de Favereau, agreed that the interest rate was too low to attract the necessary capital.[47]

Meanwhile in China, French and Belgian diplomats and businessmen were implementing the strategy of Gérard and de Vinck in negotiating supplementary conventions to make the contract more appealing to European investors. The task was made more difficult for them as Sheng continued discussions with British, German, and American investors and used these to improve his bargaining position. Negotiations took place in both Shanghai and Beijing. In the capital Gérard and de Vinck persuaded the Chinese to grant imperial approval for the railway, approval which Gérard would argue constituted an imperial guarantee.[48] In Shanghai Dufourny and Walin negotiated directly with Sheng, although de Bezaure not infrequently supported them, at times, according to himself, quite subtly and to good effect.[49] Eventually a supplementary protocol was signed in Shanghai on 27 July 1897, the last day on which it was possible to do this and exactly two months after the contract. The interest rate was increased from four to 4.4 per cent, or 5.28 per cent net. The responsibility for adjudicating tenders for equipment for the railway was taken from the Chinese director and given to the Belgian company, which thereby gained effective control of their supply. Such control of industrial markets is an essential element of economic imperialism, and it is no wonder that de Bezaure was enthusiastic:

The door of the Chinese market is now open to our industry which can find, if it wants, a remedy on the great Asiatic continent for the crisis it is undergoing in Europe. For our diplomacy too, it is a new means of influence; and for our country [it means] the certainty of having its share when the Celestial Empire finds itself carried away by the economic evolution which must inevitably occur sooner or later.[50]

De Bezaure's uncharacteristic optimism was understandable. The summer of 1897 was the time when France's economic position in China appeared to be strongest, both absolutely and relative to that of other powers. Co-operation with Belgium and also with Russia had given her the edge over her rivals in the important sector of railway construction, in a way that Article 7 of the 1885 treaty had been unable to do. For not only was a largely French-financed company going to build China's first main line, but also the only railways sanctioned in China's frontier provinces were a French line in the south and a Russian line, once again largely French-financed, in the north.

The apparent brilliance of the French position proved transitory. The Belgians, like the Russians, were happy to use French capital for their own purposes but proved to be as determined as the Russians to maintain as many of the advantages as possible for themselves. French financiers also continued to be less enthusiastic about investing in China than had the diplomats who had made it possible for them to do so. Gérard had foreseen the problem in the early stages of the negotiations. In view of the strong competition for the concession, the terms of the loan would not be particularly favourable, although the orders for French industry would be highly profitable. As the railway was hoped to secure industrial rather than financial advantages, as early as January 1897 Gérard advised Hanotaux to ensure that the funds would be available. 'The success of the industrial enterprise is ultimately dependent on the acceptance of the financial scheme,' he observed. 'It is not therefore our industrialists but our bankers who need to be persuaded.'[51]

Pressure on French bankers to invest in the railway was to come from three sources. The Chinese, most notably Zhang Zhidong, continued to negotiate with representatives of other powers, including the German Minister, and Sir Robert Hart. They indicated that their banks could provide the funds and even offered to prosecute the Belgian and French directors of the Société d'études for breach of contract.[52] Leopold himself was anxious that the project be commenced and that the French provide the capital. He told de Montholon, the French Minister in Brussels, that he would be obliged to look for another partner if the Parisian financiers were unable to find the funds quickly.[53] As well as the Chinese and Belgians, the French government attempted to arouse the enthusiasm of its own wealthy citizens in the enterprise. The chosen instrument for this official pressure was Auguste Gérard who had returned from Beijing in August. Before informing him of his new post, Hanotaux asked Gérard to spend some time speaking to financiers and industrialists in Paris in order to convince them to exploit the opportunities which had been opened for them in China. Gérard discovered that French financiers were at best indifferent and often hostile to the prospect of investing in China. He later made the acid comment, 'I made the discovery, in fact, that after not without trouble convincing the Chinese, I now had to preach to and convert the French.'[54] This evangelist for imperialism did have some influential allies in business, notably Henri Germain, president of the Crédit lyonnais, and Edmond Duval, president of Fives-Lille; but the resistance of most financiers to such daring investments was great.[55]

On 23 September 1897 a meeting of French financiers was held in Paris to discuss the matter. One of the Belgian engineers who had just returned from surveying the proposed railway, Dufourny, addressed the hesitant audience. He pointed out that it was a case of lending one hundred million francs to the Chinese company which would build and operate the railway and receive the profits. The Chinese company would pay interest of 5.3 per cent on the capital advanced. On the contentious issue of the guarantee of payment of this interest, he took the optimistic view that, although it was not a state loan guaranteed by specific Chinese taxation revenue, the imperial sanction that had been given for the contract was 'enough to dispel all concern on this issue'. All the rails, locomotives, wagons, and other equipment were to be provided by France and Belgium, not, however, in proportion to the capital provided by each country, but in proportion to the 250,000 francs advanced to support the costs of the initial survey — that is three-fifths by Belgium and only two-fifths by France.[56] Not surprisingly, the financiers were unimpressed. Villars, a director of the Banque de Paris et des Pays-Bas, expressed the feeling of the meeting when he stated that there were two main and very considerable problems: that the investment was neither secure nor lucrative. It was insecure because the contract did not state clearly that the railway would become the property of the investors if the borrowers defaulted, and it was not lucrative because the rate of 5.3 per cent was too low for such an investment. Although the meeting did not come to any firm decision, there was some support for Villars's point of view in the Direction commerciale of the Quai d'Orsay: 'It can be concluded ... that if this business can bring profits to our industry, it is, on the other hand, mediocre for the banks which would handle the loan and for the public who would bring its money'.[57]

Although there were some sceptics in the Foreign Ministry, Hanotaux and Gérard began their campaign quickly. Gérard prepared a statement on the necessity for French capital to be involved in the project, and, on his Minister's instructions, spoke to Jules Gouin, who was president of the Banque de Paris et Pays-Bas as well as a director of the heavy engineering Batignolles company.[58] Gérard argued that the Chinese had given in fact two guarantees: implicitly by granting imperial approval and, more directly, by a guarantee on the line itself. Gérard considered these guarantees adequate and suggested that the railway would be very profitable. As well as pointing out the considerable industrial advantages which would flow from the loan, Gérard concluded with a political argument: that the railway was destined to link the

French railways of southern China with the Russian network in Manchuria and Siberia. The Russian card was a strong one to play with French financiers, as the Russian empire was a highly regarded area for investment. Gérard did not hestitate to tell the financiers that Russian diplomatic support in Beijing had been instrumental in the Belgian company's success in obtaining the concession. [59]

The Russian alliance, however, was a dangerous argument to use, for Gérard was not alone in his rather wild depiction of the railway as a link between emerging French and Russian 'spheres of influence'. It was a theme to which British and German businessmen and diplomats were to rally in 1898. Indeed, a few weeks before Gérard spoke to Gouin, Gustav Detring, who was a senior German employee of the Imperial Chinese Maritime Customs and a man who did not lack political ambition, had petitioned the *Zongli Yamen* in alarmist tones:

The French really are the masters of the business and the Russians help them, and so Sheng has been able to arrange this loan with Belgium. This matter is now settled [sic] but China's danger is ever present. For instance, the Russian railway is to connect with Manchuria and French railways will connect with Longzhou. France has had her eye on Hankou for many years. The North and South are very distant but they are opposite each other. Their object is to obtain the central portion of China. At present the money and name are Belgian, but really it is France and Russia who are assisting its completion. [60]

Neither Detring nor Gérard were to make much headway with such arguments in 1897. In the case of the former this was scarcely surprising, but Gérard, as well as speaking for the Minister for Foreign Affairs, had the support of the imperialist press in Paris. Thus, in the same week that Gérard commenced his campaign, the *Quinzaine coloniale* published an apocryphal story about the alleged success of the British Hooley–Jameson syndicate in obtaining valuable railway and mining concessions all over China. The *Quinzaine* also claimed that Hooley and Jameson had been promised the Beijing–Hankou concession in the event of the failure of the Société d'études to construct the line. [61] Gouin was unimpressed, and so in October Leopold II himself visited Paris, more in his role as a 'businessman of the first order', as Gérard described him, than as King of the Belgians. Gérard later accorded to Leopold a crucial role in bringing about the success of the enterprise: 'It is the King who is really the soul of the project. It is he who, as much in Paris as in Brussels, has given himself the mission of converting the hesitant or lukewarm'.[62] However, neither Gérard, nor Leopold, nor Victor Stoclet, a director of the Belgian Société générale who visited Paris later in the month and threatened to look elsewhere for funds if the French did not provide the capital soon, was able to convince the French financiers.[63] The financiers insisted that the Chinese government would need to provide a more formal guarantee. They also wanted made more explicit the provision that the loan was to be secured on the railway itself. They further wanted the loan to be a loan to the Chinese government, rather than to Sheng's Chinese Imperial Railways so that it would be taxed in France at the low rate of one half of one per cent, which applied to government loans, instead of at the rate of ten per cent which applied to private loans. Following the failure of Stoclet's threats at a meeting on 23 October to persuade the French financiers to part with their money, Gérard admitted defeat and proposed that the French and Belgian Ministers at Beijing seek such changes to the contract. Hanotaux assented to the proposal.[64]

Within a few days telegrams had been sent from Paris and Brussels to Beijing instructing de Vinck and Georges Dubail, who was the French *Chargé d'affaires*, to obtain the required modifications. The Russian government supported the Franco-Belgian requests in Beijing, but asked in return that the Russo-Chinese Bank be given the railway's business.[65]

There can be little doubt that the French financiers would have withdrawn from the project if the diplomats had not been instructed to attempt to meet their requirements. Patriotism and a concern for France's position in East Asia were insufficient to persuade them to invest such large sums of money on what they considered to be such flimsy security. Indeed, it would appear as though the French companies were about to recall their personnel who had gone to China with the survey party. One of these French engineers, d'Orival, had been invited to become chief engineer for the southern section of the railway which was to be constructed northwards from Hankou. However, his employers, the Banque de Paris et des Pays-Bas and the Société des Batignolles, both concerns in which the Gouins were involved, had instructed him to return to France. He was only given permission to take up the Chinese appointment on the eve of his departure. Even though d'Orival was replacing a German in Sheng's employ in Hankou, and despite the heated political situation which German activity in Shandong was creating in November 1897, French capitalists were clearly unwilling to make any concessions in order to maintain their position.[66]

Their extremely hard-headed attitude had produced the desired results, for, as Hanotaux put it, French diplomacy was to seek to interpose the Chinese government between the Franco-Belgian syndicate and Sheng's railway company. In so doing, the Quai d'Orsay would solve at once the problems of taxation and security which concerned the shy investors:

The loan would be issued by the Chinese government itself which, in consequence, would have to ensure its service. It would also no longer be a loan to a railway company, taxed in France at ten per cent, but a state loan, taxed at present at half a per cent . . .[67]

In the atmosphere of extreme tension and pressure which permeated relations between China and the European powers, and between those powers themselves during the last weeks of 1897, it is not surprising that some Chinese concessions were rapidly forthcoming. The *Zongli Yamen* agreed to a formula whereby the Chinese government would contract the loan 'through the intermediary of the Imperial Railway Administration'. De Vinck optimistically commented that as a result of his and Dubail's triumph, the loan would evidently become governmental.[68] Dubail agreed and told Hanotaux that he had met the conditions demanded by French financiers, not neglecting to remind Paris that matters were coming to a head in China and that British and German pressure made a firm acceptance of the formula extremely urgent.[69]

The French financiers, however, remained suspicious and were unsatisfied by the formula. Their intransigence moved Baron d'Anethon, the Belgian Minister in Paris, to plead with Hanotaux to insist that the French financiers fulfil the agreements made with their Belgian colleagues.[70] The Belgian bankers adopted a harsher tone but agreed to seek further amendments in China. In making this concession Stoclet and Baeyens were moved to lecture de Frondeville, who was a director of the Banque Parisienne, on the necessity for these amendments to be concluded quickly and definitively.[71] Although exasperated with their French colleagues, the Belgians by this stage were too deeply committed to them to search elsewhere for the necessary capital. Above all they insisted the French state clearly and definitively what they wanted in the contract.

Early in January the French Finance Minister, Georges Cochery, told Eugène Gouin that he believed that the Chinese government would be willing to negotiate on a new basis. As a result, a series of meetings of French financiers was held in Paris during January 1898 at which evolved conditions satisfactory to both French and Belgians.[72] The financiers initially sought to interpose not merely the Chinese government, whose credit had declined over the previous six months, but the French and Belgian governments as well between themselves and the railway company. They proposed that the two European governments appoint two commissioners who would seize and operate the railway in the event of a default. The French Ministry of Finance was sympathetic to this idea, but Hanotaux refused to consider it.[73] Following the failure of this formula, the bankers decided on the conditions for their involvement at a crucial meeting on 22 January. The most notable and innovative of these was for the creation of a Belgian operating company (or *société d'exploitation*), which would be responsible for the daily administration of the railway on behalf of the Chinese Imperial Railways until the loan was repaid. The creation of the operating company was intended to ensure that the guarantee given to the shareholders could be enforced and to avoid the possibility of the railway passing into foreign hands. The company was to be established under Belgian law with two-thirds of its directors being Belgian or French, the other third Chinese. Its president and general manager were to be alternately Belgian and French and its principal employees half Belgian and half French. Any disputes between the operating company and the Chinese Imperial Railways were to be subject to the arbitration of the Ministers of France and Belgium in Beijing.[74]

As well as demanding the creation of a European-controlled operating company, the French financiers insisted that the line be constructed from its two extremities, that Chinese capital be used first, and that funds from Europe be forwarded to China gradually as required for further construction. The financiers proposed initially to lend only 30 million of the 112 million francs nominal capital of the railway, which was to be issued at a rate of 90 per cent and to return five per cent interest.[75]

These conditions were acceptable to the Belgians, whose interests they protected as much as the French. Another Belgian engineer had arrived in Shanghai at the end of January. To this man, Eugène Hubert, was given the task of forcing Sheng to accept the new conditions. The conditions themselves had been made even more specific than those proposed by the French. Stoclet told Hubert by telegram on 8 February that it was proposed initially only to construct the railway from Beijing southwards to Baoding and from Hankou northwards to Xinyang. The Chinese capital was to be used to build the former section, European capital of 1,400,000 pounds (or 35 million francs) for the latter.[76] French and Belgian diplomatic support was provided for Hubert in these negotiations.[77]

In Shanghai de Bezaure was shocked by the extent to which the financiers were demanding revision of the contract. He supported Hubert strongly, assisted by Emile Francqui, the Belgian Consul at Hankou who was then visiting Shanghai, but was nevertheless embarrassed at the demands which would prolong the negotiations on the railway project. This situation was clearly depressing him. He described it as 'this thankless business, whose history has been marked by so many vicissitudes and disappointments'.[78] Despite de Bezaure's embarrassment, Hubert's position was in fact stronger than Sheng's. The decline in China's international credit, combined with Sheng's insecurity of tenure and awareness that he had many enemies in Beijing, meant that he was in no position to resist the Franco-Belgian demands.[79] Sheng's weakness, however, did not prevent him from trading on French and Belgian insecurity, and he constantly referred to British and German offers to construct the railway. It was also during February 1898 that the Chinese government agreed to place the indemnity loan with the Hongkong and Shanghai Banking Corporation and the Deutsche-Asiatische Bank. In fact, the railway and indemnity loans were quite separate issues; but, to de Bezaure at least, the British success with the latter seemed to jeopardize the Franco-Belgian hold on the former.[80]

By the end of March Sheng and Hubert had reached agreement on a new contract drafted in Shanghai on the basis of Stoclet's telegraphic instructions. Both Dubail and de Bezaure considered that the conditions imposed by the French financiers had been met and urged Hanotaux to obtain their immediate approval.[81] This they were unwilling to give, and on 4 April they replied to Hubert's request for authorization to sign the contract with a list of further modifications. Hubert was also told that a new contract would be drafted in Paris and sent by sea to Shanghai where he was to obtain Sheng's acquiescence in it.[82] This attitude on the part of the financiers outraged de Bezaure, who considered that it indicated a lack of faith in their own representative, Hubert, as well as involving considerable political risks.[83]

Sheng's reaction to the failure of his concessions to Hubert to bring about the signing of the contract supported de Bezaure's analysis, and for two months it appeared as though the Franco-Belgian syndicate would lose the concession. Sheng's position was strengthened during April by the serious interest of the American China Development Company in the railway. Late in March Stoclet had felt sufficiently confident about his position to demand that any funds remaining after the construction of the Hankou–Xinyang section were to be used not for work between Xinyang and Baoding, but for surveys on the Hankou–Guangzhou line for which, he insisted, the Société d'études would be given preference.[84] Three weeks later a provisional agreement was signed by Sheng and the American China Development Company for the construction of the Hankou–Guangzhou railway.[85] This represented a very real check to the prospects of the French and Belgians as it specified that, in the event of the Belgians failing to build the northern railway, the concession would be transferred to the Americans. While the contract drafted in Paris was on the water somewhere between Antwerp and Shanghai, Sheng, emboldened by his American success, sent Francqui and Hubert a draft contract of his own in which he withdrew most of the concessions he had given in March.[86]

Reaction to Sheng's proposal was strong, both by governments and by financiers. The new French Minister in Beijing, Stephen Pichon, suggested that Hanotaux take up the matter with Jing Zhang, the Chinese Minister in Paris.[87] This he did in the most pressing terms, while in Beijing Pichon and de Vinck complained to Li Hongzhang and the *Zongli Yamen*.[88] Both Li and the Chinese ministers protested about the behaviour of the bankers, but Li nevertheless attempted to persuade Sheng to mollify his stance.[89]

French financiers replied to Sheng's attempt to play his American card with an ultimatum. At a meeting in Paris on 4 May they decided that they would in no way modify the fundamental conditions stipulated in the contract they had sent out to China (and which was still *en route*), and to inform Sheng that if he did not sign the document within a week of its arrival in Shanghai, negotiations would be considered ruptured.[90]

The contract arrived in Shanghai on 18 May, and the following week was one of intense diplomatic activity, both in China and in Europe. There were no direct negotiations between Sheng and the Société d'études, as relations between them were all but broken: 'Sheng poetically compares his links with the Belgian company to a waterlily which has been half pulled out but which still lives because a few strands of the stalk remain'.[91] One of the tenuous links to which Sheng's metaphor made such a colourful allusion was Li Hongzhang, to whom he very quickly sent a telegram indicating that he was willing to accept the French contract with two modifications. Firstly, the company was to agree to construct the entire railway from both ends and to complete it in three years. Secondly, he wanted the clause referring to the arbitration of the French Minister removed from the contract itself and relegated to a private letter.[92]

In Brussels Leopold approached the French and Russian Ministers and asked for their governments' support in Beijing.[93] The new French Minister was Gérard himself, who was clearly committed to the project and had, from the moment of the presentation of his credentials to the King on 7 March, devoted much of his energies to Franco-Belgian railway interests in China.[94] As requested, both French and Russian Ministers in Beijing intervened with the *Zongli Yamen*.[95] Hanotaux himself saw Jing Zhang, the Chinese Minister in Paris, to complain about the delay. The latter suggested that Sheng was probably hesitant as he had not yet received any personal reward. The banks had proposed to give Sheng a commission of four per cent on the orders; such an arrangement was not sufficiently attractive to Sheng, however, as he would already be committed before he received the funds. Hanotaux therefore asked Gérard to arrange with Stoclet for the forwarding of a down payment to Sheng as a virtual bribe.[96]

Despite all this activity, Sheng had not signed the contract within the week stipulated by the bankers, who therefore considered themselves no longer bound by any previous agreements. The Banque de Paris et des Pays-Bas, in an attempt to bluff Sheng into signing, declared that it was not convening any further meetings of the syndicate as negotiations were at an end. The bank was no doubt willing to carry out its threat, as the French financiers had consistently demanded that the contract be on their terms or not at all. They did, however, state that the syndicate would meet again if Hubert could persuade Sheng to sign the contract.[97] In this way the bankers at once carried out the threat in their ultimatum and yet allowed the possibility of a resumption of discussions.

Over the next fortnight nothing was done about the affair in Europe, although in China there was much activity. Sheng was pressured to sign, especially by the Belgian diplomats, Francqui and de Vinck. According to Pichon, Hubert became increasingly less significant as he absorbed himself in the pursuit of the pleasures which Shanghai had to offer.[98] Eventually, on 11 June, Sheng indicated his willingness to sign provided some minor changes were made, most notably to the clause referring to the arbitration of the French Minister.[99] Li Hongzhang objected to this clause, and despite the intervention of the Russian Minister, albeit a rather weak intervention in Pichon's opinion, would not be moved.[100] Despite their concern that the clause regarding French arbitration remain in the contract, the bankers at a meeting in Paris on 17 June agreed to authorize Hubert to sign.[101] Hanotaux spoke firmly, indeed brutally, to Jing Zhang on 20 June about the necessity for French arbitration.[102] According to Pichon it was this intervention on Hanotaux's part which ensured that the contract was signed.[103] Eventually de Vinck secured a face-saving formula from the *Zongli Yamen* which omitted any reference to French arbitration in the contract. The arbitrator, it was agreed, would be the minister of the country taking part of the loan.[104] This proved to be acceptable in Paris and the contracts were signed in Shanghai on 26 June 1898. At the same time Sheng signed a declaration in which he specifically stated that the arbitrator would be the French Minister.[105]

There were, in fact, two contracts signed on that day. The first was the loan contract of 29 articles by which the Chinese government would borrow 112,500,000 francs (or 4,500,000 pounds) at five per cent interest. Amortization was to occur over 20 years beginning in 1909. Initially the sections from Beijing to Baoding and from Hankou to Xinyang would be constructed, using the 13 million taels already held by the Chinese Imperial Railways and 33,100,800 francs (or 35 million francs nominal capital issued at 90 per cent) to be raised immediately in Europe. All equipment was to be supplied by the Société d'études except for that which could be furnished by the Hanyang Iron Works. The Russo-Chinese

Bank was to be the railway's banker in China. The second contract of ten articles provided for the operation of the railway for 30 years or until the amortization of the loan, by the Société d'études. The contract gave the Belgian company complete control over the railway's daily management for that period and guaranteed it 20 per cent of the net profits. If the railway's operations were not profitable, the Chinese Imperial Railways would cover the Société d'études's losses. In the case of both contracts, final arbitration in case of disputes was to be undertaken by the French Minister in Beijing, although he was described in a circuitous way to avoid any mention of France in either contract.[106]

The concession of the Beijing–Hankou railway was easily the largest and most significant yet granted to a foreign company in China proper: its only rival was the Russian Chinese Eastern Railway in Manchuria. It was a very real triumph for Belgian and French diplomacy in China; and was certainly recognized as such by the British, although not, curiously enough, in Paris. When news of the signature of the contract arrived in Paris, there were the twin distractions of a ministerial crisis which culminated on 28 June with the fall of the Méline government and the departure forever from the Quai d'Orsay of Gabriel Hanotaux, and of the revival of the Dreyfus affair. In these circumstances, and in view of the long period during which the signing of the contract was expected daily, the news from Shanghai was not accorded the importance which might have been expected. *Le Temps* dutifully published the official communiqué by the Ministry of Foreign Affairs on 27 June, but the press generally showed remarkably little interest. In an article in the *Messager de Paris*, however, J. Raubert did remind his readers that 'the event is not however without importance' and proceeded to describe that importance in fundamentally political terms: 'This railway is destined ... to connect with the Russian railway in the north and with our Tonkin railways in the south'.[107]

It was exactly this sort of political inference which other powers, such as Germany and, above all, Britain, feared. Late in May an article in *The Times* had suggested that by working through a Belgian company, Russia and France would control the main trans-Chinese railway which would bisect Britain's interests on the Yangzi.[108] A further article had argued that the French and Russians were planning to use railway loans 'for facilitating the creation of future claims of a political order within the Chinese empire'; in other words to stake a claim for a future Protectorate or annexation in the event of the collapse of the empire.[109] Whatever may have been Russia's intentions for Manchuria, there is no evidence that French officials had any such thing in mind for the plains across which the Beijing–Hankou line was to be built. Nevertheless, British reaction to the announcement of the loan was dramatic enough. The British Minister in Beijing, Sir Claude Macdonald, had been led to believe by the *Zongli Yamen* that he would be shown a copy of the contract before it was signed. The Chinese failure to do so provided the pretext for British recriminations with China. The *Zongli Yamen* blamed Li Hongzhang, who, they claimed, insisted that the contract be signed before Macdonald saw it so that the Russians, whom he believed to be firmly supporting the French and Belgians and whom he greatly feared, would not be offended.[110] This attempt to blame Li Hongzhang convinced neither Macdonald in Beijing nor Salisbury and Balfour in London. Macdonald attempted to prevent imperial ratification being given to the contracts by threatening that England would take unspecified measures in the Yangzi valley.[111] In this he failed, and the sanction was granted on 11 August. British opinion by then had been further inflamed by the publication of a partial and inaccurate English translation of the contract in the Shanghai *China Gazette*. This translation omitted mention of the Belgian Société générale and exaggerated the role of the Russo-Chinese Bank. A special meeting of the China Association, which was dominated by British interests and especially by Jardine Matheson and Company, was convened in Shanghai, and, on the basis of the *Gazette*'s translations, sent vigorous protests to London and Beijing.[112]

The response in London was strong, as Britain feared nothing more in China than the imposition of a Russo-French Protectorate:

Her Majesty's Government have come to the conclusion that the Belgian Syndicate is only a screen to conceal French proceedings and that the action of the Chinese Government in ratifying the Beijing–Hankou Concession, in spite of our request for delay, is an act of deliberate hostility.[113]

Macdonald conveyed the savage missive from Westminster to the *Zongli Yamen*, demanding the concession of a swag of railways to Britain by way of compensation. To strengthen what was in effect an ultimatum, Balfour ordered demonstrations of British naval power at the mouths of the Yangzi and the Beihe. Vice-Admiral Seymour placed the Royal Navy's China Station on a war footing.[114] Britain was willing to fight over the issue, although ironically not at the expense of the Belgians, nor of the French, nor even of the Russians whom they most feared, but of the Chinese. Under these most pressing circumstances the Chinese government agreed to apologize to Macdonald and to grant four important railway concessions to British syndicates. Significantly, the contracts for the Beijing–Hankou concession were to be the model for these British concessions.[115]

During the period of British recriminations, which lasted through July and August, French diplomats were only able to attempt to assert, quite accurately, that there were no Russian interests or capital involved. The British had ascribed a solidarity to the Franco-Russian alliance in China which, by the summer of 1898, had ceased to exist — as Pichon's complaints about the lack of energetic Russian support during the negotiations on the contract amply demonstrated. The subtleties of Franco-Russian relations, however, were lost on both Chinese and British. Indeed, the imperial sanction was only given after Li Hongzhang had arranged for de Vinck, Pichon, and their Russian colleague, Pavlov, to declare quite firmly to the *Zongli Yamen* that there was no Russian involvement.[116] The French and Belgian press also joined in the debate and in August 1898 the *Indépendence belge* published a counterblast to the English accusations. The article was reprinted in two French papers, *Les Débats* (16 August 1898) and *Le Rentier* (27 August 1898). The basis of the article's case was the face-saving fiction that the railway was to remain Chinese property and that the concession had been granted not to the Société d'études but to the Chinese Imperial Railways. The Belgian company was merely going to lend the Chinese company the required funds and operate the line until their amortization in order to guarantee their repayment. In this the article was accurate, although clearly the Belgian company would control the line. Less accurate was its assertion that Russian intervention was a 'fantastic invention' although it is true that the British press and government had greatly exaggerated the extent of Russian support for and involvement in the railway.[117]

The strong British reaction to the Belgian concession is indicative of the high imperialist passion reigning on both sides of the

Channel. The British threats and naval demonstration may have secured some valuable concessions, but they also indicate a mounting tension and atmosphere of unreality in which national prestige was as important as economic advantages. The readiness of Salisbury and Balfour to make such aggressive moves when they believed that British interests were threatened had a sobering effect on French diplomats, who, within a year, were seeking an accommodation with their British counterparts and actively promoting the concept of an Anglo-French accord in China. The Fashoda crisis, which erupted within a few weeks of Macdonald's successes in Beijing, was to repeat the lessons of the conflict over Chinese railway concessions. Then British and French forces almost came into conflict in the Sudan without the presence of a state like China to act as a buffer between them and victim of both. Thus, although the Beijing–Hankou concession may have been a triumph for France, the strength of the British reaction and the humiliation at Fashoda soon after removed much of its gloss.

4

The Implementation and Consequences of the Beijing–Hankou Concession

THE concession of the Beijing–Hankou railway to a Franco-Belgian group was in a sense the culmination of the aspirations of de Freycinet and Cogordan expressed in 1885 and those of Gérard and Hanotaux a decade later. Certainly it appeared for a short time as though it would be largely French capital which would build China's railways, even if the companies operating the railways were Russian in Manchuria and Belgian in central China. Article 7 of the 1885 treaty would be honoured, albeit indirectly. Nevertheless, much remained to be done to ensure that the construction of these railways would provide the financial and industrial advantages which French imperialists, whether in government or business, expected from their investments.

In the case of the Beijing–Hankou concession the first tasks were the organization of the company and the floating of the loan. Since the railway would operate under the name of the Chinese Imperial Railways, it was decided to retain the name of the Société d'études de chemins de fer en Chine, even though the company would be doing far more than making surveys. The first meeting of its board of directors was held in Brussels on 1 August 1898. Although it was to operate under Belgian law and to have a Belgian president, the principle of Franco-Belgian equality was affirmed in its organization. Although its president, Victor Stoclet, was a Belgian and a director of the Belgian Société générale, its vice-president, Baron Hély d'Oissel, was French and also vice-president of the French Société générale. Six of the other 12 directors were French and six Belgian.[1] The company's *secrétaire*, Ernest Felsenhart, was Belgian, but meetings were to be held alternatively in Paris and Brussels. An executive committee with wide powers to direct and manage the project (described as the *comité d'exécution*) was appointed: two of its members, Focquet and Spruyt, were Belgian and two, Ferré and Ristelhueber, French. The latter, a former Consul in Tianjin where he had been more adept at making money than diplomacy, was also the Paris *secrétaire* of the Russo-Chinese Bank, thereby reinforcing the links between the railway and its banker.[2]

It was not until April of the following year that the efforts which had gone into securing a contract satisfying the bankers were tested in the market place and the loan floated. The leading French publicist for colonial affairs, Paul Leroy-Beaulieu, had expressed some doubts as to the attractiveness of Chinese railway projects, pointing out the risks which revolt or collapse of the Chinese government or even merely slow construction could present to investors. For Leroy-Beaulieu, 'we must first see if they will easily find the money'.[3] At the time when the loan was offered for subscription, however, Leroy-Beaulieu's doubts were forgotten. *Le Temps* claimed that Hankou was about to become China's greatest centre of exportation and reminded its readers that the shareholders of the Société d'études were 'in brief, all the great Franco-Belgian banking and industrial houses'.[4] *Le Rentier* not only endorsed the financial arrangements made for the railway, but also rather fancifully anticipated that its construction would open up a new era of intercontinental rail travel:

If one casts one's eyes over a map of Europe and Asia, one is immediately struck by the exceptional, one might say unique since railways were first built, importance of the series of lines which, leaving Europe by the Transsiberian, this central line (Beijing to Hankou), connected with the Beijing–Tianjin–Niuzhuang line and the Russian system in Manchuria in the north and with the Hankou–Guangzhou line in the south — links our Indo-Chinese Empire to all Europe.[5]

In spite of the far from encouraging political situation in China and the doubts of Leroy-Beaulieu, the response to the loan when it was floated on 19 April was as enthusiastic as rhetoric in *Le Rentier*. The loan was for a total of 112,500,000 francs, issued in 225,000 shares of 500 francs. The loan was issued to the public at a rate of 96.5 per cent, or at a cost of 482.5 francs per share.[6] Initially only 133,000 of the shares were issued as it was proposed to construct the railway in stages. Requests for a total of 226,800 shares were received, 190,800 from France and 36,000 from Belgium.[7] Belgian investors received all the shares they requested, but French investors were only to receive 45 per cent of the number for which they had applied.[8] It was an extraordinarily brilliant result and showed that Chinese railway loans could appeal to French investors. The remainder of the company's capital was paid up in two further share issues, a small one of 556 shares in August 1899 and a final issue of the remaining 91,444 shares in April 1902.[9] Important as the Beijing–Hankou concession was in its own right, it was also seen as a lever for obtaining further industrial advantages in China. The two enterprises in which the syndicate was interested and which were linked more or less intimately with the Beijing–Hankou project were the Hanyang Iron Works and the Hankou–Guangzhou railway. The destiny of all these enterprises, of course, was in the hands of Sheng Xuanhuai. During June 1898, while negotiations on the Beijing–Hankou contracts were drawing to a close, Sheng had announced his intention of selling the Hanyang Iron Works. At first he had discussed the sale with Jardine Matheson and Company, but following intervention by de Vinck and Pichon he offered the Société d'études preference in any sale.[10] The new French Foreign Minister, Théophile Delcassé, was enthusiastic about the prospect of French participation in the ownership and operation of the factory, especially as it was assured of a ready market as Chinese railways were constructed. Immediately he sought out the attitude of the Banque de Paris et des Pays-Bas, and after much prevarication was told that no French bankers were interested but that the Belgian Société John Cockerill was. Reluctant to see the factory pass into Belgian ownership, Delcassé asked Hély d'Oissel to interest the French members of the Société d'études in the Hanyang works.[11] Eventually the Société d'études decided that it would attempt to assume the administration of the ironworks. The board of directors, meeting in Paris on 4 and 5 November 1898, authorized Francqui, who was carrying out business on behalf of the Société d'études as well as his consular duties, to negotiate with Sheng on the matter. The board proposed to advance three million francs at six and a half per cent to Sheng, in return for which it would receive a measure of control of the factory and 30 per cent of the profits.[12]

These negotiations never reached any conclusion, mainly because the Société d'études refused to reply to Sheng's request that 60 per cent of the bridges, rails, and steel sleepers required for the Beijing–Hankou line be furnished by Hanyang. In response, Sheng turned to Japanese and German companies. From the

Japanese he sought loans, which would enable him to develop the ironworks while retaining control, and from the German firm of Carlowitz and Company he sought and quickly obtained a loan for the development of the Pingxiang coal mines in Jiangxi as a source of coke.[13] Hély d'Oissel commented frankly that the opportunity to assume control had been lost because on one hand the Société d'études had not shown sufficient enthusiasm and on the other Sheng had discovered that he could retain control. He had decided to do so and to attempt to use the works to supply most of the more easily constructed equipment needed for the railway. The company had therefore decided to seek to limit as much as possible the supply of equipment from the Hanyang Iron Works.[14] The preservation of industrial advantages always was a high priority amongst both investors and politicians.

The extension of the Beijing–Hankou line southwards to Guangzhou would complete China's grand trunk railway and give the Chinese a rapid means of communication which would not be susceptible to interference by the foreign navies which dominated the coast. It was therefore accorded considerable importance by the Chinese, and Sheng had begun negotiations soon after his appointment as director of the Chinese Imperial Railways. His policies of avoiding dependence on one source of capital, personnel, and equipment and of using the resources of powers which professed to have no political ambitions in China led him to favour negotiations with American syndicates. In this he used the services of the Chinese minister in Washington, Wu Tingfang, who, on 14 April 1898, signed a provisional contract in that city with Thurlow Weed Barnes who represented the American China Development Company. This contract contained the provision that, if the arrangement with the Belgian syndicate to construct the Beijing–Hankou line were to be broken, the American group would be able to take over the concession.[15]

Although the contracts signed with the Société d'études on 26 June 1898 contained no such provision, a convention which Sheng signed on the same day did grant the Belgian company the same right of preference in the event of American failure to construct the Hankou–Guangzhou section as had been accorded the Americans in respect of the northern section. It appeared as though Sheng had skilfully divided his patronage between Belgians and Americans and ensured that should either fail, their place would be taken by the other and not by one of the great European powers.[16] The only problem with Sheng's strategy was that, like the Belgians, the Americans did not have sufficient capital available to construct the railway. The American China Development Company certainly looked solid. Its president, Calvin S. Brice, was a retired senator; and even after his death in New York in December, the syndicate's membership, which included individuals such as Rockefeller, Vanderbilt, and Morgan and companies such as Standard Oil, Con-Negri Iron and Steel, and the American Sugar Refining Company, inspired confidence. Nevertheless, at this time the United States was still an importer of capital and it was widely suspected on three continents that it would be unable to honour its contract with Sheng. Moreover the syndicate was divided on various issues: at the time of his death Brice was locked in litigation with Barnes over the distribution of future profits.

Pichon feared that British interests would exploit the American syndicate's lack of capital and that the good relations between Washington and London would lead them to come to an arrangement analogous with the Franco-Belgian combination. To prevent this he warned the *Zongli Yamen* of the dangers of allowing the British to control a railway from Hong Kong to the Yangzi and sought the support of Aleksandr Pavlov, the Russian *Chargé d'affaires*.[17] Neither approach was successful: the *Zongli Yamen* assured him that no British capital was involved and Pavlov stated that Russia would not involve itself in affairs south of the Yangzi.[18] Delcassé was enthusiastic about the prospect of the Société d'études taking over the American concession. He almost sent a telegram to Pichon asking him to attempt to foil the American proposals so that the Franco-Belgian right of preference could be exercised. However, he thought better of it and asked Pichon instead merely for opinions and information on the matter.[19] In the same spirit Delcassé consulted Hély d'Oissel in Paris, who stated that the Société d'études intended to take up the right of preference, if it was offered. He also pointed out that the difficult terrain over which the railway was to pass would make it a more expensive and time-consuming project than the Beijing–Hankou line.[20]

Despite Delcassé's enthusiasm, the first serious proposal for the involvement of French capital in the Hankou–Guangzhou proposal came not from Paris but from Brussels and, indirectly, from New York. Rather than admit defeat, the American conces-

Map 1   The Beijing–Hankou Railway and Connecting Lines

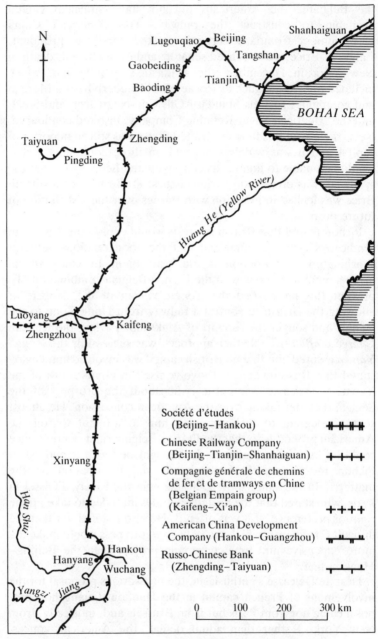

sionaires made overtures in a number of European financial capitals, including Brussels. On 31 May 1899 Raphael de Baüer, who managed the Brussels branch of the Banque de Paris et des Pays-Bas, approached Gérard about the possibility of the Société d'études co-operating with the Americans. De Baüer told Gérard that the company would probably be interested, that Leopold certainly was, but that it would not be possible to raise sufficient capital in Belgium and that French capital would probably be sought. De Baüer wanted to know the attitude of the French government to such a proposal and indicated that a number of Belgians, including Leopold, were also interested in co-operating with English capitalists in the affair. In reporting this conversation to Paris, Gérard expressed his support for the proposal which would prevent a railway leading towards the putative French sphere of influence in China from falling into the hands of an English, American, and Belgian group.[21] Delcassé agreed and told the French directors of the Société d'études that the government could only see advantages in listening to the American proposals.[22]

At this time, however, reports from London and Washington indicated that the Americans had found the capital they needed in England. It was claimed that the British and Chinese Corporation, a syndicate formed by Jardine Matheson and Company and the Hongkong and Shanghai Bank which had already obtained the Kowloon–Guangzhou concession, would help the American China Development Company to build the line.[23] Despite this news, Delcassé still held hopes for the success of the Belgian–American negotiations so that, even if British involvement could not be prevented, at least it could be reduced by the addition of French and Belgian capital.[24] The prospect of a financial consortium of British, American, French, and Belgian interests was not without supporters in the French Foreign Ministry, including, strangely, Stephen Pichon. At least at this point in time, Pichon believed that as French interests were diverse and spread throughout the empire, rather than seeking to concentrate her efforts in particular projects or in a particular sphere of influence, France should endeavour to 'internationalize' her investments where possible so that French interests became inextricably linked with those of other European powers. Provided that France received a share of the orders, profits, and management of the enterprise commensurate with her investment, Pichon strongly supported the international combination: 'A union of the Franco-Belgian and Anglo-American syndicates, in consolidating their interests could have fortunate results for our political position in China'. A month later the mercurial Pichon had changed his views on the matter of spheres of influence.[25] Yet briefly he was ahead of his contemporaries in his support for such arrangements, for it was not until the reorganization loans of the last years of Qing rule that much broader international loans became a feature of Western investment in China.

In the meantime, the rumours of an Anglo-American agreement proved to be false and the Société d'études kept its options open. It favoured an agreement with the Americans, but in the event of a definitive contract with them being signed, it was going to arrange to have the Belgian Minister in Beijing protest to the *Zongli Yamen*. This protest, however, was intended only to pressure the Americans into association with the Société d'études or to prepare the way for compensation in imitation of the powers during the spring and summer of 1898.[26] Indeed, Sheng had already anticipated this protest and had agreed to compensate the Franco-Belgian group by granting it the right to build branch lines and exploit mines along the Beijing–Hankou route in the event that definitive arrangements were made with the Americans.[27]

The whole affair was extremely slow moving, essentially because none of the parties involved was particularly anxious to invest the extremely large sums of money required and all parties feared the political consequences of acting too precipitately. Thus, in Beijing Sheng broke off negotiations with the American China Development Company in September 1899, only to re-open them in a desultory fashion the following month at the insistence of the British and American legations. In Brussels Stoclet and Gérard were convinced that the British market was in no condition to absorb another large Chinese railway loan and believed that the consortium most likely to succeed would involve both the Americans and the Société d'études. Typically, neither Gérard nor Stoclet worried about the ability of the French market to absorb the loan.[28]

Whatever difficulties may have existed in Europe to frustrate Gérard's and Stoclet's plan, there were very definite barriers to its implementation in China. Sheng and the Chinese government in general had been impressed mightily by the strength of British reaction to the Beijing–Hankou contracts of 1898. Thus, when

Sheng was asked by Jean Jadot, Hubert's replacement as manager of the Société d'études in China, about the prospects of a Franco-Belgian-American consortium, he replied that he could not agree to the involvement of the Société d'études in any way as the Chinese government had promised the British Minister not to discuss the concession with anyone other than Americans or British.[29] Sheng later confessed to Jadot his fear of possible British recriminations and his desire to avoid a repetition of the events of the summer of 1898.[30]

Gérard believed that the British had neither the right nor the power nor the will to prevent the Société d'études from being involved in the Hankou–Guangzhou project. He declared that the conclusions the British were attempting to draw from the declaration of inalienability of the Yangzi valley and the Anglo-Russian agreement on spheres of influence of 28 April 1899 were 'inadmissable'.[31] Gérard favoured an active policy which would challenge British pretensions to a sphere of influence in the Yangzi and use the railway to unite the French empire in Indo-China with the great French enterprise in the north, the Beijing–Hankou line. Such a vision was very different from Pichon's concern for the security which he felt could only be offered by international consortia. Opposing a cautious policy, Gérard advised Delcassé that:

On the contrary, it is a good opportunity to prevent the general acceptance of the opinon that, as of now, the Yangzi valley is recognized as a British sphere of influence. The state of the London market and the South African preoccupations of Her Majesty's Government are such that it is reasonable to assume that British policy will not look for further difficulties, especially in China, when everything suggests it should avoid them.

The Société d'études being ready either to claim the exercise of its rights or to co-operate with the American syndicate, Your Excellency could well appreciate just how favourable circumstances are for us to seek from the Chinese government, with the agreement of the Belgian and maybe even other governments, the concession of this line which is destined to extend to Guangzhou, that is into the provinces bordering on our Indo-Chinese empire, the great and profitable Beijing–Hankou railway.[32]

The problem with Gérard's activist position was that it ignored the reality that nobody was in a hurry to do anything. Perhaps Gérard had been talking to Leopold too much: at any rate his advice was such that no responsible French Foreign Minister could take it seriously. The Société d'études, far from 'being ready' to take on the concession, either alone or in association with the Americans, above all sought delay. At a meeting in October 1899 the directors unanimously decided that they would take their right of option if offered, provided that a survey be made first so that the cost of the railway could be determined. It was a brave decision, for Hély d'Oissel acknowledged that, in the unlikely event of Sheng offering the company the concession, the company would be greatly embarrassed, partly because it did not know the reasons why the American syndicate might withdraw, partly because it had no idea of the cost or physical difficulties involved in constructing the railway, but above all because the financial and political situations were far from brilliant. The prospect of raising perhaps up to 400 million francs was not encouraging, especially in view of the difficulties involved in securing favourable conditions for raising the half of that sum required for the Beijing–Hankou loan. The board considered that it would be much easier to raise capital when several hundred kilometres of the Beijing–Hankou line were already in profitable operation, when the Hankou–Guangzhou route had been surveyed and when the market was more likely to be sympathetic to another Chinese loan. Therefore, while the company wanted to be involved in the Hankou–Guangzhou project, its interests were best served by delay and Hély d'Oissel requested the Quai d'Orsay prevent matters from moving too quickly.[33]

Although he did not know it, Hély d'Oissel had nothing to fear and there was no danger that the concession would be concluded with any undue speed. The Americans, clearly, did not have sufficient capital; so early in 1900 their negotiator in China, the lawyer Clarence Carey, returned to New York without signing a definitive contract, claiming that he lacked the power to do so. Sheng was obliged to accept this fiction, as at least it was preferable to the Americans renouncing their concession altogether and thereby obliging Sheng to negotiate with Europeans whom he feared.[34]

This widely sought delay was seen as an opportunity by the ambitious King of the Belgians and his agents. While desultory discussions continued in China, the United States, Brussels, and Paris, the men who had brought Europe the scandal of the Congo Free State were moving in on the American China Development

Company. A loose Belgian consortium, consisting of the Banque d'Outremer, the newly established Compagnie Internationale de l'Orient, and Leopold himself, was buying shares in the American company. The raiding was being organized by Colonel Albert Thys and Emile Francqui, who had left the consular service to become an agent of the Compagnie Internationale de l'Orient.[35] Both men were veterans of Leopold's Congo venture. By July 1900 Francqui and Thys had acquired 78 of the 150 shares, each of US$10,000, of the American China Development Company. Francqui approached Gérard to see if the French would object to the purchase of more shares using French funds. Delcassé indicated that there was no objection, provided Colonel Thys's group remembered that the Société d'études had a right of preference should the Americans fail to complete their contract.[36]

Ironically, the infusion of Belgian capital emboldened the Americans to do what they had feared until that time, and on 31 August 1900, while foreign troops occupied the ruins of Beijing, Carey and Wu Tingfang signed a definitive contract in Washington. With the expiry of its option for a third of the American syndicate's shares by the British and Chinese Corporation on 21 November 1900 and the reorganization of the syndicate into a company capitalized at only US$600,000, Thys and Francqui were able to acquire control of the company. Delcassé's attitude to all this frantic activity was that in any assumption of the Americans' rights, the Belgians would first have to negotiate with the Société d'études and that it was essential that any agreement contain the principle of absolute equality of the French and Belgian elements.[37]

The affair, however, drifted beyond the control of the French Foreign Ministry when in June 1901 Leopold acquired US$400,200 of the US$600,000 of the company's capital on the account of the Congo Free State. Leopold's intention was to leave US$199,800 of the capital in American hands and offer US$99,800 to French investors, leaving the King with US$300,400 or a clear majority. He proposed to construct only the two extremities of the railway first, connecting Guangzhou and Wuchang with wealthy mining districts at Foshan and Pingxiang respectively. The capital and profits were to be divided equally between Americans, French, and Belgians, but five of the nine directorships would remain in American hands as required by the Chinese.[38] The French Minister of Finance, Caillaux, was impressed by Leopold's scheme and told him on one of his visits to Paris that he thoroughly approved of it. The French Foreign Ministry, however, found little in it to commend it. Although possibly remunerative, Leopold's proposal guaranteed no industrial privileges for France and in no way would further French interests generally. The head of the Direction commerciale at the Quai d'Orsay, Maurice Bompard, insisted that there was no point in investing in Chinese railways without obtaining either a market for industrial production, which would bring immediate benefit to the French economy; or control of the enterprise, which would make it a sound long-term investment. Leopold's scheme allowed the French neither option, and would only use French money to further Belgian, American, and above all Leopoldine interests:

We must not forget that agreement between American and Belgian elements in the syndicate will be quite easy to achieve, given that their interests are the same. For both the Hankou–Guangzhou is above all an industrial matter: to sell as many products as possible needed for its construction and to use that to seek new business which would provide outlets for their superproduction, such fatally will be their common aim. But we who ... will be, if not from the beginning at least in a very few years, the only shareholders, we have to be concerned with the enterprise's future ... and ensure its efficient operation, the source of income with which the shareholders will be paid. It follows that ... we must assure ourselves of the upper hand in the management of the enterprise.[39]

In analysing the King's schemes, Bompard had eloquently distinguished between the financial and industrial aspects of imperialist endeavours. To either a French financier or a French industrialist, Leopold's railway proposal offered little as it stood, so the opposition of the Direction commerciale was scarcely surprising. In 1898 Eugène Gouin had described the aims of the quest for the Beijing–Hankou concession as 'to develop French influence in China and open great outlets for her industry there'.[40] Gouin, like Bompard, knew what imperialism was about: it was scarcely likely that French financiers, any more than French bureaucrats, would permit themselves and their resources to be used to further the pretensions of the King of the Belgians.

Ultimately, the issue was never brought to a test, as the company only sought limited funds for the construction of the railway. Such work as was done on the line by the American China Development Company, about fifty kilometres in all near Guangzhou, was funded by the company's formation capital and by

US$2,200,000 which it had floated by bonds sold in the United States. Leopold's plans failed because Sheng Xuanhuai would not permit a Franco-Belgian monopolization of the railways for which he was responsible. Sheng was so fearful of such a monopoly that, even after Leopold sold US$120,000 of his holdings in the formation capital of the company to J. P. Morgan to restore American domination, he insisted on the revocation of the concession. Sheng achieved this aim in 1905, but only in face of strong opposition from both Leopold and Theodore Roosevelt, the American President, and at considerable cost, far greater than the value of the work done, to the Chinese Imperial Railways. This experiment of Sheng's in using the resources and personnel of unambitious nations to avoid the involvement of great powers had failed disastrously and expensively. Thereafter, he espoused openly international financial groupings rather than attempting either to use supposedly unambitious nations or play off great powers against each other.[41]

The failure of French investors to participate in the Hankou–Guangzhou project and the implementation of the contract for the construction of the Beijing–Hankou railway both reveal advantages and limitations of the same policy. Association with a militarily weak, albeit industrially powerful, nation such as Belgium could provide opportunities for French imperialism, but at the same time restricted French freedom of action and inhibited the full economic realization of a project's potential so far as French industry and investors were concerned. Thus, the history of the construction of the railway is full of unedifying incidents which illustrate all too effectively the disadvantages of operating through a small power. For, although Sheng was certainly not ignorant of the extent of French involvement (de Bezaure had told him all the details), and the behaviour of the British in July 1898 indicates that they certainly suspected, indeed exaggerated, what was occurring, the necessity to preserve the appearance of the fiction that the Société d'études was a Belgian company greatly weakened the French position in disputes with their partners and in their attempts to acquire all the advantages imperialists would expect from so brilliant a concession.

An early example of the difficulties raised was the replacement of Hubert as the managing engineer in China. Hubert's failure to do his job effectively led all involved in the project to agree that he needed to be replaced. Delcassé and Gérard were strongly of the opinion that his successor should be French; but Pichon, fearing British recriminations and anxious to placate Sheng and the *Zongli Yamen*, believed that it would be better to have another Belgian fill the position. The directors, despite Hély d'Oissel's appeal for a French manager, an appeal made at Delcassé's specific request, decided that Hubert's assistant, a Belgian named Jean Jadot, should be the new manager.[42] Although his administration was not to be without controversy, Jadot remained at his post to see the railway completed and was a fine choice. A Belgian journalist later wrote of him:

His name is not unknown in Belgium; at Tianjin, at Shanghai, at Hankou it is famous. M. Jadot is to the Beijing–Hankou railway what Colonel Thys is to the Congo railway... The rough legion of pioneers whom he leads recognize in him a clear-sighted and forceful general. M. Jadot is the confidant of viceroys; he is the friend of Sheng, the famous head of the railways and the greatest speculator in China; he deals as power to power with these magnificent satraps — and why not? For them he is the most authoritative representative of our country's industrial power; morally and materially, the most important Belgian in China.[43]

Despite Jadot's undoubted talents, the fact that a Belgian rather than a French citizen occupied such an elevated position due to Chinese susceptibilities was painful to the French Foreign Ministry. Railway builders, especially in remote parts of the world, had great prestige and influence in the late nineteenth century, as the careers of Thys and Rhodes in Africa, Whitton in Australia, and Kinder and Jadot in China, amply demonstrate. Although the French had provided most of the money, there was to be no great French railway builder in central China.

Not only did the French element have to forsake the premier position, but it was also apparent that Belgians were being favoured in appointments to the European staff. Thus, in March 1899 there were 24 Belgian employees in China but only 15 French.[44] This inequality prompted Delcassé to contact the board of the Société d'études, and he extracted a promise that in future an effort would be made to achieve an equilibrium between the French and Belgian elements.[45] The directors did in fact honour their promise, and the engineer who was appointed to take charge of operations of the railway, Henri Bouillard, was French. He had previously served on the French Nord Railway and was to leave for China early in April 1899 taking with him 22 subordinate personnel, of whom no less than 20 were to be French.[46] Delcassé continued to take an active interest in the nationality of the employees of the Société d'études in China. When Jadot indicated that he needed an assistant, Delcassé wrote to Gérard asking him to remind Stoclet of his agreement to employ equal numbers of staff and to Pichon to ask him to ensure that the company's servants in China kept the agreements it had made.[47] Further, he had Bompard write to Hély d'Oissel to insist that Jadot's assistant be French.[48] Gérard was such an enthusiastic protagonist of French rights in this matter that he cornered the Belgian Foreign Minister, de Favereau, and associates of Stoclet (who was then in Spain) at a royal garden party and asked them to ensure that French industry was accorded its rightful place both in the appointment of personnel and in the orders of material for the Beijing–Hankou railway.[49] Pichon had suggested that the engineer-secretary in China, a French citizen named Diamanti, would be an appropriate choice for Jadot's deputy, especially if he were given the right to correspond with the board in Europe.[50] Delcassé took up this suggestion and asked Gérard to put forward Diamanti's claim to Stoclet.[51] The latter was initially reluctant, but eventually agreed to Diamanti's elevation to the status of deputy. Additionally, responsibility for construction at Hankou would be given to a French engineer named Petit. As there were in June 27 French and 20 Belgian employees in China, the requirements of the French Foreign Ministry appeared to have been met.[52] Indeed, so anxious was Stoclet to conciliate his French partners, that when Petit had to resign in July 1899 following a scandal of some sort, Diamanti's status was raised to equality with Jadot by way of compensation.[53]

The first tangible indication that investment in the Société d'études would further widen French interests in China came when Bouillard and his party of almost entirely French officials took over the operation of the Lugouqiao to Baoding section of the railway on 1 October 1899. These 145 kilometres of railway had already been constructed by English engineers from Kinder's staff, who, understandably, showed some pain and disappointment at handing over their project to their French professional colleagues but national rivals. Within a few weeks French had replaced English as the language in which notices and tickets were printed and correspondence conducted. Those senior Chinese employees such as station masters who were unwilling to learn French instead of English were replaced by French speakers.[54] This policy was carried out at the request of the French Foreign Ministry, which was enthusiastic at the prospect of an enterprise offering employment to the French speaking graduates of French schools in China. Bompard had asked Villars, a director of the Banque de Paris et des Pays-Bas, to ensure that a knowledge of French was expected of all senior Chinese employees.[55] The Société d'études did as Bompard requested and even allocated a subsidy of 650 taels per month for the operation of a French school by the Marist brothers in Hankou.[56] The diffusion of the French language was in many ways the most important aspect of France's self-appointed 'civilizing mission' in the colonies and China. Whereas Germans operating in China were content to use English in their intercourse with Chinese and other Europeans, the French did so only with reluctance. This was one of the advantages of association with Belgium. As a French diplomat in Beijing was to put it ten years later:

By language every Belgian enterprise is a French enterprise, and language remains a precious means of our general influence. The future of our language is very much threatened in this country. It depends above all on how much railway development we can keep for ourselves; our schools will attract students in proportion to the jobs they can promise them.[57]

Bouillard's francophone railway was both a novelty and a success. Although the English engineers who constructed the line had been competent, the untrained and ill-paid Chinese who had administered it for Sheng before Bouillard's arrival had not. Bouillard claimed that he found there was 'the most complete arbitrariness and the most appalling disorder that one could imagine', and that Chinese merchants were relieved to be able to ship goods without paying the *fort pot-de-vin* or large bribe demanded by the previous administration. During its first month of operation the railway earned a gross revenue of 102,535 francs, a little over half coming from passenger traffic, or 8,742 francs per kilometre. Bouillard estimated working expenses at 6,500 francs per kilometre, which gave the railway an operating co-efficient of 0.743. In view of the difficulties involved, this was a fine result and Bouillard suggested that after the railway's completion the co-efficient would fall to about 0.450. At that rate it would be a very profitable railway by any standards.[58]

The regard in which Bouillard's work was held by the Quai d'Orsay was revealed when he requested that a specific French engineer be appointed to assist him in his management of the railway. The board of the Société d'études wished to choose the engineer itself, but at the request of the French Foreign Ministry it agreed to Bouillard's request. Delcassé informed Pichon that: 'My Department has been happy to recognize the distinguished services that our compatriot has already rendered to French interests in China'.[59] This support was given Bouillard in spite of Sheng's complaints that the company was employing too many French citizens in senior positions. Sheng was well aware that there was some tension between the French and Belgian elements and sought to exploit this. He told Stoclet that he did not want too many French working on the railway because they were too haughty in dealing with Chinese employees, an attitude which, of course, was not confined to the French amongst the European populations in China; and because in the unlikely event of war between Russia and China, the French, as allies of the Russians, would be able to give Russia quick access to the Yangzi.[60] Informed of Sheng's complaints, Delcassé told Hély d'Oissel to reply in a way that would categorically dismiss them.[61] Hély d'Oissel drafted Stoclet's reply which met entirely Delcassé's requirements. Sheng was informed that any complaints by Chinese against French employees would be investigated, but that as French capital had been necessary to construct the line, it was necessary that French citizens be employed and equipment ordered from French factories. Not to do so, he was told, would lead to serious discontent on the part of French capitalists and even to protests on the part of the French government.[62] This curt reply and the changed conditions in China when it arrived in July 1900 mollified Sheng.

Although the French obtained satisfaction on the issue of personnel, there were other points of dispute between the Belgian and French partners which also required resolution. The most important of these was the question of the distribution of orders for material between French and Belgian factories. The close links between French finance and heavy industry, exemplified in the connections of the members of the Gouin family, and the interest of the French government in promoting manufacturing industry, meant that the concession had always been expected to bring as many industrial as financial benefits. Yet, in the first 18 months following the signing of the contracts hardly any orders at all went to French industry. Gérard regularly collected statistics on such matters and forwarded them to Paris. An alarming report from Joseph Dautremer, the French Consul in Hankou, accused Jadot of buying rails from the Hanyang works, heavy equipment from Belgium, wooden sleepers from Sumatra and Australia, and only two girder bridges from France.[63] By late 1899 the orders were so disappointing from a French point of view that Delcassé had enquiries made as to why French industry was not being accorded its share of the spoils. The Société d'études replied that the Belgians had received all the orders because the French metallurgical industry was hardly able to cope with the orders it had received for the International Paris Exposition of 1900, the construction of the Paris metropolitan railway and other large public works. For this reason it was decided to give most of the early orders to Belgian factories while later orders would go to France.[64] Delcassé was not totally satisfied with this response and wrote to Jules Gouin, president of the Société de construction des Batignolles, enclosing Gérard's statistics and asking if the Société d'études's explanations were accurate. Delcassé further asked whether, if the explanations were true, French industrialists had made the necessary preparations for the flood of orders which could be expected to come from China in 1901 and 1902. In approaching Gouin about this delicate question, Delcassé did not hesitate to remind him that, so far as the French government was concerned, imperialism was largely a quest for markets for her national industry:

I can only regret that circumstances have not yet permitted French industry to share in the advantages involved in the construction of the Beijing–Hankou railway. In lending support to the Franco-Belgian group which sought the concession, my Department was not only considering the financial side of the affair; it seemed to it to be interesting above all because of the work it would bring to our industrial establishments.[65]

Gouin's reply indicated his broad agreement with Delcassé as to the aims for seeking concessions such as the Beijing–Hankou. He pointed out that when the concession had been awarded, French industry had been suffering from a lack of orders for several years and that as a result hoped to attract strong new markets overseas. That situation, however, had changed; not only because of the Paris Exposition and its associated works, but also because of increased investment in new locomotives and rolling stock by French railways. The buoyant domestic market of the last couple of years of the century had led to the inability of French industrialists to take export orders. Nevertheless, as the productivity of factories was improved, the French were able to supply some of the needs of the Beijing–Hankou project; and Gouin was confident that the balance between French and Belgian orders would improve. Significantly, Gouin at no stage suggested the possibility of investment in new plant to increase capacity in order to serve a growing export market. He hoped for nothing more than 'a more rational and constant utilization of the means of production'.[66] While enterprises such as the Beijing–Hankou railway were useful protection from downward turns in the cycle of the domestic market, even for Gouin they were not the basis for any great expansion of heavy industry. Whatever the aspirations of the bureaucrats at the Quai d'Orsay and imperially minded politicians, the commitment of French capitalists to industrial imperialism was never more than half-hearted.

As Gouin suggested, the balance did begin to tip in favour of French industry during 1900. Thus, in January the company ordered 20 locomotives, six heavy main line engines from the French company of Fives-Lille, and six small tank engines from two Belgian companies. However, the country with the reputation for low prices and fast delivery, if not for high quality of finish, the United States of America, received the largest order of eight locomotives which went to Rogers and Company.[67] From the beginning of 1900 Gérard monitored the orders made by the company every month and forwarded the figures to Paris. Clearly, there was no intention that the Belgians should be able to monopolize the industrial side of the enterprise any more than the personnel. In both March and April 1900 the Société d'études placed very large rolling stock orders totalling over three and a half million francs with French firms.[68] As a result, by the end of May there was something far more nearly approaching a balance between the two elements: the French had received orders totalling 6,624,752.84 francs and the Belgians a total of 8,142,794.12 francs.[69] Relatively few orders were made during the rest of 1900 so the imbalance, although less dramatic, continued to exist. Delcassé therefore approached Gouin once again about the matter and reminded him that as the Paris Exposition and the first stage of the metropolitan railway were now complete, the balance between French and Belgian orders should by then have been redressed. He further indicated, very forcefully, that Gouin and his colleagues should be doing more to ensure that it was.[70] Once again French industry was prodded by the almost neo-mercantilist zeal consistently shown by Delcassé, Hanotaux, and the officers of their department. Delcassé was also anxious to limit as much as possible the orders being placed with the Hanyang Iron Works. In this he had the full concurrence of the board of the Société d'études. Unfortunately for the French, however, the only equipment being produced at Hanyang was rails; and, under the terms of the syndicate's agreements, all rails ordered from Europe were to be manufactured in Belgium.[71]

Ultimately French industrialists were well satisfied with the orders received from the Société d'études. After the railway had been completed and international syndicates become the norm for Chinese railway concessions, the president of the Chambre syndicale des fabricants et des constructeurs de matériel pour chemins de fer et tramways, a former Minister for Colonies and engineer, Florent Guillain, referred to the 'good results' obtained by 12 large French firms which had supplied equipment for the Beijing–Hankou railway. They were sufficiently pleased to be enthusiastic about the prospect of supplying the third of the equipment required for the Hankou–Guangzhou and Hankou–Chengdu lines which France would supply as a consequence of her participation in the international loans for those railways.[72] As an experiment in international industrial co-operation the Société d'études was a success. At least in part that success was due to the insistence of the French Foreign Ministry on the importance of the industrial side of the affair, while French capitalists remained concerned above all with its financial aspects.

The support of the French government for the railway was not confined to diplomacy and putting pressure on French industrialists. During the Boxer Rebellion French troops and the French navy were to be involved not only in protecting and capturing the railway but also in its construction. At the time of the Rebellion the northern section from Lugouqiao to Baoding was in operation under the direction of Bouillard whereas the southern section starting from Hankou was still under construction. Although there was condiserable disruption in Henan, Zhang Zhidong's decision as Viceroy at Hankou to dissociate himself from the Rebellion meant that work continued on the southern section of the line. The

Europeans in Hankou, however, were understandably fearful, so Jadot confined works to the southernmost section near Hankou.[73] This involved the dismissal of Chinese labourers who, it was feared, would destroy the work they had just completed in revenge. Gérard, however, considered that the possibility of British forces moving in to protect the line was a greater danger. There was an atmosphere of panic at Hankou, despite Jadot's determination to keep working and Zhang Zhidong's assurances that his troops would protect the railway. To reassure the railway personnel the French and Russian governments insisted that the Russo-Chinese Bank not close its Hankou branch and evacuate its employees as it had planned.[74] By way of further reassurance a French gunboat was ordered to Hankou in mid-August at the company's request.[75] In October Jadot ordered that work recommence throughout the southern section of the line, possibly rather prematurely in view of the tense social conditions which, according the French Consul, still prevailed.[76]

If the situation on the southern section of the railway gave some cause for alarm in the summer of 1900, that in the north was disastrous. The northern section stretched from the heartland of the Boxer movement to Beijing and was an early victim of Boxer attacks. The railway was both a symbol and a potent economic tool of Western penetration. Its large stations, locomotive depots and godowns transformed the face of the towns through which it passed; and it brought with it not just Europeans, but scores of anglophone and francophone Chinese whose acquaintance with European ways and privileged positions made them powerful agents of Western influence. It was exactly such people and such disruptions that the Boxers resented. During May much of the line was destroyed and Bouillard and his French staff took refuge in the Beijing legations.[77] Amazingly enough, after the epicentre of Boxer activity had moved northwards to Beijing, the Chinese personnel at Baoding managed to restore over a hundred kilometres of line and maintain for a time a service of sorts from Gaobeiding to Jiangfu.[78]

The mobilization of an international force to liberate the Europeans and Chinese Christians besieged in the legations in Beijing raised the question of the recapture of the railway. After the entry of this force into Beijing of 14 August 1900 the issue became pressing. The railway, as the *Chargé d'affaires* in Brussels claimed, was the largest French enterprise in the world apart from the

Panama Canal; and one of the principle elements of French influence in East Asia. Moreover, French railway employees and missionaries were besieged in the station of Zhengding near Baoding.[79] Delcassé believed it to be crucial that French and not foreign troops retake the railway and asked Pichon if possible to order the occupation of a strategic point on the railway such as ⋆ Gaobeiding. Once Beijing was relieved, the railway was the highest priority for French forces in Zhili.[80]

Pichon's problem was that there were only eight hundred French troops in the international force of eighteen thousand. While the Japanese and Russian contingents were easily the largest with eight and nearly five thousand troops respectively, the British contingent of three thousand was also impressive.[81] Pichon did not have the resources to occupy Gaobeiding, some hundred kilometres to the south, so instead he ordered a French occupation of the bridge at Lugouqiao on 29 October. Tricolours were flown all around, prompting Sir Claude Macdonald to describe rather testily the practice of placing flags about the place as *enfantillage*. Far from being childish, of course, Pichon was protecting, symbolically as much as by force of arms, and from British as much as Boxers, France's greatest asset in China.[82]

Pichon was unable to send a French contingent as he would have wished to march along the route of the railway to Baoding, so he allowed Bouillard to leave Beijing with British troops on 13 October. Four days later they reached Baoding and to their astonishment found it already occupied by French African troops under the command of Lieutenant Colonel Drude. They had marched from Tianjin at a frantic pace and had occupied Baoding on 15 October without firing a shot. Although Drude did not waste any ammunition he was liberal in his distribution of the tricolours with which the town was covered. General Wilder and his British detachment, having nothing further to do, returned to Beijing. Bouillard was delighted to see French troops in control of the railway, largely because he was convinced that the British wished to seize the line; probably only temporarily, but for sufficient time to create great difficulties for the French management when it returned.[83] Perhaps his fears were exaggerated: at any rate Drude's forced march had prevented the possibility of their realization.

The concern of the French government for the railway was

expressed in even more positive ways than its protection. French sailors and soldiers in China, acting on orders from Paris, actually assisted in the railway's construction. The navy was first to be involved. At Hély d'Oissel's request, Delcassé asked the Minister of Marine, de Lanessan, as early as 17 August if the navy could provide assistance at Dagu, near Tianjin, for the landing of equipment arriving from Europe for use on the railway.[84] De Lanessan was at first reluctant, in view of the navy's role in supporting the expeditionary force, but Delcassé refused to accept his excuses and insisted that he give telegraphic orders to Vice-admiral Pottier at Dagu to give every assistance to the Société d'études in its attempt to land material in the crowded port.[85]

The involvement of French troops was even more direct. Prior to the Boxer Rebellion and the allied occupation of northern China, the Société d'études had sought and obtained the right to extend the railway from Lugouqiao into Beijing itself. However, the Anglo-Chinese company which operated the railway to Tianjin and Shanhaiguan also had aspirations to build the line and succeeded in preventing the decree's imperial ratification. Pichon was able to circumvent its opposition by using the exigencies of the military occupation. He suggested to General Voyon, who commanded the French forces in Beijing, that he use his troops to build the railway on the pretext that he needed rapid communication from his command post in Beijing with the French forces occupying the area around Baoding. Voyon's proposal met with no opposition from either the supreme allied commander, General von Waldersee, or from the British military authorities, so Bouillard started work immediately with the assistance of French troops. Pichon was aware that the railway was illegal, but considered that if the Chinese objected it could be claimed as part of the indemnity owed for the occupation expenses. The rival company accepted the French *fait accompli* and signed an agreement with Bouillard for the transit of its trains over the new line to the capital.[86] The railway into Beijing was opened on 16 March 1901 with considerable ceremony in the presence of Pichon, his Belgian colleague, and General Voyon, who reviewed the troops who had assisted in its construction.[87] Pichon was so impressed with Bouillard's diplomatic and technical talents in this delicate operation that he recommended his decoration.[88] Although the Société d'études promised to repay all the military's expenses, the construction of the Lugouqiao to Beijing extension was a remarkable demonstration of the extent of French official commitment to the railway.

While this commitment extended to the enterprise as a whole, the French government was always anxious to protect the interests of the French element in the company. Although French investors had provided the bulk of the capital, both the president of the company, Stoclet, and its executive manager in China, Jadot, were Belgian. The French directors occasionally used French official pressure to get their way in the internal business of the company. The most revealing example of this occurred in the wake of the Boxer Rebellion. The French directors proposed to send a mission to China to assess the damage done by the Boxers, to support the claim for an indemnity against the Chinese government and generally to see how their money was being spent in China.[89] Diamanti, Jadot's French deputy, had written to the French directors to complain that Jadot's autocratic nature was preventing him from exercising the authority of his position, so Gérard suggested that the mission also investigate the general administration of the project.[90]

Not surprisingly, Stoclet and the Belgian directors opposed the mission on the grounds that it would undermine Jadot's authority and, indeed, question the entire management of the company.[91] Delcassé, however, insisted that the mission be sent and that it must have the authority to inspect and report on the work carried out and to estimate how much more money would be required to complete the railway.[92] Gérard was inclined to conciliate the Belgians by restricting the mission's functions to sorting out the arrangements for the Boxer indemnity, but Bompard bitterly rejected such an approach, insisting that the mission have wide powers so that it could investigate the competence of Jadot's management:

What we are interested in, for it is our responsibility to a certain extent, is knowing what is being done and what will be done with our money in China. And it is even more necessary for us to make enquiries since, if you believe rumours, they are wasting money in all sorts of ways. In China they call the Beijing–Hankou the 'Chinese Panama'. Is there any truth in the rumours we hear? None, I hope, but it is our duty to assure ourselves of that... We all know that [Jadot] has spent 50 million francs. It is a matter of knowing whether if it has been well spent, if for 50 millions spent there are 50 million worth of works done or almost so.[93]

A senior bureaucrat could hardly be more solicitious about his compatriot's investments; although, as he acknowledged, his department was to a certain extent responsible for this investment having been made in the first place.

Because of Stoclet's continued hostility to the mission and Belgian feeling that it was 'useless and dangerous', it was sent not by the Société d'études, but by the French financial syndicate.[94] Since the loan had been floated this body had been dormant, but it was revived for this purpose on the initiative of Hély d'Oissel and met in Paris on 23 March 1901. The bankers decided to send a two-man mission consisting of a financial and a technical expert.[95] The financier they chose was the last manager of the Shanghai branch of the Comptoir national d'escompte, Vouillement. The choice of a suitable engineer was left up to Hély d'Oissel, who asked for Bompard's assistance in the matter. He decided on E. Bousigues, an *ingénieur en chef des ponts et chaussées*, who was employed by the Department of Public Works as the head of the technical control it exercised over France's largest railway company, the Paris–Lyon–Méditerranée (PLM). As a director of the PLM, Hély d'Oissel would have been familiar with Bousigues's work.[96] Faced with the decision of the financial syndicate, the board of the Société d'études agreed to send letters of introduction for the mission's members to Jadot, but Stoclet specifically told him that it was being sent to China 'in spite of the formally expressed opinions of the Belgian directors'.[97]

Neither Stoclet nor Bompard need have been so anxious. Vouillemont and Bousigues left Paris on 1 June and returned on 1 December 1901. Their report was optimistic and highly complimentary of Jadot's management. Bousigues estimated that the railway could be completed by the end of 1905 at a total cost of 105,450,000 francs, of which 85 million were for construction of the Zhengding to Xinyang section and 15 million for the long and difficult bridge over the Huang He. This cost, he believed, could be met without floating a further loan as he estimated the company's resources to total 105,603,413 francs, including the Boxer indemnity of 26,312,413 francs. As well as having sufficient capital to complete the railway, the company was already making considerable profits from those sections already in operation.[98] So far as the railway's European personnel were concerned, by September 1901 French dominance was assured. Of the 176 Europeans, 86 were French, 41 Italian, 37 Belgian, five Swiss, and the other seven anything from English to Ottoman.

In fact, it was necessary to float a further loan of 12,500,000 francs in 1905 to complete the railway;[99] but in general Bousigues's optimism was amply justified and the Huang He bridge was opened on 15 November 1905 thereby enabling trains to run throughout from Beijing to Hankou. The total cost was 172,500,000 francs, of which 33,750,000 had been on rolling stock.[100] As the railway's early results indicated, it proved to be a highly profitable venture. In 1908, its total receipts were 21,810,030 francs (or 9,693,347 Chinese dollars) and its working expenses 7,492,907 francs (or 3,330,181 Chinese dollars), leaving a profit to service the loan of 14,317,123 francs. This was an extremely good result and represented an operating ratio of 0.344. This was even better than Bouillard's 1899 estimate that on completion the ratio would be 0.450 and made the railway one of the most profitable in the world. Under the terms of the operating contract most of these profits, including those above the funds required to service the loan, went to the shareholders of the Société d'études. The Chinese therefore exercised the right of repurchasing the line which had been included in the contract. This was funded by a loan of five million pounds sterling bearing five per cent interest and issued at 94 per cent. Symbolic of the reorientation in French foreign policy since the turn of the century, the loan was issued jointly by the Banque de l'Indo-Chine and the Hongkong and Shanghai Banking Corporation. Thus, Belgian participation in the railway was replaced by British. The railway was intergrated more fully into the Chinese Imperial Railways and its profits beyond those required to service the loan would remain in China. Bouillard, who had been managing its operations since 1899, remained in charge; but the railway did lose its distinctly French ambience after the refinancing brought about direct ownership by the Chinese government on 1 January 1909.[101]

The Beijing–Hankou railway was one of the largest French overseas investments of its time. It was almost certainly the most · profitable. It was the nearest French capital was ever to come in exercising the pretensions to monopoly which had been written into Article 7 of the Franco-Chinese Treaty of 1885. As well as being a highly successful investment, the railway provided a large number of orders to French industry. Yet the concession was always sought more actively by French diplomats and bureaucrats than by the capitalists who had most to gain. While Ferry, de Freycinet, Hanotaux, and Delcassé and the officials of the department were enthusiastic about the prospect of French participation in Chinese railways, and worked hard to ensure that it was as great as possible, they were often frustrated by what de Bezaure once described as 'the inexplicable attitude taken by our financiers'.[102] French capitalists were reluctant imperialists, despite the rhetoric of the age and the conviction of so many senior French bureaucrats and politicians that imperialism was fundamentally an economic movement aimed at finding investments for surplus capital and markets for surplus production. Association with Belgium enabled these aims to be met in China without the necessity for a conquest and the expenses of administration of an overseas empire. Only the capitalists seemed reluctant to take this opportunity and looked to the security of government-guaranteed loans such as they could readily place in Russia. This attitude contrasted all too acutely with that of British capitalists, who were generally willing to make Chinese investments with far less support from their legations and consulates than a French investor. At a time when it appeared as though the French would lose the concession, that ardent imperialist, the Comte de Bezaure, wrote in despair from Shanghai:

China will escape from us like India escaped from us before, only this time not because of the inferiority of our arms or of our diplomacy, but because of the invincible inertia of our capitalists.[103]

He was very nearly right.

5

Banking, Industry,
and the Russian Alliance, 1895–1901

FRENCH bankers had sought to benefit from Western imperalist penetration of China long before French industrialists. As early as 1848 of all years, a branch of the Comptoire d'escompte was opened in Shanghai. This was only three years after the Indian-based Oriental Banking Corporation had become the first British bank in the field. Reorganized in 1862, the Comptoir was the only bank, foreign or Chinese, seriously to challenge the British monopoly of exchange banking in China until 1889.[1] Despite this early start, the lack of French merchant companies, such as the British Jardine Matheson and Company, Dent and Company, or Butterfield and Swire, as potential customers meant that French banking was slow to grow in China. When that growth did occur in the 1890s it was, like so many aspects of French economic endeavour in China, largely under the impulse of official patronage. It is a further testimony of the weakness of France's economic position in East Asia that much of that official patronage was Russian rather than French. This situation, despite the Franco-Russian alliance which had made it possible, resulted in French capital being used to further Russian rather than French diplomatic ends in China, an eventuality which evoked more distress on the part of French diplomats and politicians than of French capitalists.

After 1889 competition in the banking field increased, first with the establishment in that year of the Deutsche-Asiatische Bank as an agent of Wilhelmian financial Weltpolitik, followed by the Yokohama Specie Bank. The powerful Chartered Bank of India, Australia, and China, which had been operating in China since 1857, and the Hongkong and Shanghai Banking Corporation, established in 1865, were the most significant British banks, whereas French banking had not expanded beyond a second branch of the ill-starred Comptoir d'escompte at Tianjin. The collapse of the Comptoir in 1889, an event largely caused by its unsuccessful speculation in copper, almost resulted in the disappearance of French banking in China.[2] Naturally, the new German bank was considered to be anxious to exploit this situation.[3] However, the reconstitution of the collapsed bank with government assistance as the Comptoir national d'escompte ensured the continuity of a modest French presence on the Bund. The reluctance of French bankers to involve themselves more enthusiastically in Chinese affairs requires some explanation in view of France's strength as a creditor nation and the world importance of Paris as a financial centre in the late nineteenth century. Much French foreign investment at this time was within Europe, and particularly in Russia, where loans could be placed backed by the security of guarantee by the imperial government. The security of such investments, enhanced by growing Franco-Russian *entente* culminating in the alliance of 1894, made exotic ventures such as Chinese investments relatively unattractive.[4] Similarly, even French colonies had difficulties attracting investment without government guarantee, as Paul Doumer found in the case of Indo-China. Moreover, there was not a sufficiently large French colony in China, despite the existence of French concessions in Shanghai and elsewhere, to provide an adequate clientele for widespread French banking operations. And the French citizens who were in China were not, on the whole, amongst the more economically dynamic foreign residents. Thus, of the foreign industrial concerns established in China, mostly in Shanghai, by 1885 only two were French, whereas 57 were British.[5] As the directors of the Banque de l'Indo-Chine observed as late as 1897, the natural tendency of foreigners in China to deal with banks of their own nationality, the relative unimportance of French commerce, and the competitiveness of banking there combined to make China an unenticing field of activity.[6]

The real growth of French banking capital in China, together with much else, began following the Treaty of Shimonoseki of 17 April 1895 and the subsequent celebrated 'triple intervention' of Russia, France, and Germany to prevent the Japanese annexation of the Liaodong peninsula. The intervention was undertaken on the initiative of Russia, although Hanotaux quickly became an enthusiast for submitting the so-called 'friendly advice' to the Japanese.[7] The intervention led to the rapid and, from the French point of view, successful conclusion of negotiations concerning the Sino-French frontier and commercial privileges in Yunnan.

Additionally, the requirement that China pay Japan an indemnity of 200 million taels plus a further million in return for the retrocession of the Liaodong peninsula forced China to borrow extensively on the international market.

The Chinese government had anticipated the Japanese demand for an indemnity, and, no doubt aware of the low interest rates prevailing on the Paris money market, two ministers of the *Zongli Yamen* approached the French Minister in Beijing, Auguste Gérard, as early as February 1895 to enquire about the possibility of a loan of about 125 million francs from the Comptoir national d'escompte. Gérard informed his government of the Chinese interest but received no reply until 9 May, the day after the ratification of the Treaty of Shimonoseki, when Hanotaux cabled him with the information that a group of French banks 'of the first rank' was prepared to co-operate in providing the loan which the Chinese government was expected to seek on the European market. At the same time Hanotaux indicated that other powers would also be offering loans.[8]

It was the Russians who had acted most quickly. The desire of the Russian government, and in particular of its Finance Minister, Count Sergius Witte, to extend its influence in China, especially in Manchuria, prompted the Russians to follow the successful diplomatic campaign against Japan which they had organized with a financial coup.[9] Witte was anxious that Russia be the power to offer a loan to China first, so that China could pay the first instalment due to Japan. At Witte's behest, Adolph Rothstein, the president of the International Bank of St Petersburg whom Jules Cambon later described as Witte's *eminence grise*, arranged in Paris for a French syndicate led by Joseph Höttinguer to take up the loan. The French banks involved were the Banque de Paris et des Pays-Bas, MM. Höttinguer et cie, the Crédit lyonnais, the Comptoir national d'escompte, the Société générale, and the Société nationale de credit industriel et commercial. The French Rothschilds declined to be involved as they feared 'international complications' if the British were not included in the arrangement.[10]

Following Rothstein's and Höttinguer's efficient mobilization of French capital in support of the proposed loan, the Russian Minister in Beijing, Count Cassini, was able to offer to Prince Qing, the acting head of the *Zongli Yamen*, a loan of four hundred million francs (one hundred million taels or 15,820,000 pounds) repayable at an interest rate of four per cent over 36 years. During the subsequent negotiations in Beijing Gérard strongly supported Cassini and attended all the sessions, at times soothing ruptures between Cassini and the Chinese officials. Gérard later wrote of his association with Cassini: 'the joint action of the French and Russian ministers was characteristic of this period: it has been fruitful.'[11]

At that time engaged in what proved to be highly profitable negotiations over Yunnan, Gérard constantly emphasized the political nature of the loan, suggesting that it was owed to France and Russia in return for their assistance. Without a hint of irony he argued that the loan would continue the 'task of liberation' which the intervention had begun while simultaneously expressing the opinion that a speedy conclusion was desirable to maintain its political character.[12] Word of the negotiations soon spread around Beijing and Baron Schenck, the German Minister, indicated that he had strong objections to the proposal, while Sir Nicholas O'Conor, the British Minister, and Sir Robert Hart, head of the Imperial Maritime Customs, expressed some not unfounded scepticism that the loan was entirely a commercial venture on the part of the Russians when such a low interest rate was being offered. Hart advised his employers to place the loan with British bankers, but, under pressure from Japan for payment of the indemnity, the Chinese accepted the Russian offer on 6 July. The loan was to be issued at a rate of purchase of 94.125 per cent and repaid at a rate of interest of four per cent.

The ultimate success of the Russian offer was largely the result of the fact that money was available more cheaply on the Paris market than elsewhere, and hence was as much a triumph of the parsimony of the French peasant and *rentier* as of the diplomacy of the 18-month-old Franco-Russian alliance. Russian loans were well known in Paris, as much of Witte's programme for industrial development and railway construction was financed by French loans. A guarantee of repayments by the Russian government, as was accorded the new loan, almost ensured the success of any loan admitted to the Paris Bourse. The loan was issued on 19 July in Paris, St Petersburg, Geneva, Brussels, and Amsterdam and, largely due to its combination of novelty and security (repayments were guaranteed by the Russian government), was massively successful, being covered by subscriptions more than 12 times.[13]

The consortium which had financed the loan soon began to

acquire a more permanent identity. As with the loan itself, the initiative was Russian and affairs were carefully orchestrated by Witte, who, during the discussions of the arrangements for the loan, had obtained the support of the French and Russian bankers involved for the creation of a new bank. Witte asked Rothstein and Edouard Noetzlin, a director of the Banque de Paris et des Pays-Bas who was also closely involved with Fives-Lille, to supervise the project.[14] By October Noetzlin could inform Hanotaux of the arrangements: the bank was to have a capital of 24 million francs (or six million roubles) in shares of 500 francs (or 125 roubles) each. 28,000 shares were to be subscribed to in Paris by the Crédit lyonnais, the Comptoir national d'escompte, and MM. Höttinguer et cie; the remaining 20,000 in St Petersburg. The board of directors, however, was to be dominated by Russians: three members were to be nominated in Paris, two or three in St Petersburg, one in Moscow and one in Shanghai. Ultimately, however, there were to be nine directors. The three nominated in Paris were Noetzlin, who was in fact Swiss, Joseph Höttinguer and Auguste Chabrières, a businessman from Lyon. Witte ensured that the bank would remain very much his own creature by arranging for the appointment of one of his own secretaries, Pokotilov, a former Russian Consul in China, as the bank's agent in Bejing in charge of negotiations with the Chinese government. The bank's other senior official in China, Werth, was also a Russian chosen by Rothstein from the personnel of his International Bank of St Petersburg. The bank was to operate under Russian law and have its head office in St Petersburg. Noetzlin defended these arrangements on the grounds of the 'great importance' of Russian commerce in China and because Witte had promised the new bank both commercial privileges in Russia and political support in China. Despite the appearance of Russian domination Noetzlin considered that the bank would further both French and Russian ambitions:

The bank's activities must include, first of all, French and Russian import–export business in China, as well as existing and future industries the two countries may have there. The bank will also undertake local business in China and, finally, will manage financial operations between the Chinese government and our two countries as well as seek various types of concessions.[15]

The new bank, to be known as the Russo-Chinese Bank, was to have its debut in China expedited by the Comptoir national d'escompte, which transferred its premises, patronage, and personnel (including an American manager, Hincelot) at Shanghai to Witte's creation on the following Chinese New Year, 13 February 1896.[16] The transfer of the Comptoir's branch to this clearly Russian organization distressed Gérard in Beijing, despite Hanotaux's sanguine assurance that the bank's 'principal objective is to develop the commercial interests of France and Russia in China'.[17] Gérard was particularly concerned that the bank's title reflect the French interest it contained, especially in view of the importance of maintaining French prestige in Chinese eyes. He suggested that the bank be named the 'Banque de Paris et de St Petersbourg', or, if this were unacceptable to the Russians, at least the 'Banque russo-chinoise, avec sièges à St Petersbourg et à Paris'.[18]

Both the Ministries of Finance and Foreign Affairs in Paris appreciated the force of Gérard's observations but France's aristocratic ambassador in St Petersburg, the Comte de Montebello, was unsuccessful in his endeavours to extract a change of name from Witte. After de Montebello's departure, the equally aristocratic Chargé d'affaires, the Comte de Vauvineux, was equally unsuccessful in obtaining from Witte even the minor concession of including in the sub-title of the bank a reference to a Paris office. De Vauvineux commented that although the French Ministry of Finance was clearly unhappy with Witte's intransigence, the Russian Minister was unconcerned as 'until now he has, in spite of everything, always obtained from us what he wanted.'[19] At that time Witte had not even authorized the establishment of a Paris branch, nor had the board of directors solicited such a measure.[20]

Apart from the bank's name Gérard was concerned about its personnel and the scope of its activity. He pointed out that of the three senior personnel in China in May 1896, two were Russians, one of whom, Werth, was an experienced banker but a novice in Chinese affairs, while the other, Pokotilov, was a Sinologue but knew little of banking. The third man, although representing the French element was the American, Hincelot. Gérard felt that another man better able to exploit the potential of the bank should be appointed, suggesting that such a man could be more easily found in France than in Russia.[21] Hincelot's death soon after caused Gérard to return to this theme. His concern found a quick response in Paris and Hanotaux asked Finance Minister Georges Cochery to approach the bank's French board members about the appointment of a new French manager in China.[22] Gérard was also concerned about he bank's viability if it confined itself to Russian business in China, pointing out that Russian trade was virtually limited to the export of tea (mainly from Hankou and Yantai) and the import of kerosene. He argued in a way that indicates his quite unfounded optimism for French economic penetration in China and his naïveté concerning the extent of Russian military and political power in East Asia:

If the bank is going to succeed it will be as a result of the initiatives of French industry which, right now and despite numerous difficulties, is making a fair attempt in China, something of which Russian industry would at present be incapable.

He concluded his argument by pointing out that the Chinese were not ignorant that France was 'the world's greatest financial power' nor that Russia was her financial 'tributary'. For these reasons Gérard suggested that it was in the bank's commercial interest to emphasize the French aspects of its character.[23]

The failure of French banks to secure the second Chinese indemnity loan early in 1896 had also led to resentment in Paris of Witte's control of the Russo-Chinese Bank. Initially the Chinese officials sought the funds with which to pay the second instalment due to Japan from an Anglo-German syndicate; however, early in February 1896 they approached Gérard in an attempt to improve the rather onerous terms that the syndicate was offering. Those terms were an interest rate of five per cent and a rate of emission of 89.5 per cent.[24] The Minister of Finances, Paul Doumer, was quick to sound out Paris bankers who might be interested in taking up the loan. Despite a very cold reaction from Gouin of the Banque de Paris et Pays-Bas, Doumer was confident that the proposed loan could be taken up by a group of French banks, including the Russo-Chinese, provided that the Chinese government made a precise statement of the guarantees it was prepared to offer.[25] Witte, however, refused to allow the Russo-Chinese to be involved, on the grounds that the guarantees offered were inadequate.[26] The French banks were reluctant to rush in where the Russo-Chinese feared to tread and were unwilling to offer a loan at terms better than 90 per cent emission. Unimpressed, the Chinese resumed negotiations with Baron Schenck, the German Minister in Beijing. Concerned at the prospect of another French loan, Schenck rashly offered a loan at 94 per cent, an offer which the Chinese found too good to refuse. A contract was signed with the Anglo-German syndicate on 11 March. Schenck, his unnecessary generosity resented in Berlin and London, was demoted to a post in Tangiers for his pains.[27] But although the syndicate had some cause for disappointment with its terms, it had acquired the loan and the French had not. While Gérard expressed disappointment that, once again, French capital had not given French diplomacy the support it had needed in China,[28] in Paris it was Witte who was blamed for the loss of the loan. His refusal to allow the Russo-Chinese Bank to participate had done much to weaken the confidence of French banks in the proposed loan, and de Montebello transmitted French discontent with Witte's role in the affair to St Petersburg.[29]

From the point of view of French diplomacy the Russo-Chinese Bank was clearly a disappointment. In an attempt to change the bank's policies Hanotaux held lengthy discussions with Rothstein in Paris on 11 July 1896. It may have been more appropriate to have opened negotiations directly with Witte, for it was the Imperial Minister of Finance and not the bank's St Petersburg directors who determined its policy. However, Rothstein had clearly been given authority to make concessions to mollify French opinion, and Hanotaux achieved considerable success in the negotiations. His greatest disappointment was that Rothstein refused to entertain the possibility of a change in title to Banque Russo-franco-chinoise. Nevertheless, Rothstein's concessions appeared to be highly beneficial to French interests: it was agreed that the Russo-Chinese would soon open branches in Paris and Lyon, that a French manager would be appointed in China within the year, and that as the bank's Chinese operations were extended, the personnel of new branches would be Russian in northern China but French in the south. This last agreement was qualified by the admission that it would be difficult to find French bankers who could speak Russian, English, and Chinese. Rothstein further agreed that if any railways were conceded to Russia in northern China, French industry would be awarded large orders. To ensure that this would occur it was decided that the bank would employ a French engineer in China.[30]

Paul Hivonnait, a civil engineer in the French Ministry of Public Works, was seconded to the bank from 1 November 1896.[31] The

Comité des Forges and the Société d'études industrielles en Chine briefed Hivonnait as to the types and prices of equipment, notably railway equipment, which they were prepared to supply.[32] The directors of the Russo-Chinese defined his role as 'an engineer attached to their service, in order to undertake missions in China and seek industrial or other business there of interest to the Bank'.[33] In fact, he was to play a significant part in the development of French investment in China, although not as much as hoped in the proposed Transmanchurian railway for which Hanotaux had almost certainly had him in mind at the time of his appointment.

In Beijing, Gérard was encouraged by the bank's appointment of Hivonnait, and also of Wehrung, a French citizen, to replace Hincelot as its Shanghai manager.[34] However, his disappointment at the impossibility of the incorporation of the name of France in the bank's title led him to suggest that another French bank could well be tempted to enter the Chinese banking arena in competition with the Russo-Chinese.[35]

The attempts of French diplomats both in Paris and Beijing to ensure that the bank could be used in their service ignored the reality that during the second half of 1896 it was in neither of those cities but in St Petersburg that the bank's future was being determined. In the wake of the Chinese defeat in 1895 Li Hongzhang had been discredited and was fortunate in losing only his high offices and not his life. Out of favour and an embarrassment to the court, Li was sent to St Petersburg as the Emperor's representative at the coronation of Nicholas II. Travelling in a steamer of the Messageries Maritimes to Port Said, Li was met there by the president of the Russo-Chinese Bank, Prince Esper Ukhtomskii, who accompanied him in a Russian warship to Odessa and ultimately to St Petersburg. After Li's arrival on 30 April 1896, Witte quickly opened negotiations in an attempt to secure a concession for a railway across Manchuria. Such a line would be the final Far Eastern section of the Transsiberian, a railway which was very much Witte's creation, and would bring Russia's railways to the Pacific at Vladivostok. Witte also pressed Li, unsuccessfully, for the concession of a branch line from the Transmanchurian to a warm water port on the Yellow Sea.[36]

Despite the enormous bribe which Li accepted, the former Viceroy was unwilling to meet Witte's demands in full. Refusing to concede the Transmanchurian line directly to the Russian government, he agreed to its construction under a concession to the Russo-Chinese Bank.[37] The concession was easily the largest granted to foreigners in China at that time. Compared with its predecessor, the disastrous Longzhou line which Gérard had secured with such difficulty in 1895, it was a massive project which would bring a Russian political, cultural and, potentially, military presence deep into Manchuria.[38] The price Russia paid for the concession was a 15-year defensive alliance with China directed against Japan. This secret treaty, the text of which was not revealed to the public until 1910, provided that in the event of Japan attacking Chinese, Russian, or Korean territory, the two powers would assist each other with all their available forces. The proposed railway, it was argued, was necessary to enable the Russians to aid China. Additionally, the treaty provided that Russian warships could use Chinese ports for the duration of any hostilities. To the Chinese government it appeared that the concession of the railway was a relatively small price to pay for a treaty which would ensure that the débâcle of 1895 would not be repeated, and, following approval from Beijing, Li, Witte, and the Russian Foreign Minister, Lobanov, signed the treaty in Moscow on 3 June 1896.[39]

Despite its financial and diplomatic interests in these arrangements the French government was never officially informed. Gérard was told of its existence by Li Hongzhang and Prince Qing in March 1897. At that time Gérard was pressing the *Zongli Yamen* for the concession of a railway from the Vietnamese frontier along the valley of the Hong Ha to Kunming. His appeals to the precedent of the Russian Transmanchurian line were rejected and it was intimated that the latter was a 'special arrangement' occasioned by a secret treaty on Chinese defence. When the details of the treaty were leaked to him. Gérard was furious, describing it as an *arcanum imperii* from which France was excluded and suggesting that China had become virtually a Russian Protectorate.[40] Clearly the days were over when the close cooperation between the French and Russian ministers had prompted an English journalist to describe Gérard and Cassini as 'the Siamese twins of Peking diplomacy'.[41] Gérard's cynicism as to Russian aspirations for the Russo-Chinese Bank was fed further by the revelation by the bank's Tianjin manager, Stazzeff, that Witte had originally intended the bank to be Russo-Chinese in capital as well as name and that it was only because the Chinese were unwilling to be involved that Witte turned to French financiers.[42]

The sequel to the secret alliance was the signature on 8 September 1896 of the contract between the Chinese government and the Russo-Chinese Bank for the construction of the line. According to some British sources, there was considerable opposition to the concession in Beijing, overcome partly by Russian threats, partly by Li's generous distribution of some of the funds he had acquired during his travels across Europe and America. The contract maintained the fiction that the railway was a Chinese enterprise: the bank was to establish a company known as the Chinese Eastern Railway, at the head of which would be a Chinese president; the company would be granted considerable taxation, legal, and customs privileges and was promised the assistance of Chinese authorities in a wide range of matters. The company was also given the right to exploit mines and establish industrial and commerical concerns along the line. The concession was for 80 years, although the Chinese government had the right of repurchase after 36 years, albeit on fairly tough terms. In one important respect it was impossible to disguise the company's Russian quality: although the Chinese had adopted the standard gauge of 4 feet 8½ inches (1,435 mm) for all railway construction, the Chinese Eastern Railway was to be built to the Russian gauge of five feet (1,524 mm).[43]

The Russo-Chinese Bank was to subscribe the entire five million roubles capital of the new company. Originally, it was intended that the Russian government would take three and a half million roubles and the remaining one and a half million would be offered to the bank's French shareholders. However, probably because Witte considered French involvement too risky the public sale of those shares held on 29 December 1896 was a carefully orchestrated deception and the French were excluded.[44] Neither the Chinese government nor French capitalists, nor for that matter private Russian investors held any financial interest in the company. Like the Russo-Chinese Bank itself, the Chinese Eastern Railway was to be a creature of the Russian government, and particularly of Witte.

The use which the Russians had made of the bank in furthering their political, economic, and military position in Manchuria excited Gérard's admiration, but also left him disappointed that the bank did not seem to be able to produce the same impressive results for France. This disappointment was felt especially keenly by Gérard, for it had been he who in July 1895 had extracted for France the first foreign railway concession from the Chinese government. By early 1897 it was Russia in Manchuria rather than France in Guangxi and Yunnan which appeared to have gained most from Gérard's early success. More galling still was the fact that the institution which had made possible the Russian advance was funded largely from Paris. Gérard, taking the long view, commented that Witte was continuing the policy first implemented by General Muraviev and Admiral Poutiatine of patiently absorbing much of East Asia into the Russian empire:

Russian diplomacy, skilful at changing according to different circumstances, has understood that the time has come to substitute for generals and admirals, the engineer and the banker, who are now the true conquerors of China.[45]

Gérard's early hopes that the Russo-Chinese Bank would be a powerful agent of French economic penetration in China were rapidly replaced with a deep suspicion of the institution. This attitude was revealed in January 1897 when Hanotaux enquired by telegraph as to the possibility of the Russo-Chinese Bank co-operating with the Franco-Belgian consortium which was hoping to construct the Beijing–Hankou railway. Gérard's advice was against even informing the bank of the proposal, for fear that it would attempt to prevent French involvement and form a Russo-Belgian consortium instead.[46] Although he hoped that pressure from the French government on the bank's three Paris directors might induce it to be more considerate of French interests, Gérard was already arguing early in January 1897 that as Russo-Chinese co-operation was increasing, France was in danger of becoming nothing more than an annoying third party in the enterprise. If this occurred, he considered that French financial and economic interests in East Asia would need to find another means of representation.[47] It was a theme to which Gérard frequently turned. He was profoundly disturbed that French involvement was not indicated in the name of the bank, and in mid-January argued that:

If this name is not restored ... in the title of the Russo-Chinese Bank, then the creation of a French bank in China, that is to say in a country so profoundly concerned with forms, appearances, labels, and ceremonial, appears to me to be a measure as urgent as it is necessary.[48]

Hanotaux forwarded Gérard's advice to the Finance Minster Cochery and indicated his support of this suggestion.[49]

By early March Gérard was unequivocal: if France was to achieve in Yunnan what Russia had in Manchuria it was essential to establish a French bank in China:

The instrument, the lever of our penetration in southern China must be a French bank. The more I reflect on the circumstances which preceded and made possible the Russo-Chinese contract of 8 September last, the more convinced I am that in China's present state it is by a bank, a bank clearly French, that we will obtain in Guangxi, Guangdong, and Yunnan the results that Russia has just obtained in Manchuria. The role played by the Russo-Chinese Bank in the formation of the Chinese Eastern Railway Company and the absolute secrecy the Russian government kept towards us about these negotiations suffice to prove that the Russo-Chinese Bank is above all a Russian institution and what we need to achieve our own ends is an institution which is primarily, indeed rigorously French.[50]

Gérard's argument had one serious flaw in that it ignored the fact that the Russians were aided in their Manchurian negotiations not just by a bank, but also by a bribe and, most importantly, by a secret treaty which placed considerable military burdens on them. Nevertheless, it attracted considerable attention in Paris and ultimately on his initiative a new policy towards French banking in East Asia emerged.

During 1897 the tendency towards the exclusion of French interests from the bank became accentuated. In May of that year Prince Ukhtomskii, who as well as being a friend of both Witte and the Tsar was president of the Russo-Chinese Bank and a director of the Chinese Eastern Railway, visited China, allegedly to present gifts from St Petersburg to the Guangxu Emperor. Ukhtomskii's mission included Baron Ziegler de Schaffhaussen who was an engineer and managing director of the Chinese Eastern Railway. Despite rumours current in Shanghai that Ukhtomskii intended to make a Russian loan available to China to cover the entire indemnity due to Japan, the mission accomplished little in material terms, even though the fact that the telegraph line from Shanghai to Irkutsk was left at its sole disposition for two days indicates that serious business was intended.[51] Symbolically, however, the mission had considerable importance and Ukhtomskii's visit to Beijing painfully emphasized to Gérard the extent to which the Russo-Chinese Bank reflected far more its name than the source of its capital. While in Beijing, Ukhtomskii opened the bank's branch there. The affair would have been entirely Sino-Russian, in language, toasts, and music, had not Gérard rather desperately asked Ukhtomskii for an invitation to speak so that he could remind his audience, in French, of the role of French financiers in establishing the bank and of the circumstances of its foundation.[52]

While clearly distressed at these developments, the French Ministry of Finance believed that, as the Russo-Chinese Bank was the only bank operating in China in which French interests were involved, it would be best to use diplomatic pressure to persuade the Russians to give France 'a reasonable share' in its management. The Ministry of Finance adopted this view partly because no other French bank appeared willing to enter into the competitive Chinese banking area.[53] In contrast, Gérard, following his return to Paris, claimed that an urgent priority of French policy in China had to be the creation of the Franco-Chinese bank which he had earlier advocated.[54] Moreover, Hanotaux's objections to the Russo-Chinese Bank leading the French element in a proposed loan to China in July 1897 indicate clearly that he had already rallied to Gérard's views as to the futility of attempting to use the bank to further French interests in China.[55] This tension over the bank's behaviour indicates that the Franco-Russian Alliance did not always work effectively for the French in China. Although the French enjoyed Russian support in their struggle for concessions in southern China and the Russians needed French funds, in many ways it was an uneasy partnership.

In these circumstances it was scarcely surprising that Hanotaux sought to give French banking the independent presence in China that it had possessed prior to 1896. The reluctance of French banks to open branches in China and consultations with Finance Minister Cochery led Hanotaux to the conclusion that the bank which could most easily be installed in Shanghai was the Banque de l'Indo-Chine. Although owned privately by a consortium of French banks, its official status as the bank issuing currency in Indo-China meant that the government had more control over it than over any other bank. Thus, on Hanotaux's initiative, the Ministers of Finance and Colonies agreed in November 1897 to oblige the Banque de l'Indo-Chine to open a branch in Shanghai.[56]

As anticipated, the bank's directors were unwilling to take this step. Although they were anxious to open a branch in Hong Kong, which was the entrepôt for most of Indo-China's trade, the directors considered that the opening of a branch in Shanghai, as well as being a 'very heavy burden' on the bank, would not achieve the government's aims. This was because there were already French interests represented in the Russo-Chinese Bank; because the Banque de l'Indo-Chine did not possess the necessary expertise either in financing large industrial projects or servicing the tea and silk trades; and above all, because any French bank was ill equipped to succeed in a highly competitive market:

Moreover the competition between various banks is very lively in China and, as a result, the potential profits are minimal. French commerce there is at present less significant than that of certain foreign countries; and merchants naturally tend to deal with Banks of their own nationality.[57]

Cochery and Hanotaux were unmoved by these arguments and the directors deferred to the government's wishes. Ultimately, a decree of the Colonial Ministry authorized the opening of a Shanghai branch of the bank on 12 March 1898.[58]

Witte and Rothstein endeavoured to prevent this French initiative. Rothstein travelled to Paris in December 1897 and made no attempt to hide his opposition to the proposal. However, in the face of French determination, he negotiated an agreement with Cochery which provided for the division of China into two 'spheres of influence' so far as French and Russian banking were concerned. The Banque de l'Indo-Chine was to open its branch in Shanghai but to confine its activities to the south of that city. Similarly, the Russo-Chinese Bank was to confine its activities to the north except for its branch in Hankou which had already been opened. The two banks were generally to assist each other in major projects and to represent each other, except in Hong Kong where the Russo-Chinese was to conserve its freedom to make arrangements with any bank. Additionally, the agreement provided for the creation of two new directorships of the Russo-Chinese, one held by a Russian, the other by a French citizen whose election was to be ratified by the French Ministry of Finance.[59]

These arrangements were accepted both by Hanotaux and by Delnormandie, the president of the Banque de l'Indo-Chine. Delnormandie, however, was unable to persuade a majority of his directors to join him until the bank's statutes were modified to allow it to operate effectively in China. Initially the bank's board favoured the establishment of a subsidiary to operate in China.

The government opposed this arrangement, as the subsidiary, unlike the bank itself, would not be under the control of the government. Ultimately the directors did as the government wished.[60] Gérard's and Hanotaux's ambition of a uniquely French bank operating in China in close consultation with the government had been achieved, even if only because the Banque de l'Indo-Chine's statutes placed it more or less under government control. Indeed, far from being an expression of a dynamic, imperial capitalism, the arrival of the Banque de l'Indo-Chine in Shanghai occurred as a result of official pressure and in spite of the timidity of French capitalists.

The entry of the Banque de l'Indo-Chine into the Chinese scene did not, however, mark the end of Franco-Russian rivalry over control of the Russo-Chinese Bank. If anything that rivalry intensified rather than abated in the ensuing months. The initiative was taken, as usual, by the Russians, who, after all, had appeared to concede most in agreeing to the expansion of a rival French bank in China. During June and July 1898 Witte managed to bring the bank under Russian financial, as well as political, domination. Originally 30,000 of the bank's 48,000 shares were held in France and 18,000 in Russia. Witte arranged for the Russian State Bank to purchase 6,000 of the French shares, which meant that French and Russian holdings were equal. He then proposed to increase the capital of the bank by issuing 12,000 new shares, all of which would be held by the Russian State Bank. Under this proposal 36,000 of the 60,000 shares would be in Russian hands, including 18,000 held by the State Bank, leaving 24,000 in French hands. Additionally, Witte arranged for the bank's statutes to be modified so that the election of directors would need, henceforth, to be confirmed by himself as Imperial Finance Minister.[61] Despite the refusal of Chabrières, Höttinguer, and Noetzlin to sign the minutes on the grounds that the proceedings had been irregular, these proposals were adopted by a meeting of the bank's board in St Petersburg in July.[62]

Although in this way the bank became more truly a Russian institution from mid-1898, the French government determined that through the new French director it would maintain close scrutiny, if not control, of the bank's affairs. The Méline government fell late in June 1898, and with it Gabriel Hanotaux, who was never to return to the Ministry. His successor at the Quai d'Orsay, however, Théophile Delcassé, was equally concerned for the maintenance of French influence in the Russo-Chinese Bank. It was decided, therefore, that the new French director should be Maurice Verstraete, the French commercial agent in St Petersburg. As a French diplomat, Verstraete would, of course, regularly report to the Ministry in Paris. Witte, through his agent in Paris, Arthur Raffalovich, objected to the appointment, largely on the grounds that he would prefer to see a financier rather than a diplomat selected.[63] On this point Delcassé was firm: there were already two French financiers on the board (three if Noetzlin, the Swiss, was included) and there were other non-financiers on the board, including its president, Prince Ukhtomskii.[64] Witte deferred to Delcassé's insistence and approved Verstraete's appointment.[65]

Despite Verstraete's appointment, relations with the Russo-Chinese Bank were often difficult. In 1900 it attempted to establish a branch in Hong Kong, in violation of the agreement of April 1898. Delcassé was willing to allow the bank to do this if in return its directors agreed to allow the Banque de l'Indo-Chine to share in the lucrative business of the Beijing–Hankou railway.[66] Rothstein refused to be involved in such an agreement and Noetzlin agreed that such a sacrifice on the part of the Russo-Chinese was too great to make for the sake of a Hong Kong branch.[67]

During the early years of the new century the Russo-Chinese Bank was plagued by mismanagement, according to Verstraete largely because of Rothstein's 'infatuation with himself' and 'mania for total power'.[68] While Verstraete suggested that Rothstein's German nationality may have provoked his hostile attitude to France, the French ambassador in St Petersburg, the Comte de Montebello, described him as 'our confirmed adversary' as well as 'the fatal man for the Russo-Chinese'.[69] Moreover, the bank continued to take over French assets, and in 1903 was ceded the Calcutta and San Francisco branches of the Comptoir national d'éscompte, the very institution with whose Shanghai branch it had begun operations.[70]

Two events were to transform the bank from the aggressive agent of Russian imperialism into an ordinary commercial institution. In December 1904 Rothstein died.[71] A month later the Russian base at Port Arthur fell to Japanese troops after a long siege. This was the same base Thévenet had built for Li Hongzhang's Beiyang fleet nearly twenty years before. It had been held by Russia since 1898. By the end of May 1905 Russia's Baltic fleet, brought to the Yellow Sea in the hope of retrieving something from the defeat, lay at the bottom of the Straits of Tsushima. With that fleet sank Witte's East Asian policy in which the bank had played such a conspicuous part:

Times have now changed; the Russo-Chinese Bank has changed its direction, it is changing its aspect and taking a different orientation from what it had before the war ... If it has been the auxiliary of M. Witte, the instrument of Russian policy on the China coast, it cannot be that now and in the future: the failure of the Far Eastern policy is manifest, there is little chance of it being renewed and no longer is there need for such a financial institution ... We can think, therefore, that henceforth the bank will confine itself to proper banking operations.[72]

The analysis of the French *Chargé d'affaires* in St Petersburg was accurate, and one of the first acts of new-found humility of the bank's directors was their approach to the Paris Bourse for additional funds.[73] Despite its dramatic, if transitory, success for Russian ambitions in East Asia the bank had been a constant source of irritation and offence for French diplomats. From their point of view, if not from that of French financiers, the bank was a failure. Nevertheless, there could be little cause for joy in Paris at the circumstances which brought about the limitation of the Russo-Chinese Bank's more aggressive activities.

It had been the hope of French diplomats that the establishment of the Russo-Chinese Bank would give French financiers and industrialists great assistance in developing their operations in China. The unexpected tendency of the Russians to seek to monopolize the benefits conferred by the bank's resources meant that these expectations were at times frustrated in both the financial and industrial spheres.

China's need for capital had been the impetus for the bank's foundation and that need did not come to an end with the conclusion of the two indemnity loans of 1895 and 1896. Under the terms of the Treaty of Shimonoseki, China still owed Japan 100 million taels which was scheduled to be paid in six equal and annual instalments from 7 May 1897 to 8 May 1902. Additionally, China was to pay interest on the sum outstanding at the rate of five per cent and the expenses of the Japanese occupation of Weihaiwei. Late in 1896 Gérard analysed China's revenue as best he could and concluded that, to repay this debt, China would be obliged to borrow further funds every year, and suggested that

China's budgetary interests would be better served by borrowing sufficient funds to repay the entire indemnity. As well as saving the Chinese money, immediate repayment would remove the Japanese from Weihaiwei and with them a tangible reminder of China's recent humiliation. With these considerations in mind Li Hongzhang approached Gérard in December 1896 and the Hongkong and Shanghai Bank (which continued to work with the Deutsche-Asiatische Bank) in February 1897 about the possibility of a third loan of about 16 million pounds or 400 million francs to repay the indemnity.[74] Gérard was typically enthusiastic about the proposed loan, particularly if it were to be a uniquely French affair:

The fact that French savings would have another 400 million francs invested in China would be enough, together with the 250 million francs of the first loan, without counting all the other reasons, to assure us a supreme mastery, like a blank cheque . . . The first effect, if the French market accepts this loan alone, would be to give our financial and economic interests in China more scope and independence. As for the risks involved, I estimate that, given China's resources and her financial probity and our possible means of recovering funds, the risks would be reduced to the minimum.[75]

The reality was less brilliant. The Anglo-German loan of April 1896 had not enjoyed very great success on the London and Berlin markets, leaving a large number of shares in the hands of the underwriting syndicate.[76] The success of any further Chinese appeal to those markets seemed doubtful unless repayments were covered by very secure guarantees. The two previous loans had been guaranteed by the revenues of the Imperial Maritime Customs. The balance of those revenues would be insufficient to cover the repayments for a third loan, so further security had to be obtained.

Moreover, Gérard's enthusiasm for a uniquely French loan was not shared in Paris. The head of the Direction commerciale, Maurice Bompard, expressed with some passion his belief that France should receive industrial privileges in return for risking capital in a loan:

The placement of a Chinese loan in France must not be considered a favour to us. On the contrary it is for us to stipulate the advantages we want in return for opening our market to a Chinese loan. In my opinion these advantages must be industrial; in fact in this regard China has treated us with a rare offhandedness. Germany has gained important industrial advantages from her intervention in China's favour; . . . England, which has done nothing for China is as well treated as Germany . . . Only France is deprived of profits. We have obtained *nothing* since we saved her [*sic*]. Our situation in that country, which should rival Russia, is really painful; it would become ridiculous if we went now giving our money to China for the pleasure of obliging her.[77]

British and German bankers were only slightly more enthusiastic about lending China the money than Bompard, and, in order to obtain access to the Paris capital market, sought to include the Russo-Chinese Bank in the consortium. The negotiations, however, soon collapsed.[78]

Li Hongzhang again attempted to raise the loan from two sources in September 1897. Early in the month he approached Dubail, the French *Chargé d'affaires* in Beijing, about the possibility of a uniquely French loan, while his agent in Shanghai, Sheng Xuanhuai, signed a preliminary agreement to borrow the money from the British Hooley–Jameson syndicate. Hooley and Jameson proved to be little more than adventurers and on 10 October both Sheng in Shanghai and Li in Beijing asked de Bezaure, the French Consul-General, and Dubail if the French could finance the loan.[79] Following further discussions with Li Hongzhang, Dubail informed Paris that he believed that France could obtain whatever industrial advantages she wished if her bankers could supply the funds which Li so desperately wanted. This was the kind of opportunity to develop the market for French industrial products in China which the Quai d'Orsay had been seeking. Bompard suggested that in return for supplying the funds, France should demand that French officials participate in the administration of the taxes, notably customs and the salt tax, which would provide the guarantees for the proposed loan; that railways from Beijing to Guangzhou via Hankou, Shanghai to Nanjing, and Laocai to Kunming as well as mines in Yunnan be conceded to French companies; and that ships and arms be ordered from French manufacturers.[80] Possibly Bompard's ambitions reflected the rather overheated and rapacious atmosphere of November 1897, but Hanotaux also favoured consistently an entirely French loan if it were possible. In view of the risks of the loan, all those willing to underwrite it demanded considerable compensation. Cochery as Finance Minister, however, could see great problems with such a loan. He would prefer such a risky loan to be subscribed internationally, so that if the Chinese defaulted any action would be international. Additionally, he was concerned at the possibility of 400 million francs leaving the country at a time when interest rates were increasing and the franc had weakened on the international market.[81]

By December Hanotaux accepted Cochery's arguments as to the desirability of an international loan. At that time the Disconto Gesellschaft of Berlin, the Hongkong and Shanghai Bank, and the Comptoir national d'escompte were discussing the formation of a syndicate, and Hanotaux was prepared to support this combination, although he would prefer to see the Banque de l'Indo-Chine rather than the Comptoir at the head of the French element.[82] However, affairs rapidly drifted out of the hands of French policy makers altogether, as the German acquisition of a long term lease of Qingdao early in January prompted the British government, contrary to its previous policies, to offer a direct loan of 12 million pounds, subject to certain conditions. At the same time rumours were current that Witte had similar intentions.[83] In these circumstances the French were unable to compete and Hanotaux's emphasis shifted from securing French participation in the loan to preventing concessions to Britain which would be inimical to French interests.

The British conditions for a direct government loan included the opening of Dalian on the Liaodong peninsula; the opening of Nanning, and of a treaty port in Hunan or Hubei; a declaration that no part of the Yangzi valley was to be alienated to any third power; and the right to construct a railway from Burma through Yunnan.[84] The politics of loans was drifting into the politics of spheres of influence, and Hanotaux's dual concerns were the protection of French influence in south-western China and the prevention of other powers establishing a monopoly of influence elsewhere. He instructed Dubail to make the French government's attitude quite clear to Chinese officials:

You let them know verbally that: Nanning is in the zone where we have special interests and must be left out of any possible agreement with a third power. For the Yangzi Jiang, no advantage can be given to England. For Yunnan and Sichuan railways; we would consider the concession for the line from Tonkin to Yunnan and Sichuan granted *ipso facto* by a concession for a British line from Burma.[85]

In response to Hanotaux's instructions Dubail delivered a fero-
cious *note verbale* to the *Zongli Yamen* on 30 January, warning the
Chinese that any arrangement with the British concerning Nanning would bring severe French counter-claims.[86] At the same
time the Russians issued a similar warning concerning Dalian and
offered to lend the Chinese the money themselves.[87]

Due partly at least to the parallel, although not co-ordinated
French and Russian campaigns, the Chinese decided not to accept
the British government's offer.[88] An attempt, which can hardly
have been serious, was made to raise the funds in China, while Li
Hongzhang approached Dubail about the possibility of a French
loan of 100 million francs. This sum would only enable the Chinese
to meet the instalment due to Japan in May (about 61 million
francs) and would not bring the Chinese the benefits of borrowing
the larger sum.[89] The French bankers consulted about the loan,
however, were unwilling even to take on this sum.[90] Li's approach
to Dubail may well have been a feint, as negotiations were also
proceeding between the Chinese and the original syndicate of the
Hongkong and Deutsche-Asiatische Banks. These negotiations
came to a successful conclusion during late February, and on 22
March 1898 the third great indemnity loan of 16 million pounds
was offered to the public in London and Berlin. There was no
French participation in the four and a half per cent loan which was
issued to the Chinese at the heavily discounted rate of 85 per cent,
to the syndicate at 88 and to the public at 90. The loan was far
from a spectacular success: in Berlin only 35 per cent of the
German half and in London only 27 per cent of the British half
were subscribed.[91] By October it was being quoted at 86.2 per cent
in Berlin, at which price the syndicate was selling its shares at a
loss.[92]

Bompard's analysis that a Chinese loan was of itself nothing to
pursue with great vigour was, in this case at least, vindicated by the
reception the loan was given by British and German investors. It
is clear from the long negotiations with bankers in Paris and
Brussels that French investors would have been as unenthusiastic
as their British and German colleagues had they been given access
to subscriptions on the Paris Bourse. Nevertheless, European
governments at times went to considerable lengths to obtain the
loan, especially during the later stages of negotiation in 1898.
None went further than the British government, in direct conflict

with its general policy of non-intervention in financial affairs. Hanotaux probably would have liked to have done so as well, but was prevented by the antipathy of French bankers and the relative weakness of France's financial position at the time. The offer of a British government loan was almost certainly prompted by concern that Russia would make such an offer first and by a desire to create a Chinese obligation to Britain which would provide some insurance against further encroachments on China on the model of the German seizure of Qingdao. To all of this France remained something of an onlooker, not threatened by a possible Russian loan, but anxious to ensure that a British loan did not involve Chinese concessions at the expense of what Paris considered to be French interests. In this French diplomacy was reasonably successful, but ultimately no power, apart from Japan, achieved any real benefits from the third indemnity loan. Indeed, even had a French loan been made it is doubtful whether the industrial advantages anticipated by Gérard and demanded by Bompard would have been realized.

Just as industrial advantages were seen by Bompard as the main benefits of lending funds to China, so orders to French industry were part of the price which Witte paid for the French capital used in the first indemnity loan and in establishing the Russo-Chinese Bank. The secondment of the French engineer, Paul Hivonnait, to the service of the new bank was intended to ensure that the anticipated orders were placed. Although employed by the bank, Hivonnait would also represent a group of industrialists headed by the Schneider organization. As well as Schneider's le Creusot works, the artillery industries of the Mediterranean and le Havre and a number of regional industrialists' organizations, the Société des forges et chantiers de la Méditerranée, the Société des chantiers et ateliers de la Gironde, and the Société des ateliers et chantiers de la Loire were united in the group.[93]

Although Hivonnait was to be based in Shanghai, his first destination was St Petersburg where he went in January 1897 in an attempt to secure orders for French companies from the Chinese Eastern Railway. The Railway's purchases committee consisted of two engineers, one of whom was the great Russian railway builder, Kerbedz, who was also the committee's chairman, and two directors of the Russo-Chinese Bank, one of whom was Rothstein. In addition, all important orders had to be approved by Witte himself. In principle, as many orders as possible were to be given to

Russian industry, but its capacity was limited and many orders would have to go to foreign companies. Armed with letters of introduction from Hanotaux and supported by Noetzlin, who visited the Russian capital in February, Hivonnait was given an encouraging reception by both Rothstein and Witte. Kerbedz, however, showed a certain hostility towards French industry, largely because its prices were normally considerably higher than those of English or German manufacturers. Indeed, the high prices which the domestically protected French manufacturers charged were to be a source of great difficulty for Hivonnait in St Petersburg, despite his claims that French products were superior in quality.

During his stay in St Petersburg, Hivonnait attempted to promote the use on the Chinese Eastern Railway of French products of four main types: tugs and barges for the transport of railway equipment on the Ussuri River, rails, telegraphic wire, and locomotives and rolling stock. French prices for such industrial products were higher than those charged by British and German competitors, and very few orders went to France. Satre et cie. of Lyon supplied a dredge for work on the Ussuri River, but the large orders for tugs and barges went to British ship-builders. None of the rails or wire came from France, but three French companies, Fives-Lille, the Société alsacienne, and the Société franco-belge, did supply a total of 50 locomotives in 1899. Their price of 70,000 francs each was 13 per cent more than the German price, but it did involve a considerable discount on the prices French locomotive builders normally received.[94] This order, worth three and a half million francs, went to French industry not because it was competitive, but, according to Witte, because the Tsar had declared that 'political sympathy would in this case be more important than economic interests'.[95] These 50 locomotives were the only large order placed in France for the Chinese Eastern Railway. Hivonnait later observed that 'the very high prices generally demanded by our metallurgical industry have unfortunately not allowed us to obtain further orders'.[96]

From St Petersburg Hivonnait travelled to Shanghai via New York and San Francisco. On arrival he quickly found himself involved in a new railway project. Following represensations from two British syndicates, the Chinese government decided to construct a railway from Zhengding, on the Beijing–Hankou line in Zhili, to Taiyuan in Shanxi. Chinese officials approached French finance to build the railway, an approach no doubt taken in order to provide some counterweight to growing British intersts in Shanxi. The *Chargé d'affaires*, Dubail, believing rightly that there was no chance of interesting French banks, 'who appear to rebel at the idea of exporting their funds to China', advised the manager of the Russo-Chinese Bank, Pokotilov. In doing so Dubail stated that he believed any contract should include provision that material be obtained from French factories where possible. Pokotilov was enthusiastic and Hivonnait was almost immediately sent to Shanxi to begin surveys of the route.[97] In St Petersburg Rothstein was also enthusiastic. At the time he was fulminating against the French for their involvement with Belgium in the recently conceded Beijing–Hankou line. His discontent was probably largely motivated by his desire to see French capital enter China only through the Russo-Chinese Bank: at any rate he had no objections to the Shanxi scheme.[98]

Hivonnait worked quickly and in Novemeber submitted a report to Pokotilov recommending the construction of a 246-kilometre metre gauge line, the cost of which he estimated at 25 million francs. He considered that the line would tap a large traffic, especially as there were potentially rich coal and iron mines in the area. If desired, the line could be built in two stages, the first of which, from Zhengding to Pingding where there were fine anthracite mines, would cost about 13 million francs.[99] Following this favourable report Pokotilov signed a tentative contract with the line's Chinese concessionaire in December 1897.[100]

This contract brought the Russo-Chinese Bank into direct competition with the Anglo-Italian Peking Syndicate. The syndicate's representative in Beijing, Angelo Luzzati, had been acquiring concessions in Shanxi by the liberal distribution of bribes estimated at 300,000 francs and with the strong support of the British Minister, Sir Claude Macdonald, and the Italian *Chargé d'affaires*. Although initially a rather shaky affair, during 1898 the Peking Syndicate acquired the financial backing of the London financiers, Carl Meyer and Lord Rothschild. For Macdonald the syndicate's claims in Shanxi were a means of advancing British interests in an area which, since the Beijing–Hankou railway had been conceded to a Franco-Belgian group, appeared to be falling under French and Russian domination.[101] Pokotilov sought to protect the Russo-Chinese Bank's interests by using similar techniques to Luzzati, distributing 50,000 taels worth of shares in the proposed railway and enlisting the support of the French and Russian legations. Although there were strong elements of the politics of spheres of influence in this competition, it was ultimately resolved commercially. On 26 April 1898 Pokotilov and Luzzati met in Li Hongzhang's office and signed an agreement by which the Russo-Chinese Bank and the Peking Syndicate respected each other's concessions.[102] The syndicate retained its mining concessions and gained the right to build branch railways to serve its mines, while the bank retained the concessions of the main railway line and some mines near Taiyuan and agreed to charge fair rates for the transport of the syndicate's minerals on its railway.[103] Thus, on 21 May 1898 two contracts were signed and approved by the *Zongli Yamen* in Beijing. The first conceded the exploitation of coal, oil, and iron deposits in about a quarter of Shanxi to the Peking Syndicate; the second was for the construction of the Zhengding–Taiyuan railway by the Russo-Chinese Bank.[104]

Simultaneously with the negotiations in Beijing which led to the granting of the concession, attempts were being made in Paris and St Petersburg to raise the capital necessary to implement the proposal. Once again French financiers were reluctant to involve themselves in what they regarded as an area of high risk. Hanotaux persuaded a five-member industrial group to send a party of engineers to survey the line in detail. By January 1899 they had reported favourably, but potential investors remained unwilling to sink their capital in the project.

If the future of the proposed railway was looking bleak in Paris by mid-1900, conditions in Shanxi were even worse. The social unrest spreading through northern China lead to increasingly anti-foreign official attitudes. The Shanxi Board of Trade, despite the payments its members received from the bank, was uncooperative and refused even to accept Pokotilov's telegrams about the matter.[105] Asked for support, Pichon conveyed Pokotilov's complaints to the *Zongli Yamen*, which replied that as there was no mention of 'French merchants' constructing the line in the contract, dispute was to be expected.[106]

Faced with its difficulties in obtaining French industrial and financial support, the bank had proposed to construct the first 25 kilometres from Zhengding itself, using standard gauge equipment from the Beijing–Hankou line. However, the Boxer Rebellion made even that project impossible as the Beijing–Hankou line was itself largely destroyed in the Rebellion.

The Shanxi railway was finally constructed by a Franco-Russian consortium organized by the Russo-Chinese Bank, but under very different conditions from those envisaged in the contract of 1898. In October 1902 Sheng Xuanhuai, then Director of the Chinese Imperial Railways, and Wehrung, the manager of he bank's Shanghai branch, signed two contracts for the construction and operation of the railway. The line was to be built at a cost of 48 million francs, 40 million of which were to be raised on the Paris capital market. The interest rate was five per cent, the rate of emission 90 per cent, and the loan was guaranteed by the Chinese government. The new metre gauge line was to be part of the Chinese Imperial Railways, but operated by the consortium. The railway was completed in three years and rapidly became as profitable as Hivonnait had predicted.[107]

Ultimately, then, French industry did benefit from the association of French capital with the Russo-Chinese Bank. Those benefits, however, were neither as rapid nor as great as could reasonably have been expected in 1895 and 1896. The 50 locomotives ordered for the Chinese Eastern Railway were a very small part of the industrial products needed for Russia's penetration of Manchuria. Similarly, the Shanxi railway was not a large venture and its construction did not begin until 1903. French diplomats tended to blame the Russians for these very modest results: however, the conservatism of French bankers and the high prices charged by protected French industry were probably more significant factors behind the French failure to achieve a larger share in China's industrial development through association with the Russo-Chinese Bank. The relative failure of the French to gain the economic advantages, in particular the industrial advantages, they expected from association with Russia in Chinese affairs is significant. Combined with the tension between the allies caused by the operations of the Russo-Chinese Bank, it suggests that some revision is needed of Langer's assessments that 'the Franco-Russian Alliance reached its finest bloom in the Far East in the years 1896–1897' and that 'the French found and gave support in all they did, while the Russians, through threats, bribery and intrigue all but succeeded in realizing the full program outlined by Witte'.[108]

Despite this qualification, the first indemnity loan, the Russo-Chinese Bank, and the industrial markets which were secured through the bank were very large outlets for French capital and significant outlets for her industrial production. In a very real sense the Franco-Russian Alliance had contributed to the emergence of further French economic interests outside of what could be regarded as the French sphere of influence adjacent to Tonkin. Thus, an alliance conceived in order to give France European security helped to develop the diffusion of French investment in China. Ironically, it was to be this very diffusion of investment that was ultimately to lead France and Russia to adopt contrary policies on the issue of spheres of influence in China.

Part II

Economic Penetration versus Territorial Control: France and South-western China

Part II

Economic Penetration
versus Territorial Control:
France and Southwest China

6

Early Attempts at Penetration
and the Longzhou Railway, 1885–1899

WHILE the post-1885 French financial and industrial offensive was under way around the shores of the Bohai Sea, those Chinese provinces adjacent to Vietnam, the allure of which had prompted Ferry into war, remained relatively neglected by diplomats and investors, and virtually impenetrable for merchants. There were considerable obstacles, both political and technological, to the realization of Ferry's aim of using the new Protectorate as a bridgehead to provide access to a new market for French goods and new avenues for the investment of French surplus capital within the Chinese empire. The greatest initial problem facing the French was the pacification of the newly 'protected' population in Tonkin. The Black Flags and Vietnamese anti-colonialists were to wage guerrilla war against the French army of occupation for most of the remainder of the century before the pacification of Tonkin was at least nominally complete. The activities of these guerrillas, whom the French described as 'pirates', were concentrated along the Hong Ha, where they made commerce risky and expensive, and along the Chinese border, which was not clearly defined in many places until after 1895. Attacks on frontier towns such as Langson were common until after the appointment of the francophile General Su Yuanchuan as the commander of the military forces in Guangxi. Until Su's arrival the Chinese authorities gave support to the anti-French guerrillas and in August 1892 Chinese regular soldiers were amongst prisoners taken during a pirate raid near Langson.[1] Attacks, however, were not confined to the more remote parts of the Protectorate, and as late as 15 December 1897 four or five hundred 'pirates' attacked Haiphong, burning many buildings and killing a French book-keeper.[2] Indeed, the concrete symbol *par excellence* of imperial civilization and exploitation, the railway, was introduced into Tonkin for military rather than commercial reasons. The guerrillas, reluctant to accept the innovation for either reason, attacked a train on this tiny 60-centimetre gauge line from Phulangthuong to Langson in September 1894, resulting in considerable loss of life.

If the insecurity of the political and military situation was far from conducive to the development of European trade with China through Tonkin, the inadequacies of the infrastructure necessary for such a trade were equally discouraging. For although gunboats and patrols by the Foreign Legion could remove one of the barriers to the navigation of the Hong Ha, they could not alter the fact that the river was only navigable by steamers to the frontier town of Laocai during the season of high water, from June to October. It was not until July 1890 that the especially constructed and optimistically named *Yunnan* became the first steamer to reach Laocai. More than three years were to pass before the Société des correspondances fluviales was established to exploit the route and the new company's directors began work on improving the channel (at the government's expense) to enable them to operate the service throughout the year.[3] Even then the junks charged about half the rate of the small French steamers, consequently the bulk of the trade remained in Vietnamese and Chinese hands, despite the slowness and numerous irregularities.[4]

Significantly, it had not been French commercial interests which had established the possibility of steam navigation on the Hong Ha throughout the year, but rather French military imperatives. Following an appeal for reinforcements from Colonel Pennequin, commander of the fourth military district of Tonkin around Laocai, Léon Escande, commander of the gunboat, *Moulun*, made three voyages to Laocai between March 1893 and June 1894, the first of these at the season of lowest waters. The voyages were not completed without difficulty, but Escande concluded that navigation was possible all year for steamers drawing no more than one metre of water.[5] This modest success suggested to Escande a vision of a fleet of French steamers penetrating towards the heart of China, a vision which he considered could be appropriate compensation for the collapse of Dupleix's work in India.[6] Escande, like Ferry, saw the results of his work providing for France the mercantilist privileges necessary for her survival in an age of increasingly vicious industrial competition:

Surpluses in industrial production, the direct result of the use of machines, have driven all the civilized powers, especially during the last quarter century, to seek new outlets for the sale and exchange of their manufactured products... The original aim of the conquest of Tonkin was to ensure that France had the shortest and easiest means of penetration for European commerce into this immense agglomeration of population which lives, so to speak, isolated from the rest of the world in the centre of Asia.[7]

Unfortunately for Escande's vision, however, continuing difficulties of steam navigation during low waters precluded the development of an efficient regular service, and, to compound the problem, during the wet season; when such navigation was possible, rain made the roads on the plateau of Yunnan impassable, bringing commerce to a halt.[8] Thus, ten years after the conquest of Tonkin, the technology of industrial Europe had made singularly little impression on trade across the Sino-Vietnamese frontier.

Although the eventual opening (in 1889) as treaty 'ports' for foreign trade of Mengzi in Yunnan and Longzhou in Guangxi did lead to the development of a modest trade between the new French Protectorate and the provinces of south-western China, the pattern of exchanges that developed was not at all what the French negotiators had anticipated when they secured this right by the convention of 26 June 1887. The first French Consul at Mengzi, Emile Rocher, found that although Chinese merchants began importing goods with enthusiasm from Tonkin into Yunnan, not one French merchant took advantage of the new route. These Chinese merchants continued to obtain goods of British, Indian, and German origin in Guangzhou or Hong Kong, so the development of the trade was of no benefit to French industry whatever.[9] This use of Hong Kong rather than Hanoi as a source of European goods was reinforced by the fact that tin, Yunnan's major export, could be sold readily on the international market in the British colony, but not in Indo-China.[10] Despite the difficulties of navigation along the Hong Ha, the trade, still entirely in Chinese hands, continued to grow rapidly, largely as a result of the security provided by the French military presence in Vietnam. But fully 87 per cent of imports were consigned from Hong Kong, and the largest item amongst these was cotton goods manufactured in India. Although some Indo-Chinese products (notably cotton, wool, and timber) found a market in Yunnan, not one article produced in metropolitan France was imported into China through Mengzi throughout 1892.[11] Moreover, despite its growth and potential, the trade through Laocai and Mengzi represented only a small portion of that of the provinces of Yunnan, western Guizhou, and southern Sichuan. It was suffering from competition by rival routes, largely because transport along the Hong Ha remained so primitive and so expensive. The opening of Chongqing to foreign trade in March 1891 enabled goods to be shipped by steamer to Yichang, thence by a remarkable system of human-drawn junks through the Yangzi gorges to the newly opened port, whence they could be distributed readily to much of Sichuan, Guizhou, and Yunnan. Besides the immediate competition of the Yangzi route to Chongqing, Rocher signalled the longer term danger of British penetration from Burma. Whereas the Hong Ha route remained unimproved by its French administrators, the English had very quickly constructed a railway parallel to the perfectly navigable Irrawaddy from Rangoon to Mandalay, following their conquest of upper Burma in 1886, and were planning extension towards the Sino-Burmese frontier.[12]

If the nature of the trade through Laocai and Mengzi was disappointing for mercantilist-minded French officials, that through the other inland treaty port opened in 1889, Longzhou in Guangxi, was the cause of considerable anxiety. Essentially trade across the frontier was of a purely local nature. During the first few months following the port's opening, not one European article was imported into Guangxi through Vietnam, and it was reported that French merchants in the Protectorate were not making any serious efforts to exploit the new opportunity offered them. This situation prompted the Quai d'Orsay to urge the minister responsible for colonies to take measures to stimulate the trade.[13] However, the disadvantages of the French route into Guangxi were such that much more than official encouragement was needed to overcome them. In 1891 the freight from Haiphong via Phulangthuong and Langson to Longzhou cost about eight piastres per picul (about sixty kilograms). In contrast, the rate from Guangzhou to Longzhou direct via the Xi Jiang (or West River) was only half a piastre. As there were 17 likin (*lijin*) stations along the riverine route, it tended to be used, however, only for shipments of considerable bulk but little value. The most favoured route was that from the Guangdong port of Beihai (Pakhoi), by porter to the Xi Jiang at Nanning, which was itself a much more substantial commercial entrepôt than Longzhou, thence by water to Longzhou. Freight by this route cost three piastres per picul.

Pierre Bons d'Anty, the port's first Consul, estimated that if freight rates from Haiphong could be reduced to a level competitive with those from Beihai, the French port would capture about half of the latter's trade, or some 13 million francs worth per annum, about nine million of which would consist of foreign imports.[14] This compared with a Sino-Vietnamese trade via Longzhou estimated at about two million francs in 1892, two-thirds of which would have been contraband opium.[15] Indeed, far from being a centre of importation of European goods into China, Longzhou became one for their exportation into Vietnam. By early 1893 an eighth of the port's exports were goods of European origin destined for Vietnam's Caobang province, and the nadir was reached in September 1893 when small quantities of European red wine and mustard were exported from China into the French Protectorate.[16]

The opening of Longzhou was clearly having exactly the opposite effect from that intended, and on a number of occasions the Quai d'Orsay encouraged the government of the Protectorate to complete the railway to Langson as soon as possible in order to increase the competitiveness of the French route.[17] The 100-kilometre line from Phulangthuong to Langson was to be completed in July 1894. It, however, was being constructed essentially for military purposes, and its narrow gauge and light contruction made it unsuitable for carrying a very large commercial traffic. Certainly it was a very fragile means of penetration. Nevertheless, the Indo-Chinese customs, determined to encourage trade along the new route, simplified transit procedures on goods entering the Protectorate bound for Longzhou, and the Hanoi-based entrepreneurs, Marty et d'Abbadie, who controlled most of the riverine and coastal shipping in Tonkin, began quoting through freight rates from Hong Kong into Guangxi via Haiphong and Longzhou.[18]

Despite these efforts, the value of trade recorded at the Longzhou customs office actually declined by two-thirds in 1895 as compared with 1894. The Chinese Customs Commissioner there (an American) blamed difficulties with navigation, and the complications of the French customs procedures were modified.[19] A French journalist who visited Longzhou in 1895 was disappointed to find a quiet town with a population of only 20,000. Expecting 'a market of the first order, a vast entrepôt for goods destined for Guangxi and Yunnan', he found instead a tropical backwater, and, comparing it with other Chinese cities, concluded that 'Longzhou looks like a ghost town'.[20]

In view of these problems of pacification and communication in the new Protectorate, it is perhaps not surprising that its potential uses as a means of penetration into China were neglected for nearly a decade after its conquest. Additionally, Victor-Gabriel Lemaire, who was the French Minister in Beijing from July 1887 to March 1894, was preoccupied with the fortunes of the Syndicat de la mission de l'industrie française en Chine and hence his diplomatic activity tended to be exercised more forcefully in matters concerning French interests in northern China. The most pressing concerns of the French colonial administration in Indo-China remained pacification and frontier delimitation, both necessary prerequisites to the development of international overland trade. France's difficulties in Indo-China seemed greater on the Siamese than on the Chinese border, and the rather heavy-handed manner in which the French obtained the frontier they demanded from the Siamese court almost provoked war with England in July 1893.[21] English objections were mollified by agreements signed in Paris at the time of the crisis and in November 1893 which provided for the creation of a buffer state between the French and British possessions on the upper Mekong.[22]

Despite the attention given to these more urgent matters by French officials in Beijing, Saigon, and Paris, Ferry's original aims of 1883 remained central to French policy. Given the speed with which the English had constructed the railway from Rangoon to Mandalay and Bhamo,[23] and Salisbury's interest in Holt Hallet's scheme for a Burma to Yunnan railway (an interest dating back to 1866),[24] French policy makers could hardly ignore the fact that there was what one English traveller described as 'a race for Yunnan' in progress.[25] In case they were tempted to do so, a plethora of French publicists ensured that they could not. The most notable of these were Jules Ferry himself, whose *Le Tonkin et la Mère-patrie* articulately reiterated in 1890 the mercantilist vision which had led to his downfall,[26] and Prince Henri d'Orléans, who, after travelling in the area, suggested that in possessing Tonkin, France 'held all the trumps ... [and] may win the game with the products of [her] national industry in the great markets of China'.[27] A truer liberal than many members of the French bourgeoisie, the Prince argued that all that was necessary to achieve this vision was more initiative on the part of French merchants and less reliance on protective duties, reminding his readers that Tonkin was taken 'in the first place, in order to gain access to China' and lamenting that 'this primary object seems to me to have been quite lost sight of'.[28] A former associate of Paul Bert during his residency in Tonkin, Joseph Chailley-Bert, joined d'Orléans in praising the British example and urging the French to use their Indo-Chinese possessions for commercial activity rather than leave them as a preserve for inefficient functionaries.[29] The Consul in Mengzi from 1889 to 1893, Emile Rocher, had published a two-volume book describing the riches and potential of Yunnan as early as 1879,[30] while Louis Pichon was probably the most wildly optimistic French publicist discussing the possibilities of trade with China during the early 1890s. He considered that the establishment at Mengzi of a French commercial establishment, offering a wide variety of French goods for sale and distribution throughout Yunnan, would have spectacular results:

Our commerce would soon undergo a colossal growth which would recall the era of the merchant princes of Shanghai and Hong Kong... We would find in Yunnan, as in the first Chinese ports opened to Western trade, millions of people famished for want of European products... And all this trade is ours if we want it, because we have the shortest and hence the cheapest route for the penetration of China.[31]

Despite the developing awareness of the fact that Yunnan was not the eldorado it had been described as earlier in the century, such propaganda about the dazzling prospects for trade was reflected by British publicists, who sought to see the province penetrated either from Chongqing or Burma.[32]

The climate of opinion whch publicists endeavoured to create, the gradual pacification of Tonkin, and the desire to secure a more impressive French than British presence in south-western China, all combined to turn the attention of French diplomats to the area by 1894. Additionally, during April that year Auguste Gérard arrived in Beijing to begin a forceful, at times even aggressive, term as French Minister there, while a month later Gabriel Hanotaux, an equally committed imperialist and admirer of Ferry, was called from his desk at the Quai d'Orsay to become Foreign Minister.[33] Hanotaux was concerned at the apparent inability of French industrialists to find very extensive markets in China. In his commercial instructions to Gérard he pointed out that, alone amongst the Western powers, France had an unfavourable balance of trade with China, importing in 1892 135,212,231 francs worth of Chinese products, mostly silk, and exporting French products with a value of only 3,168,507 francs. Gérard's primary task was to improve France's export potential in China, although Hanotaux acknowledged that the disastrous histories of the Comptoir d'escompte, the Thévenet syndicate, and inferior bridges and dredges delivered to the Chinese would make this task a difficult one. Given the failure of Lemaire's policy to secure industrial concessions in northern China and the bitterness that his execution of that policy had created, it is scarcely surprising that the new team turned to a new field of activity. As the political heirs of Ferry and at once the rivals and admirers of the success achieved by British imperialists and British capital, Hanotaux and Gérard naturally turned to the Chinese provinces adjacent to the Protectorate which Ferry had conquered.

Thus, Hanotaux sent instructions to Gérard in September 1894 which required the new Minister to negotiate a definitive Sino-Vietnamese frontier which would eliminate any buffer state between British and French possessions east of the Mekong. Two months later Gérard was instructed to renegotiate the commercial conventions of 1886 and 1887 in such a way as to extend the opportunities for French trade and investment.[34] Negotiations proceeded slowly, as the Chinese were aware that the French frontier demands would lead to British objections, and considered the commercial ones a violation of her sovereignty. Gérard opened discussion with Prince Qing and other members of the *Zongli Yamen* on 24 September 1894 and the negotiations continued throughout what were for China the disastrous months of the war with Japan. Initially, the *Zongli Yamen*, 'faithful to their tactic of procrastination' as Gérard acidly, if rather insensitively commented, argued that the issue of the definition of the frontier should be settled first, as both parties agreed that there were errors in the Tianjin agreements, before proceeding the discuss the commercial question. Gérard, however, insisted that the two issues be discussed together. In October 1894 Prince Gong (1833–98), who had been dismissed from the *Zongli Yamen* at the time of China's defeat by France ten years earlier, was restored to office. In December he agreed to Gérard's demands, and in March 1895 was presented with a draft which provided for the creation of three French consulates in Guangdong and Yunnan; the linking of China's and Indo-China's telegraphs; the definition of conditions

under which trade would be conducted between Indo-China and
China, including provision for the importation of Vietnamese salt,
the troublesome commodity of 1873, into Yunnan, Guangxi, and
Guangdong; and an agreement that China consult French
engineers first when developing mines in those three provinces.
The *Zongli Yamen*'s reply, received a month later, was favourable
to much of the draft, but indicated strong opposition to the
proposed arrangements for goods to pass through Vietnam *en
route* from Yunnan to other Chinese provinces, to the importation
of salt, and to French mining privileges.[35]

After some prevarication, Prince Qing, acting during Prince
Gong's illness, resumed discussions on 9 May 1895, the day after
the ratification of the Treaty of Shimonoseki. In the new atmosphere created by China's defeat and the intervention of France,
Russia, and Germany on her behalf, agreement was reached fairly
quickly. The French were obliged to renounce their claims to the
salt and opium trades, as the former was an imperial monopoly
and the latter regulated by the Anglo-Chinese Chefoo Convention. In return for French acquiescence in these matters, the
Zongli Yamen agreed to the extension of Vietnamese railways into
China, subject to further negotiation between the two governments. Gérard gained partial satisfaction on the mines' issue, and
complete satisfaction on the customs arrangements for the transit
of Chinese goods through Vietnam. The newly claimed French
railway privilege was the result of agitation by the heavy engineering company of Fives-Lille. Late in 1893 the company made
surveys, at the request of the Resident-General, de Lanessan, for
the conversion of the Phulangthuong railway from 60-centimetre
to metre gauge and for the line's extension southwards to Hanoi
and northwards to the Chinese frontier. Fives-Lille and the Indo-
Chinese government signed a contract providing for the execution
of these works on 19 February 1895. During these negotiations the
company became aware of the potential for further extension into
China, so two of its senior directors, Sébastien Jean-Baptiste
Krantz, a senator and former minister, and his brother-in-law,
Edmond Duval, presented their case for such a project to
Alexandre Ribot, who was then premier. Ribot was impressed
with their arguments that the line would put an end to piracy in the
region and ensure French preponderance in southern China, so
sent the directors to the equally impressed Hanotaux with his
blessing.[36] It was a fateful decision. It meant, in effect, that the

direction in which France was going to attempt to penetrate China from Tonkin had been determined, not by the Department of Foreign Affairs, whose Consuls knew something about Chinese trade and conditions, nor even by well-informed company directors acting on the basis of reports as to the proposed railway's traffic, but by a company which was seeking only to extend its contract with the colonial government and thereby sell more rails and other materials. It was to be a very bad and wasteful decision.

Thus on 20 June 1895 Gérard and Prince Qing signed two documents in the *Zongli Yamen*. The first was a convention for the delineation of the frontier in a manner that favoured France at the expense of China and England; and the second a commercial convention containing, besides the provisions outlined above, añ agreement that a system of mixed Sino-French police would be established along the frontier. Probably the most notable innovation in the convention was Article 5, which provided for the extension of French railways into China.[37] There was, however, a bizarre delay of some two hours that day, while the British Minister, Sir Nicholas O'Conor, having arrived at the *Zongli Yamen* shortly before Gérard, attempted to intimidate or cajole Qing into refusing to sign the conventions on the ground that they were incompatible with Article 5 of the Anglo-Chinese convention relative to Burma of 1 March 1894. The incident marked the nadir of British influence in Beijing, and led to O'Conor's recall soon afterwards.

British resentment of the Sino-French convention of 20 June 1895 was considerable. O'Conor claimed that Gérard and Cassini had threatened to withhold payment of the indemnity loan if China did not agree to the convention. The London Chamber of Commerce believed that it afforded a 'favourable opportunity' to demand the opening of the Xi Jiang to Western trade. This view was supported by Beauclerk, the *Chargé d'affaires* replacing O'Conor in Beijing, who suggested that 'the best counter to French machinations in this area would consist in the opening of the Xi Jiang'.[38] Both O'Conor and the India Office wanted to obtain from China a 'specific assurance of equal mining and railway privileges' in south-western China, while the India Office also wanted rectification of the Sino-Burmese frontier in Britain's favour, not so much because of the intrinsic value of any of the territory claimed as to protect possible future railway routes from Burma into Yunnan.[39]

Negotiations on the British claims for 'compensation' continued through late 1895 and all of 1896. The *Zongli Yamen*, bowing to the relentless British pressure, presented a memorial to the Throne urging the opening of the Xi Jiang as early as 20 December 1895. It was another year, however, before the issue was resolved. Ultimately Prince Qing agreed to minor cessions of territory; to construct future railways in Yunnan to connect with Burmese lines, railways which nobody seriously expected to be built in the fore-seeable future; and to open the ports of Wuzhou and Sanshui on the lower Xi Jiang to foreign trade. Salisbury had expected Nanning to be opened as well and observed that 'a special importance attaches to Nanning in view of French designs'. He ultimately settled for Prince Qing's written promise that 'they will open Nanning immediately the French should be permitted to extend their Longzhou line to Bose'.[40] This Anglo-Chinese convention was signed on 4 February 1897.

As yet unaware that British counter-claims would rather tarnish his achievement, Gérard believed that the signing of the Sino-French convention of 20 June 1895 was highly significant. He was convinced that it would enable the realization of France's long-held mercantilist ambitions:

The work begun in 1885 with the Treaty of Tianjin was thus completed ten years later in the most propitious circumstances, not only for the preservation of the security of our Indo-Chinese possessions from any new threats, but also making Indo-China the most direct and shortest means of penetration into the south and west of the great empire. The dream of the first explorers, Garnier, Doudart de Lagrée, Dupuis, had become a reality. The Hong Ha was going to be the great means of communication between Indo-China and the hitherto almost inaccessible provinces of southern and western China.[41]

It was not long before French capital sought to exploit the opportunities for investment which the new convention seemed to offer. Within a fortnight of the signature of the convention Fives-Lille presented a submission to Hanotaux for the construction of a railway from the border town of Langson, scene of the notorious retreat which had prompted Ferry's fall in March 1885, to Long-zhou, the treaty port on the upper reaches of one of the Xi Jiang's tributaries in Guangxi. The company also sent an engineer, Antoine Grille, to Beijing with full powers to negotiate a contract. He arrived there only three months to the day after the convention's signature.[42] The alacrity with which Fives-Lille embraced the project is attributable partly to the company's involvement in the rebuilding and extension of the Phulangthuong–Langson line, of which the Longzhou line would be a further extension; partly to the fact that the company had suffered a loss on the notorious dredges it had supplied the Chinese government and was anxious to recoup it; and partly to the decline in French domestic railway construction and hence in the demand for Fives-Lille's products by its protected captive market.[43] They were singularly poor reasons for building a railway.

Despite the terms of Article 5 of the convention recently concluded and of Article 7 of the 1885 treaty, and despite growing Chinese interest at the time in railway construction, French diplomats in both Beijing and Paris recognized that the proposal of Fives-Lille would at best be received coolly. Gérard opened negotiations on 9 September 1895 and suggested that Fives-Lille be permitted to construct and operate the proposed railway with full proprietorial rights and the option to sell it to any other French organization. With this ambitious claim began what Gérard later described as 'the most laborious, the most painful negotiations I have ever had to undertake'. The reply of the Zongli Yamen was predictably 'a polite but categorical refusal'. Gérard was informed that China had not yet prepared her own plans for a national rail network and that there could not be any thought of constructing a line penetrating the empire from the frontier until after that system's completion; that Article 5 of the 20 June convention did not commit China to any dates or conditions for railway construction; and that, in any event, the proposal of Fives-Lille was inadmissable as it violated Chinese sovereignty in a number of ways.[44] The Chinese ministers, were, of course, quite right on all these points, but the China of 1895 was not always able to rely on the rectitude of its claims. Grille had full powers to negotiate a modified contract; Marcelin Berthelot, the new Foreign Minister, authorized any arrangements which would conciliate the Chinese while ensuring that the French retained control of the line; and Gérard was anxious, as he put it, to 'open the breach' for railway construction in China by foreign capital.[45]

Thus, at the beginning of November 1895 Gérard spoke to Prince Qing in urgent, if not coercive tones, about the obligations imposed on China by the 20 June convention. In reply the Zongli Yamen suggested that China would herself consider building the line to connect with the Vietnamese system. Under this proposal the governor of Guangxi would be responsible for surveying and planning the line and the Chinese government would undertake its construction and operation with the assistance of a French company nominated by the French government. Gérard decided to negotiate on this basis and presented a second draft contract for discussion in December. This provided for China to retain the ownership of the line, but to cede to Fives-Lille its construction and operation.[46] Gérard considered it vital that the Chinese line be operated by Fives-Lille as that company was to be ceded the line's Vietnamese section, and he considered that continuity of operation was essential, both to fulfil the terms of Article 5 and to ensure that the new line became the means of penetration into China that it was intended to be. Initially, the *Zongli Yamen* prevaricated, suggesting that nothing could be discussed until the Guangxi authorities had completed surveys, but Gérard used pressing language, and, at his request, Berthelot told Jing Zhang, the Chinese Minister in Paris, that the French government was at the limit of its concessions.[47]

The Chinese negotiators indicated their approval in principle of the Longzhou extension, and on 20 March 1896 an Imperial Decree authorized the railway's construction. At the same time the commander of the military forces in Guangxi, General Su Yuanchuan, was appointed the line's director-general. However, when the terms of the decree were communicated to Gérard at the end of the month, it became obvious that the Chinese intended merely to employ a team of French engineers and purchase French material with which to build the line, thereby retaining complete control.[48] Meanwhile, Gérard had been informed in February that the Chinese government had decided to open the lower Xi Jiang to foreign trade, thereby making the railway a far less attractive financial proposition.[49] The terms of the Imperial Decree and the decision to open the Xi Jiang had a demoralizing effect both on the Colonial Ministry and on the concessionary company. The former, which two months earlier had been encouraging the prolongation of the proposed railway from Longzhou to Bose, near the Guangxi–Yunnan frontier, feared that the trade of Guangxi and Yunnan which it had hoped to attract to Haiphong, would be diverted away from the French Protectorate.[50] Additionally, the terms the contract of 29 February 1896 between Five-Lille and the Indo-Chinese government made the concession of the Vietnamese section of the line conditional upon the construction of its Chinese section.[51] Fives-Lille's managing-director, Edmond Duval, already worried about the railway's potential profitability, told the premier, Léon Bourgeois, that as his company was not guaranteed the Vietnamese concession, it reserved the right to play no part in the operation of the Longzhou extension and to content itself with its construction alone.[52]

However, just as it appeared as though the project might be abandoned, the cabinet headed by the anti-imperialist future Nobel Peace Prize laureate Léon Bourgeois fell on 29 April. The aggressive Hanotaux returned to the Quai d'Orsay, and an enthusiastic imperialist, André Lebon, became the new Minister for Colonies. Lebon quickly reassured Duval that his Vietnamese concession was not at risk,[53] while Hanotaux threatened Jing Zhang with Gérard's withdrawal from Beijing unless the proposed contract (as modified following French concessions) was accepted. This threat was followed by another as Hanotaux indicated that France would adopt a generally less sympathetic policy towards China. Hanotaux's strong language was repeated in Beijing by Gérard.[54] All this bluster proved to be effective, for on 25 May Jing Zhang informed Hanotaux that the Chinese government would agree to the signature of the contract provided that the line was built to the standard gauge of 4 feet 8½ inches (1,435 mm) and that any reference to its ultimate extension to Bose was omitted.[55]

The question of the railway's gauge was important for both the French and the Chinese. China had decided to construct all her railways to the standard gauge, and this provision had been incorporated in an Imperial Decree. The French railways in Vietnam, however, although initially built to a gauge of 60 centimetres, were being converted to and extended as metre gauge lines. If the Longzhou railway was to be means of penetration into China it was highly desirable (although not crucial, as cheap labour enabled goods to be transshipped without great expense) that Vietnamese rolling stock be able to run on it. Moreover, the terrain around the frontier and Longzhou was fairly rugged, and a narrow gauge railway, able to negotiate much sharper curves than a standard gauge line, would be considerably cheaper to construct. Duval, ever anxious to dictate terms, sent a long note to Hanotaux in which he stated that the potential traffic probably would not justify the expense of a standard gauge line, and that if the Chinese continued to insist on that gauge, the company would have to abandon the project.[56]

Duval, however, did allow his agent in Beijing, Grille, more room to manoeuvre than his note to Hanotaux suggested. So, after some delay while Grille recovered from burns inflicted on him by wayward fireworks illuminated at the Russian Legation to celebrate the Tsar's coronation, a compromise was reached.[57] The contract which Grille and Zhou Wen, the *Zongli Yamen's* secretary-general, signed on 5 June 1896 omitted reference to the line's further extension and provided that the question of gauge would be settled after further surveys. The company was to construct and exploit the line for a period of 36 years with the option of renewal. The rates and administration of the line were to be identical with those of the Vietnamese railways but the risks and ultimate ownership were to be China's.[58] This contract was the first between a foreign company and the Chinese government which provided for foreign investment and control of a Chinese railway. Gérard considered the event a highly significant one:

I believe ... that the signature of the contract of 5 June 1896 by Fives-Lille and the Imperial Government was a landmark in the opening of China, and an exclusively French one.[59]

Hanotaux's aggressive policy appeared to have succeeded where that of Berthelot, who had been more conciliatory towards China and more sensitive to British feelings, had failed. Hanotaux's triumph was to prove a hollow one, but in the summer of 1896 French prestige and power in China was growing. At Gérard's insistence, the *Zongli Yamen* ordered General Su to visit Hanoi to celebrate Bastille Day as evidence of the amelioration of Sino-French relations. While there, apart from becoming a commander of the Légion d'honneur, Su discussed the new railway and the operation of the joint policy on the frontier with the French Governor-General, Rousseau. In contrast with his predecessors, Su appeared to be so francophile that Gérard described him as 'a force, an auxiliary, on whom the Government of the Republic now has, I think, the right to rely'.[60]

Unfortunately for the expectations of the interested capitalists and those diplomats who believed that the line would be an effective means of French economic penetration into Guangxi, the railway was not built until a time when investment by French capitalists in the area, like the French Protectorate itself, had ceased to exist. This enormous delay until the 1950s was partly the result of a Chinese desire to resist such penetration from Vietnam. The decisive factor, however, was the lack of conviction as to the proposed line's benefits held by many French observers. Additionally, the engineers Fives-Lille sent to Longzhou to get the project under way appeared to have had an almost unique propensity for souring relations both with the Chinese and their compatriots, and the bitterness they created at least delayed the commencement of work even if it was not responsible for its abandonment.

The most eloquent and effective opponent of the railway was exceptionally well placed both to develop a well argued and documented case in support of his opinion and to make this view well known in official circles in Paris. He was Auguste François, a former offical in Indo-China, who had recently been appointed French Consul in Longzhou. Significantly, François travelled to his new post from Guangzhou via the Xi Jiang rather than from Haiphong and through Langson. During this voyage he was told, both by Chinese merchants and by French missionaries such as the redoubtable père Renault,[61] that if the railway was extended to Nanning as Fives-Lille proposed, trade would be centralized on that town and hence benefit Beihai, Guangzhou, and Hong Kong at the expense of Haiphong. Clearly, Longzhou would only be an interim terminus, as it was no great entrepôt, so François suggested that it should be built towards Bose near the Yunnan border, where it could intercept traffic descending from the Yunnan plateau to Nanning and direct it towards Haiphong. As for Fives-Lille's instructions to their resident engineer that he press for extension to Nanning, François considered them ridiculous,[62] and concluded in his first dispatch to Hanotaux from Longzhou: 'The Hanoi–Nanning line would in reality only be an Anglo-Chinese enterprise.'[63] Hanotaux took François's criticisms of the Nanning proposal sufficiently seriously to consult Lebon about the Colonial Ministry's opinion of an extension to Bose.[64]

Very different advice, however, came from Beijing. The concession had been acquired largely as a result of the energy of Auguste Gérard, who predicted a brilliant future for the line. In total disregard of the realities of topography, distance, and Chinese politics, he quixotically argued that the Longzhou railway, modestly extended thorugh Nanning and Bose to some point on the

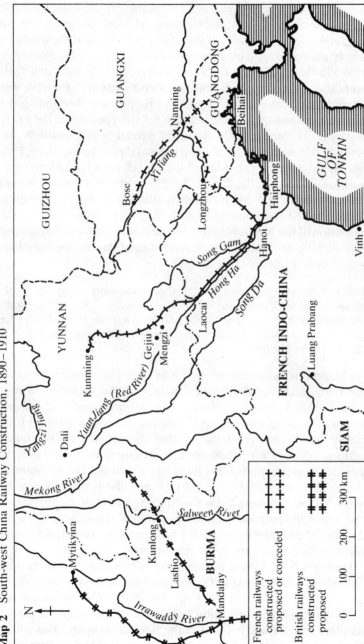

Map 2 South-west China Railway Construction, 1890–1910

Yangzi, a distance of over 1,500 kilometres, would drain the trade of large sections of Guangxi, Guizhou, Yunnan, and even Sichuan to Haiphong. 'The effects of the opening of the Xi Jiang would be amply exorcised, and even more [achieved]', Gérard suggested.[65]

While the officials at the Quai d'Orsay merely indicated their scepticism as to the practicality of Gérard's vision,[66] François was horrified to hear that his superior in Beijing was demanding the Nanning extension as compensation for the opening of the lower Xi Jiang to foreign trade. François agreed with Gérard in describing this move as 'a fatal coup [which] threatens our Hong Ha', but far from seeing extension of the French railway to Nanning as a means of averting the danger, he considered that the line 'would only aggravate and complement the disastrous measure which is turning the Xi Jiang into an English river'. If François believed the extension of the line to Nanning would benefit only the English, he was equally confident that leaving Longzhou as the terminus would benefit chiefly the Chinese:

This will be an opportunity for the French company to lay about 100 kilometres of track across sterile rocks and deserts... The railway [from the frontier to Longzhou] will really only have a strategic significance for the Chinese and will complete the system of fortifications General Su is developing on our frontier.

He based this judgement on his knowledge of the country and on the opinion of the American manager of the Chinese customs stationed at Longzhou, Morse, who claimed that the railway would only direct traffic from the Beihai and Xi Jiang routes if its freight was a quarter of the existing charges and if the French customs officials left goods alone and charged no transit duty in Tonkin. François argued that if the line was to achieve its commercial aims, it should be extended to Bose, first because that town was sufficiently remote from the Xi Jiang to escape its attraction, and secondly because General Su, in his capacity as Chinese director of the railway, intended reopening the road from Longzhou to Bose to attract traffic for his line.[67]

During 1897 François became even more firmly convinced of the railway's uselessness so far as French interests were concerned. Su was determined that it would conform to Chinese standards of construction and operation, which meant not only the use of standard gauge track but also of the English language, and young Chinese in Longzhou were studying English rather than French to improve their prospects for employment on the line. Moreover, Fives-Lille had instructed Laurence Chapron, their resident engineer, to begin construction at the Longzhou end which required, ironically, that the necessary material be imported via Guangzhou and the Xi Jiang.[68] Worse still, those materials were to be largely of Belgian or American origin. Hanotaux urged Lebon to exempt foreign railway materials from transit duty so that they could be imported into Guangxi via Haiphong, but the latter replied that the Méline tariff stipulated that 80 per cent of the import duty be paid on goods in transit and that the law provided for no exemptions.[69] Apart from the problems posed by competition from Beihai and the Xi Jiang, the railway's chances of success were severely limited by the inflexible exigencies of protecting national industry.

As an alternative to the Longzhou railway, which he believed should be abandoned before any real damage was done, François suggested building a line more towards the north, passing through Caobang before crossing the frontier and climbing towards the Yunnan plateau. In support of this proposal he cited Chapron's conviction that such a line would be more useful than the one on which he was working, and the opinion of the new Governor-General, Paul Doumer, that the Longzhou line should be constructed only because of the political cost of not doing so. Additionally, he reported that a French merchant named Jacquet, who traded in fruit from Bose to Tonkin, had stated that he was the only person who would benefit from the Longzhou line, and that other merchants would find it cheaper to ship via Nanning and Guangzhou or Beihai. Jacquet laconically concluded: 'Besides, the Tonkin customs will always ensure that Beihai keeps its monopoly on the transport of goods into this region'.[70]

François's criticisms extended beyond the role the proposed railway would have to the contract which Gérard and Grille had negotiated in Beijing. He considered that it was really nothing more than an agreement to build the railway at some undefined point in the future at a cost and under conditions yet to be determined, and that it gave the Chinese as many opportunities as they wished for prevarication. Gérard and Grille believed they held the trump card in the provision in the contract's eighth article for the formation, in the event of a dispute between Fives-Lille and the railway's Chinese administration, of a French-dominated adjudication panel. François, however, was sceptical of this clause's applicability as it referred to 'possible differences', which could be interpreted as excluding difficulties in *establishing* the railway.[71]

Gérard and the directors of Fives-Lille were far from pleased with the sentiments François was expressing with such conviction and support and in such quantity. Gérard complained to Hanotaux about the 'doctoral tone' of François's dispatches and about his attempts to denigrate the value of the contract;[72] and took the opportunity to remind François that his function was only to ensure that the government's decisions were carried out, in this case that the terms of the contract were faithfully executed.[73] Duval, on behalf of Fives-Lille, reminded Hanotaux that it was his ministry's responsibility to look after the company's interests in China, and requested that he either give orders to François to do so, or exert pressure on the *Zongli Yamen* to ensure that Su and his subordinates carried out the terms of the contract. He further indicated that unless the situation at Longzhou improved the company would have to withdraw from the project.[74] Hanotaux obediently wired Gérard's successor, Dubail, and François, asking the former to approach the *Zongli Yamen* and recalling the latter.[75]

If François's relations with his superiors in Beijing and Paris were at times difficult during his stay in Longzhou, the local situation confronting him was frequently chaotic. Fives-Lille's engineer, Laurence Chapron, arrived in Guangxi early in November 1896, and immediately seemed to establish the worst of relations with General Su and the other five members of the Chinese commission responsible for the railway. Initially these problems arose from rivalry between Chapron and Georges Betrand, whom Su had employed as a technical adviser following his short career as an agent of the Thévenet syndicate in Tianjin. Bertrand spoke Chinese, but Chapron refused to conduct any business with Su through his rival, and began making allegations about Su's relationship with Bertrand's wife. Additionally, Chapron seemed to spend more time in Vietnam than Guangxi, telling bawdy stories about the Su ménage to French officers and recruiting subordinates of the worst kind. While one of these gentlemen upset the prefect at Longzhou with his noisy parties and trading in Vietnamese girls (with whom, it appeared, he cohabited before selling them), Bertrand was threatening to sue Chapron for defamation. Su and his commissioners were constantly asking Chapron for estimates of the cost of the surveys he was desultorily undertaking, but the engineer refused to discuss the matter. François considered that Chapron's replacement was the only means of improving relations between the company and the Chinese commission and within the French community.[76] In this matter, at least, his advice was followed quickly, and early in June 1897 Antoine Grille arrived in Longzhou from Beijing to take over Chapron's duties until a new engineer was appointed. That the anticipated improvement in relations did not occur was largely Grille's fault. Having come from Beijing, where he had enjoyed the social and diplomatic life of the French legation and mixed with men such as Gérard and Li Hongzhang, he found Longzhou boring and provincial. No doubt it was, but it was not necessary for him to treat Su and the commission with disdain and at times contempt. Additionally, he acted entirely for the benefit of his employers, and thus alienated both the Chinese, by asking considerably inflated sums for the expenses of the surveys, and François, by suggesting to Su that Fives-Lille construct defences against a possible French invasion.[77] Su was well aware that Grille's proposals for a defence system in Guangxi had been made solely in the interests of increasing expenses, and had been informed by one of Grille's former colleagues of the intention to present a bill for survey expesnses which would be nearly five times the actual costs.[78] It was in these circumstances that François was recalled from Longzhou.

While little work was being done but much bitterness created in Guangxi, the railway's future was being discussed in Paris. Ironically, the initiative for these discussions came from Duval, whose company was to suffer from the decisions that ultimately were made. Anxious to lay as many rails across China as possible, he proposed that a thorough survey be undertaken of the resources and topography of southern China so that a network of lines could be planned.[79] Hanotaux rallied to the idea, as he was thinking about possible claims France could make against China in compensation for the opening of the Xi Jiang and for allowing the extension of a railway from Burma into Yunnan. He asked Lebon if the Colonial Ministry would wish to undertake such a survey.[80] Both Lebon and Duval, however, prejudged the results of the proposed survey by agreeing that the line must be extended to Nanning, while Hanotaux entertained a different vision of Tonkin's commercial future:

The aim we must have in mind for the future of our colony in Tonkin is undoubtedly Yunnan. Therefore the means of penetration leaving our Indo-Chinese colony must point towards Kunming.[81]

At any rate Lebon quickly decided to send a technical mission to make the survey as soon as the commercial mission which had been organized in 1895 by the Lyon Chamber of Commerce returned from China and furnished the information it had gathered. The new mission was to consist mainly of engineers and topographers, and its independence would be guaranteed by the provision of funds by the Colonial Ministry rather than financially interested companies.[82]

The Ministry for Colonies' mission was not formed until October and by then the debate about the future of the Longzhou railway had taken a direction which was quite irksome to the directors of Fives-Lille. Hanotaux took the disparaging comments François made about the line's prospects seriously, and consulted Lebon and Gérard on the matter.[83] Maurice Bompard, head of the Direction commerciale of the Foreign Ministry and something of an economic rationalist, supported François consistently. He first agreed that the contract Grille had signed in Beijing was of no great worth,[84] and soon after expressed scepticism as to the benefit to French economic interests of the Longzhou railway at all. André Lebon, however, despite his concern that the mission be disinterested, suggested an extension from Longzhou to Bose, via Nanning if the direct route was too difficult, then through southwestern Guizhou into Sichuan, terminating on the banks of the Yangzi either at Luzhou or Chongqing. Such a railway would provide the shortest route from Sichuan to the sea, and thus, Lebon argued, Haiphong would be able to 'defy all competition' for the trade of this wealthy province. He rejected the proposals to build railways direct to Kunming either along the Hong Ha or via Caobang on the grounds of topographical difficulties.[85]

It was this grandiose plan which provoked Bompard's denunciation of the Longzhou railway and any extensions to it. The basis of his argument was the theory of a hinterland. He believed that geography imposed certain constraints on French ambitions and that Lebon, in seeking to extend French penetration into Sichuan and the upper Yangzi valley, was ignoring those constraints. Shanghai and Guangzhou had their own natural hinterlands, and no railway could alter them very substantially. To attempt to divert trade from the natural orbit of those two great ports was, in Bompard's opinion, to chase chimeras. He considered that as Sichuan, northern Guizhou, and generally all provinces bordering the Yangzi were naturally tied to Shanghai, and southern Guizhou to Guangzhou, the possible hinterland of Tonkin was reduced 'purely and simply to Yunnan not even including its northern portion'. Moreover, Yunnan could choose other outlets for her trade besides the French route via Tonkin, namely an English route via Burma and an Anglo-Chinese route via the Xi Jiang and Guangzhou. Bompard asked:

Can this last outlet, which I have described as Anglo-Chinese, possibly be turned into a Franco-Chinese outlet by a railway from Tonkin to Bose? That is a very doubtful matter, but what is not at all doubtful, for the *Direction commerciale* at least, is that if this railway has to go through Nanning, it will not achieve the desired result but very much the opposite. A railway from Bose to Nanning, even before its extension . . . towards Beihai or Wuzhou would be an English line and a branch to Longzhou would add to its traffic rather than deviate it. From this point of view the *Direction commerciale* does not regard the construction of the proposed railway from Langson to Longzhou without anxiety; it wonders if this line, far from breaking into Guangzhou's natural hinterland will not end up extending it further until it includes even a province of Tonkin. Such would be in any case the inevitable result of its extension to Nanning.[86]

Bompard's appeal for a sober assessment of the potential for French penetration in China was a timely one, for on 3 June the lower Xi Jiang had been opened for foreign trade under the terms of the Anglo-Chinese convention of 4 February 1897. Predictably a British steamer became the first foreign vessel to arrive at Wuzhou to exploit the traffic.[87] The direct result of British demands for 'compensation' for the Sino-French convention of 20 June 1895, the opening of the Xi Jiang was a bitter blow to French commercial aspirations in Guangxi. As early as 1887 a French Consul in Guangzhou had observed that it was in French interests to allow the Chinese to be as restrictive as they wished in the Guangzhou area.[88] Sir Claude Macdonald, the British Minister in Beijing, had claimed the opening of the river as compensation for the commercial and territorial advantages France had secured in the convention providing for this concession on 4 February 1897. However, only two ports, Sanshui and Wuzhou, both on the lower river were opened, and Li had admitted to Gérard that the British request that the river be opened as far as Nanning was refused in deference to French opinion.[89] While Bompard's response was to suggest that France turn elsewhere for commercial penetration, and the *Quinzaine coloniale* considered that a more liberal and honestly enforced tariff structure in Vietnam might lessen the impact on French trade;[90] Gérard believed that the most appropriate compensation to claim would be the extension of the Longzhou railway to Nanning, Bose, and into Yunnan.[91] Hanotaux's initial reaction was to make undefined threats to the unfortunate Jing Zhang which were, however, so strongly worded that the Chinese Minister was afraid to transmit them to the *Zongli Yamen*.[92] In determining specific demands to be made, Hanotaux generally agreed with Gérard, except that he questioned the latter's assertion that a railway via Bose was 'the only possible means of penetration into Yunnan' and preferred a formula which would allow the exact route of the railway to be determined once more information had become available.[93]

Gérard's negotiations with the *Zongli Yamen* on these points were prolonged and difficult. He did receive some support from Aleksandr Pavlov, the Russian *Chargé d'affaires*, but was reminded both by Li Hongzhang and Pavlov that the concession to Russia of the Chinese Eastern Railway across Manchuria was 'the result and payment for a secret agreement of another kind' and that France could not expect such advantages without paying a similar price.[94] Agreement was eventually reached on 12 June 1897. The first and most important paragraph of the note extracted from the Chinese provided that once Fives-Lille had completed the railway as far as Longzhou, it could construct an extension towards Nanning and Bose. The second paragraph restated France's mining privileges in the three southern provinces, even though the Anglo-French Convention of 15 January 1896 obliged both powers to share any privileges in Yunnan. The third was a Chinese agreement to improve the navigability of the upper Hong Ha and the roads from the river to Mengzi and Kunming. The ultimate construction by China of a railway to Kunming, either via Bose or the Hong Ha valley was also anticipated in this last paragraph.[95]

While the provision which Gérard obtained for an extension of the Longzhou railway towards Nanning would bring considerable benefit to the shareholders of Five-Lille, there was growing doubt in Paris as to the extent to which French interests as a whole would benefit. Late in June 1897, Hanotaux was impressing upon Lebon the dangers a railway to Nanning might present to French commerce.[96] By February 1898, when Fives-Lille's directors asked their engineer to begin surveys of the Longzhou–Nanning section, the Minister for Colonies had rallied to François's arguments and opposed the undertaking of such work.[97] Lebon was anxious above all, that the mission formed under his Ministry's aegis would determine the pattern of future railway construction by French industry in southern China. The mission was established early in October 1897 under the leadership of a civil engineer, Charles Marie Guillemoto (1857–1907). Its members embarked later that month.[98] Guillemoto and his lieutenants were expected to examine not just the potential for trade, but also the most effective means of directing that trade towards Haiphong. In its aims it was as ambitious as the Lyon mission and as well funded, although this time by public funds largely from the Ministry for Colonies. Guillemoto was given very complete documentation of the debate about the Longzhou railway, supplemented by a verbal declaration from Gérard, who, back in Paris after three years in Beijing, urged the mission not to bother investigating the possibilities of the Hong Ha valley, but to build towards Nanning, thence towards Bose and on to Sichuan in one direction, and north through Hunan to Hankou in another. Despite Gérard's extremely ambitious plans and prejudice against the Hong Ha route, the mission was instructed to examine, in order of priority, the potential of various proposed railways: firstly, direct to Kunming, via either the Hong Ha or Song Gam valley; secondly, from Longzhou to Bose direct; thirdly, from Longzhou to Bose via Nanning; fourthly, from Bose into Sichuan; and fifthly, from Nanning to Guilin in northern Guangxi.[99]

Initially, it was intended that the mission would operate as a single unit. However, insurgency around Bose and its suppression by General Ma made the Yunnan–Guangxi border area inaccessible, so it was decided in Hanoi to split the mission into two groups. The first group, under Guillemoto, was to explore Yunnan and Sichuan; the second under his deputy, Wiart, was to examine the possibilities for extending the Longzhou railway to Nanning, thence through Guangxi and Hunan to Hankou. It was at Hankou that the two groups were to be reunited. During its members' stay in Hanoi, the mission's programme became influenced by Governor-General Doumer's military and territorial ambitions in

Yunnan. Both groups, particularly the one to investigate Yunnan, were liberally reinforced with military officers, and far from being a disinterested exploration, the mission rapidly became a vehicle for the furtherance of Doumer's expansionary vision. Intoxicated by Doumer's plans, after less than a month in Hanoi Guillemoto was writing about the aims of the Yunnan group in these terms: 'The mission must extend its reconnaissance as far as Chengdu, the capital of Sichuan, and proceed to a study of the economy of this rich province which is a principal objective.'[100] The military reinforcements of Wiart's Guangxi group were more modest than those of Guillemoto's group but Wiart was to be joined by Grille in Longzhou, an addition which served only to provoke General Su's hostility.[101] Its aims, however, were just as ambitious and Wiart was instructed to determine the point on a future Haiphong to Hankou railway from which traffic would flow in one direction to the Yangzi and in the other into Vietnam.[102]

While Guillemoto and Wiart were examining further possibilities for railway construction in southern China, French diplomacy was actively involved in protecting such commercial prospects as the Longzhou line might possess. In January 1898 the British pressed for the opening of Nanning as a treaty port, a move which Dubail successfully countered with vague but strongly worded threats.[103] A rumour circulating in Hong Kong early in 1898 to the effect that a British syndicate was endeavouring to secure the concession of a railway from Beihai to Nanning provoked an anxious reaction from Lebon. He believed, probably quite rightly, that this short and relatively inexpensive line (about 125 kilometres in length and estimated to cost about 25 million French francs) would quickly monopolize the Guangxi trade and destroy any chances of success which the French policy of attempting to draw this trade to Haiphong might have.[104] The Colonial Ministry, therefore, suggested that the concession should be secured for a French company which would not be obliged to construct the line within any fixed period.[105] Fortunately for the Ministry for Colonies' plan, on 21 April a French missionary, père Berthelot, had been killed in Guangxi. Stephen Pichon, the new Minister in Beijing, seized the opportunity to demand, beyond the usual reparations such as execution of the murderers, punishment of local officials, construction of a chapel, and an indemnity for the dead man's family, the concession of a railway from Beihai to the Xi Jiang. Pichon asked that the contract be similar to that for the

Longzhou railway, which had certainly allowed plenty of scope for delay, and that the Chinese agree not to permit any other than a French or Franco-Chinese company to build other railways from Beihai. The *Zongli Yamen* acceded to both requests with a quite remarkable and totally unprecedented alacrity.[106] Of all the concessions China was forced into granting in 1898, probably none was more bizarre than that of the Beihai railway. For the French sought the concession, not for the sake of the diplomatic success as *The Times* claimed, nor even in anticipation of a French annexation of Guangdong as far east as the newly acquired base at Guangzhouwan as the *Hong Kong Daily Press* feared, but solely to prevent someone else from obtaining it.[107] On 28 June Hanotaux was replaced by Théophile Delcassé, and the new Minister's briefing on the Beihai concession was succinctly summarized as follows: 'The Minister knows that this concession has above all a preventative character and that there is no intention, at the moment, of constructing the line which has been conceded.'[108]

As it happened, the Hong Kong rumour was false, even though the British had been discussing such a proposal as long before as 1889.[109] Nevertheless, at the request of the Ministry for Colonies, the Longzhou railway was further protected by an agreement Pichon secured from the *Zongli Yamen* which bound the Chinese not to allow any railway to be built on the Guangdong coast between the Vietnamese frontier and Guangzhouwan without prior French consent.[110]

Although its future traffic was being jealously protected during 1898, the railway to Longzhou was itself no nearer to construction. Grille's successor as Fives-Lille's resident engineer, Vallière, attempted to conciliate Su Yuanchuan by forwarding to him the estimates of the cost of the line. Su, however, considered them exorbitant, and not even Doumer, who paid Su a subsidy for the maintenance of his troops and whom he visited in Hanoi to celebrate the European new year, could convince him to accept them. Su was encouraged in his conviction not only by Bertrand, but also by Provison, a former employee of Fives-Lille, and by Morse, the customs commissioner, who claimed that a British or American company could do the job much more cheaply.[111] Su therefore offered to pay only 9,140,000 francs for the construction of the line, about half of Vallière's estimate of 18,138,690 francs. In these circumstances the directors of Fives-Lille desperately sought scapegoats and found them conveniently in François and

Doumer, the former for his 'sad campaign', 'calumnies', 'outrageous stories', and even 'detestable intrigues' against the project, the latter for encouraging Su's pride by the flattering welcomes he was given in Hanoi. Additionally, Krantz and Duval complained that the instructions of the Guillemoto mission were evidence of the government's lack of support for Fives-Lille's railway. (Yet it had been at Duval's suggestion that the mission was established.) Su, they claimed, was attempting to drive Vallière from his position as he had Chapron and Grille. In response, and to strengthen Vallière's position, the company indicated its resolve to proceed with work by ordering surveys to begin on the Longzhou to Nanning extension.[112] This gesture, however, was never carried out, as Guillien, the new French Consul in Longzhou, pointed out to Vallière that such action would be illegal under the terms of the agreement of 12 June 1897 until the railway to Longzhou had been completed.

Confronted with Su's refusal to negotiate a higher price for the line, Fives-Lille decided to convene the arbitration commission provided for in Article 8 of the contract to resolve differences between the company and the Chinese commission. Nominated early in May 1898, its three members were all French engineers involved in constructing public works in Vietnam who had been selected by Doumer on behalf of Su and by Vallière himself for the company. However, just as the arbitrators were about to meet, the directors of Fives-Lille in Paris ordered Vallière to refuse to proceed with arbitration. Their grounds for forcing their resident engineer into such a humiliating volte-face were, first, that the hostility François had engendered towards Fives-Lille in Vietnam might affect the arbitrators' judgement; and second, that engineers employed by the state could not understand the enormous costs an enterprise in such a difficult and inaccessible region would occasion for a private company. In effect, Fives-Lille did not want its costings evaluated by any professional engineers whose interests were not bound up with those of the company. Instead of arbitration by experts, Fives-Lille wanted the issue settled by negotiation between the French Minister at Beijing and the *Zongli Yamen*, or, as failing that, by the appointment of new arbitrators.[113] The Quai d'Orsay, however, whether under Hanotaux or Delcassé, was little disposed to help Fives-Lille, whose record in China had been little short of disastrous since its involvement had begun with the supply of dredges to the Shandong government in 1890. Krantz's requests (written on the elegant senatorial notepaper to which he was entitled) for a 'resolute intervention' by Delcassé to overcome the company's problems met only with restatements of the Department's resolve to leave the matter to arbitration.[114] When Vallière went to Beijing to enlist Pichon's support, he was only castigated for the sins of his employers, whose 'deplorable' behaviour was said to be gravely prejudicing French policy and the interests of French industry in China. Pichon suggested that the company accept the arbitrators Doumer had offered.[115]

Even the financial press was expressing opposition to Fives-Lille's Chinese railway in late 1898. The editor of the *Texte de l'Echo des Mines et de la Metallurgie*, Francis Laur, denounced the line for the same reasons that François had. He stated the problem in the classic terms, which by then had become something of a cliché:

The point to resolve from the point of view of Tonkin is this: how do we bring through our colony . . . the important Yunnan trade. That was the aim pursued by all the conquerors of Tonkin, Francis Garnier, Dupuis, etc. It is the dream of our occupation, for we believe that it was to trade with China that we took Tonkin; without that it would have been an historic piece of deception.[116]

The railway being constructed by Fives-Lille, however, far from draining traffic towards Haiphong would instead direct it towards the Xi Jiang and the English-dominated ports of Guangzhou and Hong Kong. Laur argued that the line be abandoned and that Fives-Lille content itself with an indemnity:

Finally there is a railway for the Chinese and the English, but built by some Frenchmen . . . more or less Belgian, with American equipment and materials.

This is how we are beginning to undertake public works in China!

And then? If neither Tonkin, nor France, nor even China — as Marshall Son [sic] claims . . . — have any interest in this railway, who then has?

Fives-Lille obviously.

You can let us think that in that case, with all due respect, it is not nearly enough.[117]

Derided by the press and, more importantly, denied the assistance of French diplomacy, Krantz informed Delcassé that the company had decided to abandon the enterprise and would negotiate in Beijing for an indemnity.[118] Krantz's move was probably largely bluff. At any rate, during his visit to Paris in December 1898, Doumer persuaded Fives-Lille to accept the arbitration of a reconstituted commission, two of the three members of which were to come directly from Paris.[119]

After a month's inspection, the new arbitrators handed down their decision in Hanoi on 8 May 1899. Their decision was very favourable to the company, for their estimate of the line's cost was even higher than that which Vallière had given to Su a year and a half earlier.[120] Such a decision was clearly unsatisfactory both to Su and to the Chinese government in Beijing which had ordered him to keep expenses to a maximum of 3,200,000 taels (11,200,000 francs). Constructive negotiations soon began in Beijing. The arbitrator who had been nominated by Fives-Lille, Blondel, became the company's representative in Beijing and agreed to build the line in the most economical way. The Chinese, in the interests of economy, decided to allow the railway to be built to the Vietnamese metre gauge rather than to the Chinese standard gauge. These concessions, together with certain financial advantages which Fives-Lille secured, were incorporated in a new contract signed in Beijing on 15 September 1899.[121] The company was obliged to attempt to keep expenses to below the figure of 11,200,000 francs, and Blondel succeeded in reducing them to 15,300,0000, an achievement which seemed to please Su.[122] However, in the transformed political situation of March 1900, Su was unable to obtain approval from Beijing for the construction of the line at that price and in the same month the Chinese government ceased to pay Fives-Lille the monthly instalments with which the line was being financed. Duval indicated that unless negotiations on the price were opened, the company would once again abandon the project.[123] In the circumstances of mid-1900 negotiations in Beijing were, of course, impossible. Moreover, if the situation for foreigners was not as bad in Guangxi as in Yunnan or northern China, there was nevertheless considerable social unrest and anti-foreign feeling which Su's unpaid troops lacked both the will and means to curb.[124] Fives-Lille thus carried out its threat to withdraw without having laid a single rail and decided to press for an indemnity.[125] The company sought to have its claim considered as part of the general Boxer indemnity, and although Delcassé initially believed that its losses in Guangxi were not eligible for such reparation, the Boxer Indemnity Commission ultimately awarded it some 550,000 francs of the 265 million allocated to France.[126]

After Fives-Lille's departure there were some suggestions that the Indo-Chinese government build the line, and Gaston Doumergue while Minister for Colonies was certainly enthusiastic about such a proposal.[127] However, the fundamental problem, that the line would in no way benefit French commerce, remained to prevent this occurring. The conclusions of the Guillemoto mission, despite their importance for developments in Yunnan, were of no help whatever in deciding how best to make use of the Longzhou concession. Guillemoto suggested that it should be used as the starting point for a railway to Hankou, an enterprise which he believed would be practicable provided capital invested in it was covered by a French government guarantee.[128] His proposal was received with derision in the Quai d'Orsay. Delcassé pointed out that parliament was reluctant enough to give such guarantees for colonial railways and that it was out of the question for a purely Chinese line.[129] Pichon preceded his discussion of Guillemoto's proposal by stating that there was no chance of the French obtaining the concession and cited the violent English reaction to the Beijing–Hankou concession as evidence of this. Moreover, he was extremely sceptical as to whether the line would be profitable. Potentially the most profitable trunk line in China, he believed, was that from Beijing to Hankou, and, after initial enthusiasm had subsided, the capital required for its construction had only been subscribed with difficulty. In rejecting Guillemoto's proposal, Pichon expressed agreement with Sir Claude Macdonald, who had recently confessed that while strongly pressing the claims of English syndicates on the Zongli Yamen, he was not rash enough to buy shares in them himself.[130]

The disappointing results of the Guillemoto mission, so far as its investigations in Guangxi were concerned, led to the conclusion at the Quai d'Orsay that there was nothing to be done about the Longzhou railway. The railway would never benefit French commerce and the choice of it as the first means of penetration into China had been an 'unfortunate' one. The only reason for the construction of the line was that not to do so would involve the loss of French prestige and influence in the eyes of the Chinese.[131] In view of these conclusions about the railway's worth, the Boxer Rebellion provided a convenient excuse for withdrawal without undue loss of face. Later observers agreed with this damning analysis, particularly after the opening of Nanning as a treaty port in 1907.[132] Even as a means of encouraging purely local commerce the line would have failed as Guangxi, like Tonkin, had an agricultural surplus and produced few commodities for which there was a market in Indo-China.[133]

Fives-Lille's determination to extend their Vietnamese concession into China, then, ultimately led to considerable difficulties for French diplomacy. Gérard's initial triumph in securing the first foreign railway concession in China soon became an embarrassment. The flaw in the French programme of penetration seems to have developed at the very beginning when a private company, whose interests were best served in the laying of as many miles of track as possible, was allowed to determine the direction that penetration would take. No attempts were made to gauge the potential volume of traffic nor the directions in which that traffic would flow along the proposed railway. The company itself selected Longzhou, it would appear, simply because the Vietnamese railway it was working on pointed vaguely in that direction.

The premature pressure which Hanotaux and Gérard brought to bear on the Chinese to grant the concession to Fives-Lille was perhaps the result of their enthusiasm at being presented with a proposal which a French company was willing and, apparently, able to undertake in southern China. However, Su Yuanchuan's French adviser, Georges Bertrand, ascribed more sinister motives to the French diplomats:

[Auguste Gérard], intimately tied to Fives-Lille's agent M. Grille, whom he lodged and fed at the Legation, had only obtained this concession from the Chinese government to satisfy the wishes of Grille's relations; and M. François did not accept that this company should profit from the situation it had been able to create in high places, to treat as it liked the local Chinese authorities who had to pay the costs of the line's construction.[134]

Whatever the truth of Bertrand's allegations, his patron, General Su, certainly shared his conviction that the personal interests of French officials were involved in the railway. Su told François that Prince Gong believed that Gérard was financially interested in Grille's operations, and moreover that Li Hongzhang and Prince Qing had written to him to indicate that negotiations with Fives-Lille had to be successful as Gérard was directly interested.[135] The

Chairman of Fives-Lille's Board of Directors, Krantz, assured Hanotaux that Gérard had received no personal reward for his services.[136] Certainly, these allegations did not prevent Hanotaux from using Gérard as the Quai d'Orsay's agent to convince French investors to place their capital in China after his return to Paris. However, even if Gérard was not financially involved in Fives-Lille's success, he did appear to identify himself closely with Grille's success. During has stay in Longzhou Grille often referred to his close links with Gérard ('as close as brotherhood') and attempted to use them to force his views on François and his assistants.[137] And the interests which Grille was attempting to advance, in Longzhou as in Beijing, were those of his employers rather than those of French commerce. But perhaps in that he was not being absurdly narrow. When confronted with François's arguments as to the deleterious effects of the railway he was building on French commerce, his reply was pessimistic and perfectly honest: 'French merchants? Do any of them exist? Everyone knows that the trade here will be English. The railways in this area will only carry English merchandise . . .'[138] In this way the cynical representative of the company responsible for the first French railway of penetration into southern China dismissed those dreams of Garnier and Ferry, which had been the basis of the French imperial venture in Tonkin.

7

French Initiatives in Yunnan, 1895–1898

ALTHOUGH the effects on French commerce of the opening of Longzhou had been almost disastrous, trade through Mengzi, which had been opened simultaneously, offered more encouraging results. For even though very few of the goods being imported into China via Haiphong and Mengzi were French,[1] at least the trade was growing rapidly and in the direction intended. From 888,372 francs in 1889, its value increased to 8,455,925 francs in 1893 and 13,809,164 francs in 1898.[2] The development of trade through Mengzi was due to a number of factors, the most important of which was its geographical location. Situated high on the Yunnan plateau yet separated from the head of navigation of the Hong Ha only by the short if very steep road of ten thousand steps, it commanded what was easily the shortest route from Kunming, the capital and largest city of Yunnan, to the sea. Indeed, Mengzi was almost as favoured geographically as Longzhou was cursed, and its natural advantages were to be a significant factor in British and French decision-making during the late 1890s. Another factor which contributed to the growth of trade through Mengzi was the suppression of transit passes and imposition of heavy *lijin* dues on the Xi Jiang route by the Guangdong government. Even though transit duties were quite high and often dishonestly administered in Tonkin, they remained much lower than the *lijin* dues on the Xi Jiang, and British merchants sending goods to Yunnan from Guangzhou and Hong Kong found it cheaper to ship via Haiphong than along the Xi Jiang.[3] The opening of the lower Xi Jiang did not greatly alter this situation so far as Yunnan was concerned, and in any case after 1897 one of the greatest disadvantages of the route via the Hong Ha and Mengzi, the danger of pillage or extortion from riverine pirates in Tonkin, had been removed by the French military forces. These same French military officers, particularly Colonel Pennequin who commanded the Third and Fourth Military Territories adjacent to the Yunnan border, had also succeeded in persuading the French customs officials to be a little less zealous in the execution of their duties, a measure which further encouraged trade.[4]

However, despite the progress which the French imperial cause was making in the commercial penetration of Yunnan via the Hong Ha route, there remained a number of problems which prevented fulfilment of the expectations of its pioneers and the early advocates of its potential.

These problems were the familiar ones of imperialist endeavour: what was to be exported and to whom; how the transportation of these exports was to be effected; and the efforts of competitors, actual or potential. In a liberal state such as the Third Republic the organization of trade was theoretically the domain of private capital. So, although the economic liberalism of the France of the 1890s was far from unadulterated, it was a privately organized, rather than an officially sponsored, mission which most seriously addressed itself to the problems of France's poor export trade to China.[5] It was late in April 1895, about a week after the signing of the Treaty of Shimonoseki and while Gérard was pressuring the *Zongli Yamen* to agree to the extension of Vietnamese railways into Guangxi, that the Lyon Chamber of Commerce resolved to send a mission to East Asia to evaluate the potential for French trade in the area. Other Chambers of Commerce, including those of the northern industrial cities of Lille and Roubaix, of the ports of Marseilles and Bordeaux, and of Lyon's neighbour, Roanne, also contributed to the mission, but it was very much a Lyonnais initiative. Lyon's interest in China was attributable to the city's silk industry, which was dependent on imports of Chinese raw silk. Indeed, it was this demand of the looms of Lyon for raw silk which was largely responsible for France's unfavourable balance of trade with the Chinese empire.[6]

The mission had first been suggested by Frédéric Haas, an enthusiastic Consul in China who throughout the decade constantly promoted all kinds of schemes for French penetration into Sichuan, which he believed to be part of Tonkin's natural hinterland, including a plan for a French concession in Chongqing.[7] Speaking in support of the mission at Lyon in April 1895, Ulysse Pila, head of an important Lyon family of traders and officials in East Asia, naturally alluded to the potential of Tonkin and the Hong Ha route as a means of opening markets in Yunnan. Additionally, however, echoing Haas, he suggested that the trade of Sichuan, the largest and most populous province of China, should be a further end to which French commercial ambitions direct themselves.[8]

Given the enthusiasm of Haas and Pila and the province's importance as a producer of silk, it is scarcely surprising that the mission organized by the Lyon Chamber of Commerce in 1895 spent more time in Sichuan than in any other Chinese province.[9] The conclusions of the mission's members as to the potential of French trade were restrained and certainly offered little encouragement to grandiose imperial visions.[10] Far from advocating the construction of a railway from Haiphong into Sichuan, the mission believed that the Yangzi was the province's natural means of access. Thus, it drew attention to the difficulties of navigation on the Yangzi between Yichang and Chongqing and suggested that a hydrographic study of the upper river to discover how it could be opened to steam navigation would enhance the province's mercantile prospects.[11] So far as the opportunities for penetration into the provinces immediately adjacent to Tonkin were concerned, the mission was equally conservative. Although its report welcomed Fives-Lille's construction of the Longzhou railway, it doubted whether the line would ever carry much traffic and suggested instead that trade with Yunnan be developed by building a railway along the Hong Ha route.[12]

In terms of investment in China, the only tangible result of the mission itself confirmed this faith in the Hong Ha as a means of penetration into Yunnan. For, in March 1898, the Compagnie Lyonnaise Indo-Chinoise was founded in Lyon with a capital of 1,250,000 francs. The company's aims were very broad, but it was conceived essentially as a trading operation, based in Hanoi with branches in Mengzi and Hong Kong. As well as entering the transit trade between Hong Kong and Yunnan, the new organization also hoped to sell in China the products of various industries which French capital established in Vietnam as a result of the mission's reports.[13] The initial proposal for the company alluded to the aims of the conquest of Tonkin and the fact that they were not being realized:

The troubles of conquest and pacification have made us lose sight of the aim of the first brave advocates of Tonkin: the penetration of China, especially the province of Yunnan. No French or foreign trading house yet exists in this province. Nevertheless it imports foreign merchandise; it even imports them from Tonkin, via the Hong Ha, which proves the economy of this route. But it imports them from the British colony of Hong Kong, and it is there also that it sends or exchanges its products.[14]

The company's primary aim was to suppress the intermediary role of Hong Kong, which implied creating Haiphong as a rival entrepôt.[15] The fact that Haiphong was riddled with customs officials whereas Hong Kong was a free port was seen as an advantage rather than as a difficulty for this scheme. Its promoters considered that the duty of 24 per cent on foreign imports via Haiphong would enable French cotton to compete with the Indian cottons which constituted the bulk of foreign imports into Yunnan.[16] Although they were right in anticipating the development of the Hong Ha route and the growth of Haiphong, their hopes for French cottons were to prove wildly optimistic. Louis Rabaud, who had been a member of the mission in 1895 and 1896, arrived in Mengzi in April 1899 to establish a branch of the company. He found, however, that he was unable to sell French products there. He confessed his difficulties to the Consul, Maurice Dejean de la Bâtie, who rather typically took a broad view of the problem:

it is absolutely essential that French industry be less routine, take more bold initiatives, in a word modernize itself to create a serious outlet in Chinese markets. This is why M. Rabaud does not count on, indeed cannot count on, French merchandise to make a reasonable amount of business in Mengzi. He mainly sells foreign products there.[17]

While private enterprise was concerned with a search for markets and supplies of silk for the looms of Lyon, the French government was more interested in establishing routes along which this commerce could be established. Yet this search for feasible commercial routes and the diplomatic and engineering efforts put into making them viable was from one important point of view to very little effect. For there was only a small market in China for French goods and the prospects were not good for one developing. A major result of the French imperial effort, even in the putative hinterland of Tonkin, was essentially to allow British and Indian goods to reach Chinese markets more expeditiously. French diplomats and traders had long been aware of the problem, and frequently made suggestions as to how it could be cured, although few were willing to admit that fundamentally French products were uncompetitive because the French economy was protected by tariffs.

As early as 1886 the then Resident-General at Hanoi and later Minister for the Marine, de Lanessan, suggested that the cultivation of opium in Tonkin would provide a commodity which could be imported into China, thereby rivalling the Bengali product whose shipment was a British monopoly.[18] De Lanessan's suggestion, however, was backward rather than forward looking. Certainly the great English taipans had made their fortunes from opium, but modern imperialism demanded the export of manufactured goods, to the benefit of the capitalists and workers of the metropolitan power. Gauthier, the French Consul at Beihai in 1890, was all too well aware of this and suggested that it was not enough for the French to confine themselves to the modest profits obtained from shipping British and German goods. He urged French industry to commit itself to the manufacture of the cotton cloth and thread that was the basis of European exportation to China, so that France could exploit the advantages of the Tonkin route to conquer Chinese markets.[19]

From 1888 French tariffs were applied in Indo-China, where they were to be as formidable a barrier to the development of trade with China as the rocks and sand banks in the Hong Ha. A year after their introduction, Emile Rocher, then stationed at Mengzi, complained that despite his consular status he had been forced to pay transit duty on his property which he had shipped from France. He described the reputation of the French customs amongst Yunnan merchants as 'detestable' and suggested that their operations were arbitrary and characterized by pettiness, trickery, and excessive formality. Under such circumstances he suggested that very few merchants would use the route through Tonkin.[20] A decade later a French publicist and ardent imperialist, Charles Saglio, claimed that the proposed railway from Haiphong to Kunming might not be used by foreign merchants because the reputation of the French customs was even worse than that of the Chinese *lijin*, or internal customs stations. He quoted with evident agreement the opinion of an English Consul that the operations of the French customs in Tonkin appeared to differ little from the Chinese practice of squeeze.[21] One of Rocher's successors at Mengzi who was to play a major role in the railway's construction, Dejean de la Bâtie, made a neat distinction between the development of the transit of goods across Tonkin and the development of French commerce. Writing only a week before the concession of the railway was eventually secured, he observed that it would greatly augment the former but do nothing for the latter. It was not enough, he argued, to build a railway penetrating China from Tonkin:

French industry must adapt its production to the needs of Chinese customers and our superannuated trading methods must give way to more modern methods like those of our competitors.[22]

Such opinions were not confined to consuls and publicists. The French ambassador in London, Paul Cambon, held similar views, and in 1899 expressed the opinion that a policy of active commercial penetration into China via Tonkin was incompatible with protectionism. He regretted that protection involved the erection of barriers to trade between Tonkin and China, especially since 'the principal aim of our conquest of Tonkin has been the opening of communications with central China'. He further argued, this time less accurately, that the reason for constructing a railway from Tonkin into China was to establish trade between Europe and China through Haiphong, not to provide a market of rails and rolling stock for the French metallurgical and heavy engineering industries.[23] In fact, French-financed railways were themselves to provide a growing market for French products in China from the late 1890s.

On 27 June 1901 a future premier and president of the Republic, Gaston Doumergue, sought some justification for the government guarantee of interest on capital to be invested in the proposed railway from Tonkin into Yunnan in the Chamber of Deputies. He asked rhetorically: 'What advantage will be derived from the railway? I'm searching for your economic interest, I can't find it.' In a considered, detailed, and well-argued response an enthusiastic protagonist of imperialist endeavour, Raymond Lefèvre, could only reply by referring to the hundred million francs worth of orders from the French metallurgical industry. Beyond that he could only write in general terms of the access it would provide to a Chinese market for 'manufactured items coming from Europe'.[24] Lefèvre was well aware of the economic basis of late nineteenth-century imperialism and had begun his book with the assertion that:

For the European nations which everywhere are seeking outlets for their industry and new markets ... the Chinese crisis is an opportunity to intervene in the business of the Celestial Empire.[25]

Despite this acute analysis he at no stage suggested that large quantities of French goods would travel up the railway from Haiphong to millions of Chinese consumers. Indeed, in view of the mineral wealth of Yunnan, he realistically predicted that most of the traffic would flow the other way: 'For us it is a question of making the mineral riches of Yunnan, Guizhou, Guangxi, and Guangdong come down to their natural outlet, Tonkin.'[26] Such business could well be profitable, but it hardly contributed to the welfare of French domestic industry.

In ignoring the possibility of France developing a significant export trade to China, even with the benefit of the Tonkinese tariff of 24 per cent applied to foreign goods, Lefèvre was being realistic. The aspirations of the founders of the Compagnie Lyonnaise Indo-Chinoise to replace Indian with French cottons in the Yunnan import trade were not realized. French trade improved little from the early years after the opening of Mengzi to foreign trade when, for the first two years from May 1889, *not one item* of French merchandise was imported into China by that route.[27] In 1901 the total value of imports into China via Mengzi was 3,034,910 taels. (The tael was then worth 3.73 francs.) Most of these imports were cotton goods, of which no less than 2,086,657 taels worth came from India. Japanese cotton imports were worth 103,662 taels and, for the first time, a small quantity of Tonkinese cotton worth 3,817 taels was imported.[28] No French cotton was imported via Mengzi. In a very modest way the French were experiencing what the British already had: the mills of Ahmedabad and Bombay and now Hanoi were able to produce cotton cloth and thread more cheaply than those of Manchester or Rouen. Lefèvre's judgement was further vindicated in 1903 when the first direct shipments of tin from Yunnan to France via Haiphong were made. This trade, amounting to 3,000 tons in 1903, bypassed Hong Kong and is a rare early example of the successful implementation of the policy of using Tonkin to drain the alleged riches of Yunnan.[29] The scale of trade between Tonkin and Yunnan should not, however, be exaggerated. Until late in 1903, when railway construction began to have some impact on the local economy, the only French business in Mengzi was an office of the Société Lyonnaise Indo-Chinoise; and its main operations were exporting opium which it sold to the Indo-Chinese government and importing cotton thread manufactured by the Société cotonnière d'Haiphong.[30] This was hardly an impressive result.

While the French government could not force its citizens to export goods into China via Haiphong, it did devote a great deal of energy to providing the means for them to do so. The Sino-French Commercial Convention of 20 June 1895 was an example of this concern.[31] The advantages which it seemed to accord to French railway builders and miners in Yunnan, Guangxi, and Guangdong were, however, to be compromised in a subsequent Anglo-French convention which was signed in London on 15 January 1896. The government of Léon Bourgeois, which was in power from November 1895 to April 1896, was less imperially minded than most other French cabinets of the period. It was notable not only for the London convention but also for the presence of Paul Doumer as its Minister for Finances and Doumer's unsuccessful attempt to introduce an income tax. Doumer's presence is ironic as he was later, as Governor-General of Indo-China and as an ardent imperialist, to attempt to overturn the convention by force. During 1893 there had been an Anglo-French crisis over French incursions into Siam. The frontiers between Cambodia and Siam; and between British, French, Chinese, and Siamese territory on the upper Mekong were ill-defined and as such a constant source of conflict, most dangerously between the two European powers. Both the Foreign Minister in the Bourgeois cabinet, Marcelin Berthelot, who was a senator with a scientific background, and Baron de Courcel, the Ambassador in London, were anxious to remove this source of tension. The British were co-operative as their main interests in both Siam and Yunnan were commercial. As a result they were quite amenable to giving the French a favourable territorial settlement. Under the terms of the convention the left bank of the Mekong became French territory, including an area which had been occupied by British troops. Siam was divided into British and French zones of influence but maintained its independence and hence British commercial domination there was secured. Although basically concerned with the frontiers between the Indian and Indo-Chinese empires of the two powers, the British extracted from de Courcel and Berthelot an agreement that any commercial privileges extracted by either power from China in Yunnan or Sichuan would be common to both. This had the effect of weakening the pretensions to a French railway and mining monopoly contained in the convention of 20 June 1895, but also gave the French access to any benefits that the British might gain from their convention with China of 1 March 1894 which had delineated the Sino-Burman frontier.[32] The Franco-British convention, which was as favourable to French territorial interests as it was to British commercial interests, was condemned by imperialists on both sides of the Channel.[33] It was to form the basis of both future rivalry and co-operation between the two powers in Yunnan.

Despite the efforts of Fives-Lille to build the Longzhou railway, at this time the British were well ahead of the French in penetrating Yunnan by rail. Immediately after the conquest of northern Burma in 1885, the British had built a railway parallel with the Irrawaddy from Rangoon to Mandalay. In 1892 Salisbury had obtained a report from Holt Hallet, the enthusiastic propagandist for a British route into Yunnan on the extension of that railway to Kunlong Ferry on the border between Burma and Yunnan. Surveys were carried out over the next three years and construction of the 270 miles of railway authorized on 16 October 1895. The aim of the railway was the penetration of Yunnan, and many British manufacturers were anxious to see it extended into China to forestall any French efforts to monopolize the trade of the area. On 30 June 1896 a meeting of the Chambers of Commerce of the Empire unanimously passed a resolution 'that connection by railway of a seaport in Burma with S.-W. China is greatly required in order to open out to the trade of the empire our new territories in the basin of the Mekong, and to enable manufacturers of the empire to compete with those of France in Northern Siam and in S.-W. China'. On the same day an expansive Salisbury told a deputation from the meeting that he regarded the proposed railway as 'a great benefit to the world'.[34] Salisbury was not merely cultivating his electorate, for on 4 February 1897 Sir Claude Macdonald and Li Hongzhang signed a conventioin in Beijing which contained the provision that if railways were constructed in Yunnan, the Chinese government 'agrees to connect them with the Burmese lines'.[35]

This concession on China's part to British interests greatly worried Hanotaux who sought to secure permission to construct a comparable French railway as soon as possible. The problem was that there was no agreement as to the route such a railway should take. Hanotaux's initial response to the apparent British coup was therefore to canvass French opinion on the merits of various means of penetration from Tonkin into Yunnan. Gérard's opinion that extension of the Longzhou railway to Nanning, Bose, and ultimately into Yunnan was well known but by then regarded with some scepticism.[36] Hanotaux also sought information from a former military attaché at the Beijing legation, Commander d'Amade, and from the Direction commerciale of his own department. D'Amade forwarded a 50-page report on the various means of access, while a M. Liébert prepared a summary of the evidence sent by various French consuls in the area. Emile Rocher, typically, prepared a submission of his own based on his long experience in Yunnan. He considered the proposals of Fives-Lille and Gérard for extension of the Longzhou line to Nanning and Bose as 'prejudicial to the interests of our colony', especially as Bose was some seven hundred very mountainous kilometres from Kunming and as, in any case, small junks could sail from Bose direct to Guangzhou at very cheap rates. He believed that a railway leaving the upper Hong Ha (called in China the Yuan Jiang, which translates as Red River) at the town of Yuanjiang, some two hundred kilometres into Yunnan from the Tonkin frontier, and climbing directly to Kunming would be the shortest and cheapest option. He also pointed out that such a line would not attract traffic unless the French customs régime in Tonkin were made more liberal, an observation which attracted a sympathetic response from Bompard, who wrote in the margin: 'That is the difficult yet essential point.'[37]

Liébert's note generally agreed with Rocher's conclusions. It recognized that there were two possible routes into Yunnan, one following the Hong Ha, the other via Guangxi and the upper Xi Jiang. The report dwelt at length on matters such as the navigability or otherwise of the Hong Ha and the effects on Sino-French trade of the Longzhou railway. Liébert ultimately rejected the prolongation of the Longzhou line as a reasonable option, as such a route would not lead directly into Yunnan, but would serve primarily Guangxi and Guizhou, where the commercial prospects were very limited. He reminded his Minister:

it is not a matter of favouring Chinese or foreign traders, but of obtaining the best possible situation from the particularly favourable geographical position of our Tonkin for the penetration of Yunnan, which must remain our principal objective. It was besides to this end that the conquest of the colony was largely directed.[38]

This approach, with its appeal to both history and geography, led him to favour the Hong Ha route. He believed it would be best to improve the navigability of the Hong Ha by blasting and dredging as far as Laocai and then build a railway from there to Kunming, following Rocher's suggested route via Yuanjiang.[39]

Although his conclusions differed in detail, Commander d'Amade generally agreed with the Direction commerciale. His report compared the four main routes into Yunnan, all of which led to Kunming. They were the imperial road overland from the Yangzi, the road from Burma via Dali, from Guangzhou via the Xi Jiang to Bose and thence by road, and from Haiphong by the Hong Ha and road via Mengzi. D'Amade was convinced of the superiority of the Hong Ha route, largely because it was the shortest. He rejected the possibility of the Longzhou railway as a means of penetration even of Guangxi, as Longzhou 'is located out of all the commercial currents of the province'.[40] The conclusion of his report was directed towards the best means of drawing the trade of Yunnan to Haiphong. He agreed with Liébert in asserting that 'the principal *raison d'être* of our colony lies in its commercial links with China'.[41] To achieve this aim he suggested the construction of a railway from Hanoi along the valley of the Clear River (or Song Gam), a northern tributary of the Hong Ha, across the frontier via Kaihwa (modern Wenshan) and Mengzi to Kunming. He justified this longer variant of the Hong Ha route on the grounds that the Song Gam valley was less steeply graded and more densely populated than the Hong Ha. To maintain connection between Mengzi and the Hong Ha, he rather illogically favoured construction of a rack railway which would only divert traffic from his proposed through line to Hanoi via the valley of the Song Gam.[42]

Both reports recognized that it was unrealistic to expect to channel the trade of Sichuan through Haiphong. The presence of the Yangzi was sufficient to prevent the development of such a trade, despite the early aspirations of Dupuis and Garnier. Even though as late as 1900 that enthusiastic imperialist, Prince Henri d'Orléans, hoped that a railway could capture the trade of the upper Yangzi, Sichuan, and Yunnan for Haiphong, [43] the Ministry of Foreign Affairs was under no such illusions. As early as 1892 the Consul at Hankou, Joseph Dautremer, had warned of the dangers of 'having too many illusions about the Hong Ha route', and suggested that a better way of penetrating Sichuan would be to establish a line of French steamers on the Yangzi.[44] A year later Maurice Dejean de la Bâtie, then at Fuzhou but later to play a leading role in Yunnan, came to much the same kind of conclusion: the question of the best route to Sichuan had already been resolved by European merchants in favour of the Yangzi. Nobody in China based their commercial decisions on national sentiment; if the Hong Ha route were superior in terms of cost, time, or security all merchants would use it. None did. Dejean de la Bâtie went even further: the Hong Ha certainly will not serve Sichuan, but could the French even expect to find great markets in the immediate area it could serve when they had failed to do so elsewhere in China?

It must be frankly said, it is strange that certain chambers of commerce are so preoccupied with knowing if the Tonkin route is really the best means of penetrating Sichuan, when nineteen ports in China are open to us just as they are to the English, the Americans, the Germans, etc., and we do not manage to sell more than three or four million francs worth of merchandise there, the great bulk of it, moreover, sold to the Europeans living in these ports.

He suggested that French merchants would do better to establish themselves in the treaty ports before hoping to conquer the more remote parts of the empire. He cruelly pointed out that although the Lyon Chamber of Commerce was enthusiastic about the prospects of millions of francs of commerce to be done via the Hong Ha, it was in Shanghai that some of its members lived, worked and made money, notably the arch propagandist Ulysse Pila himself.[45] Not for the last time, Dejean de la Bâtie's conclusions could remind French imperialists of the extent to which their aspirations were based on illusions. The lack of energy of French enterprise in the treaty posts hardly suggested a brilliant future for penetration from Tonkin. As Dautremer had pointed out, in 1892 there was not one French citizen in business at Hankou, a crucial centre for trade with Sichuan. As he put it: 'Now, if the French want to become traders, they must realize that the first condition [for success] is to expatriate yourself'.[46]

By 1896 the myth of the enormous wealth of Yunnan was being recognized as such. Writing in Le Temps Marcel Monier suggested that its wealth had been exaggerated and that the province was far from being the Ophir or eldorado it had once been considered. Although the small scale mining of copper and tin already being carried out by the Chinese indicated that the province possessed rich mineral deposits, their exploitation by modern methods would be difficult for legal, political, and technical reasons. Apart from its minerals, the province was one of the least populated and poorest in China.[47] This was a far cry from Rocher's description of Yunnan in 1897 as 'one of the richest provinces of the Chinese empire'.[48]

Thus, in attempting to meet the British challenge Hanotaux was confronted by a contradiction. On one side, there was overwhelming evidence from both within and outside his own department that such penetration of Chinese markets as could be achieved from Tonkin would be via the Hong Ha route to Kunming. On the other, however, there was the reality of the concession to Fives-Lille of the Longzhou railway and Gérard's commitment to that railway. In these circumstances Hanotaux decided to postpone taking a decision. At this time proposals to send a mission to investigate future railways from Tonkin into China were current in Paris. In January Duval of Fives-Lille had asked Hanotaux to send a mission to China whose aim would be to examine the possibilities for extending the Longzhou railway into 'a great system which must be extended across the southern provinces of China'.[49] Ultimately Duval's suggestion was acted upon. Hanotaux and the Minister for Colonies, André Lebon, agreed to send an exploratory mission to investigate the best means of penetration from Tonkin into China. the mission was to be funded by the Ministry of Colonies, although an interpreter would be provided by the Quai d'Orsay.[50] The decision to send the mission was approved by the cabinet on 7 July and, led by Charles Marie Guillemoto, its members departed in October 1897.[51]

It was the dispatch of the Guillemoto mission which gave Hanotaux the opportunity to avoid committing French policy too deeply to the Longzhou railway. He was able to use it as a reason for ordering Gérard to seek a formula from the Chinese which would anticipate construction of a railway into Yunnan by a route to be selected after Guillemoto's report had been completed.[52] Thus the result of Gérard's negotiations with the *Zongli Yamen* for compensation for the Anglo-Chinese Agreement of 4 February 1897 was a strangely vague document. Normally such vagueness would have been more in the interests of China than of France, but on this occasion it was prompted by the very real French confusion about the best course to follow. The agreement which Gérard and the Chinese signed on 12 June 1897 covered various aspects of French penetration including the Longzhou railway, mining rights, and improvements to roads and to the navigability of the upper Hong Ha. However, while it anticipated the construction of a railway from Haiphong to Kunming, it left open the question of the route to be followed.[53]

Events beyond French control ran faster than the Guillemoto mission could work. Guillemoto's report was available by August 1898, which was quite early considering that he had only arrived in Haiphong in November of the previous year. By that time, however, the Chinese government had already agreed to the concession of a railway from the point where the frontier crossed the Hong Ha to Kunming. This concession was the result not of the work of the Guillemoto mission, nor of the energy of French capitalists, nor even of the ambition of the Governor-General of Indo-China, but of the scramble for such advantages by the European powers following the German seizure of Qingdao in November 1897.

Ever since the triple intervention after the Treaty of Shimonoseki it had been anticipated that Germany would attempt to extract a territorial reward, for, despite her considerable commercial and naval presence in East Asia, she had no base from which to operate. German diplomats made no secret of this aspiration as they sought to prepare the other powers for their action. Thus, in an uncomfortable metaphor which Gérard found rather too Prussian, the German Minister in Beijing, Baron von Heyking, told him in February 1897 that 'bayonets ... are good for most things, but not for sitting on'.[54] Less predictable than the German move, however, was the effect which Chinese acquiescence would have on the other European powers' ambitions. From the time of the German occupation of Qingdao on 14 November 1897 until the Chinese acceptance of Germany's demands for a lease of the port and its surrounding area on 6 March 1898, there was remarkably little effort on the part of the powers to assist China to resist the German demands, even though the interests of Britain and Russia were directly involved. However, once Germany had secured her base, the other powers, most rapidly and devastatingly Russia, sought compensation. Three weeks after the concession to Germany, China agreed to lease Port Arthur and Dalian to Russia, thereby losing control over the Liaodong peninsula and hence reversing the effects of the triple intervention. In response to the Russian move, the British, who had nothing to gain from a scramble for such territorial acquisitions, were obliged to demand a similar lease of Weihaiwei as a naval base opposite the Russian base at Port Arthur.[55] Like Britain, France had diverse interests all over China and a base in the south, so she also had little to gain from imitating the Germans or Russians. Nevertheless, like Britain, she could scarcely fail to do so. Thus, while Britain leased a naval base which she did not want and never seriously fortified, France leased an absurdly useless coaling station on the coast of Guangdong less than four hundred kilometres from her own port of Haiphong.[56]

The lease of a coaling station at Guangzhouwan was only part of the claims France made on China under the terms of the most favoured nation clause. Other powers had linked their claims for leased territories with railway concessions and declaratioins of inalienability of various parts of the empire in order to preserve or extend their political and economic infiuence. France was no exception. The most important of these declarations was made on 11 February 1898 when, in response to British pressure, the Chinese agreed that no part of the Yangzi region, a vague term which embraced most of China, would be leased, mortgaged, or ceded tó any third power.[57] It was in response to privileges granted to Britain and Germany that Hanotaux resolved to demand 'equality of treatment' in accordance with the treaties. It was certainly true that France was the last power to press her claims on China, behaviour which was quite a reversal from her form during Gérard's tenure as Minister in Beijing. It was not until 7 March 1898 that Hanotaux instructed the *Chargé d'affaires*, Pierre Dubail, 'to claim certain compensations', well after other powers had not only demanded certain privileges, but had been granted some of them as well. This far from typical reticence and the absence of a French naval demonstration or ultimatum was appreciated by the Chinese, and the *Zongli Yamen* even asked Jing Zhang, their Minister in Paris, to convey this sentiment to Hanotaux.[58]

Compared with the Russian and German demands, the French, like the British demands were moderate. Hanotaux asked Dubail to obtain a declaration of the inalienability of Yunnan, Guangxi, and Guangdong; the appointment of a French citizen to head the Chinese postal service; the definitive concession of a railway to Kunming; and the lease of a coaling station on the southern coast of China.[59]

Except in Hong Kong, where the *Daily Press* asserted that

British commercial interests in Guangdong and Hainan must be defended from French interference even at the price of war, reaction to the French demands was calm and even favourable.[60] The press in Britain and Germany expressed some reserves, but no opposition,[61] and the legations in Beijing did not intervene or attempt to prevent the Chinese from acceding to the demands.[62] Thus, by letters exchanged between the *Zongli Yamen* and Dubail on 4, 9, and 10 April 1898 the French acquired almost complete satisfaction. Only on the relatively minor question of the postal service did the Chinese not comply. They pointed out that the only Western-style postal services in China were carried out by junior members of the maritime customs and that China was not yet able to establish an independent postal service. They agreed, however, that when such a service was established, the recommendations of the French government concerning personnel would be taken into account.[63] More importantly, the lease of Guangzhouwan was accorded and the inalienability of Guangdong, Guangxi, and Yunnan proclaimed. Most important was the concession of the railway from Laocai to Kunming which was granted to a French company to be named by the French government at a later date. This was the first railway concession which had been granted to a foreign company rather than to a joint Sino-European company. A further advantage was that the Chinese government was responsible for providing the land required for the railway.[64]

The significance of the concession goes beyond the generous terms on which it was made, for it represents the clear determination of the Foreign Ministry that the railway into Yunnan was to be built via the Hong Ha route, despite the advice of Gérard and the interests of Fives-Lille. After April 1898 the fate of the Longzhou railway was sealed. Yet the decision in favour of the Hong Ha route was made in great haste. Initially, the demand for a concession had been couched in very loose terms which merely specified that its terminus was to be Kunming and left the point at which it would cross the frontier undefined. Clearly, an indirect line via Longzhou, Nanning, and Bose could be permitted under these terms. However, the Chinese government wanted to know where the railway would pass into Chinese territory. Within a few days Hanotaux replied that, even though technical surveys were not yet complete, it would pass through Laocai or a point nearby recognized as that most favourably sited in the Hong Ha basin.[65] He had decided that the shortest route was the best, despite the work already undertaken by Fives-Lille in another direction. Thus, when informed, inaccurately as it happened, that the concession of a railway from Tonkin to Kunming would lead to British demands for a railway concession from Burma to Kunming, Hanotaux was able to express total unconcern and confidence as the French route was easily the shorter.[66] In deciding to obtain the concession for a railway through Laocai, Hanotaux was selecting the best route, but at the same time he was offering unconscious support and encouragement to the territorial ambitions in Yunnan of the Governor-General of Indo-China.

Almost from the time of his arrival in Hanoi after his appointment as Governor-General in May 1896, Paul Doumer was interested in extending French influence into Yunnan. He saw the project of penetration as an essentially colonial enterprise and undertook it without consultation of the French Foreign Ministry and often in direct opposition to its desire to sustain the territorial integrity of the Chinese empire. Soon after taking up his position in Hanoi, Doumer sent a military expedition into Yunnan. Commanded by a Captain Auriac of the cavalry, its ostensible aim was to study the potential of Yunnan as a source of remount horses for the cavalry in Indo-China. His report, however, was more concerned with the human than equine inhabitants of the province and suggested that, at least in lower Yunnan, they were weary of the alleged exactions of the Chinese gentry and ardently desired to come under the protection of France.[67]

It was not long before another French officer set about expanding on the theme which Auriac had sketched. Colonel Pennequin had headed the Sino-French commission which had delineated the frontiers of Tonkin and Yunnan and was commander of the Third and Fourth Military Territories in Tonkin. As a result of the good relations he had established with the fairly minor Chinese authorities with whom he had been working, he believed that they would welcome the formation of a Yunnan army of 20,000 men trained to European standards by French officers. Doumer believed this proposal to be a fine means for extending French influence in the province and in July 1897 supported Pennequin's desire to go to Kunming and present it to the Viceroy.[68] The activities of both Auriac and Pennequin anticipate the attitude of Doumer to Yunnan throughout the period of his administration of Indo-China. Both his belief that Yunnan should be at least a quasi-Protectorate if not a part of Indo-China and his conviction that military officers were the best men to accomplish the penetration of the province are typical of Doumer's attitudes. Equally typical was the reaction from the Quai d'Orsay, which consistently believed that economic penetration was far more effective and far less dangerous than military involvement. Informed of Pennequin's proposal to go to Kunming, Hanotaux asked André Lebon, the Minister for Colonies, to ensure that it be cancelled, especially as its activities could prejudice the success of the Guillemoto mission.[69] At Lebon's request, however, he did ask Dubail in Beijing to approach the Imperial Government about the possibility of such a mission. The Chinese, surprisingly, had no objections provided that they did not have to pay for the mission's expenses and that it took no initiatives, so ultimately Hanotaux, rather reluctantly, did authorize the commencement of negotiations about the mission.[70]

These negotiations never took place, because Doumer very quickly authorized Pennequin 'to survey the construction of extensions to roads in the military territories with the aim of commercial penetration from Tonkin into Yunnan and especially from Laocai to Mengzi'.[71] This road-building penetration was not at all what Hanotaux had authorized and once again he demanded the mission's cancellation.[72] Doumer, however, defended it as being merely a frontier matter which need not involve matters of foreign policy.[73] Thus, in December Pennequin and five other officers travelled from the frontier to Mengzi. Their reception there was frigid and the *daotai* absolutely rejected their proposal to build a new road from the frontier using troops from Indo-China as the labour. Pennequin's abrupt arrival, without any notification from the *Zongli Yamen* in Beijing had embarrassed the French Consul, Maurice Dejean de la Bâtie, and made the Chinese authorities suspicious of an invasion and hostile to the French. Clearly, Pennequin wished to establish a zone of French military influence in Yunnan reaching at least to Mengzi. To Dejean de la Bâtie this was folly, for, if the Chinese empire were to disintegrate, France would undoubtedly receive most of Yunnan in any case; but in view of the diffusion of her interests all over China, France should seek to maintain the integrity of the empire.[74]

This very typically rational and correct attitude on the Consul's part did not accord with the zeal or aspirations of the colonel. Pennequin made these aspirations clear in his attempts to enrol Dejean de la Bâtie as an agent in his scheme. He advised the Consul not to be too scrupulous about consulting Beijing or Paris and reminded him that if Pavie had acted according to the rules and not broken treaties, Laos would not have come under French rule. When he returned to Tonkin, Pennequin left four officers under Commander Gosselin in Mengzi. He acknowledged that their position was false, but saw that as an opportunity. He asked Dejean de la Bâtie to exploit it:

Tell any story to the *daotai* ... The most absurd pretexts are often accepted ... You have here a superb opportunity to distinguish yourself by seeking out all possible means for extending our influence in China. We must create interests here ... Obviously all that will cause you much work and embarrassment, but it is only by creating incidents that we can intervene.[75]

Neither Dejean de la Bâtie nor Angoulvant, the Consul at Hekou, just across the frontier from Laocai, accepted Pennequin's arguments or programme. Angoulvant told Dubail that he had 'committed a great error and spoiled things, even compromising the success of the Guillemoto mission'. He had aroused fears of invasion amongst the Chinese by his imprudence.[76] While understanding the ambition of successful young military officers, Angoulvant was shocked by Pennequin's open espousal of expansionist policies, all the more so because Doumer seemed to support him. The true aim of the expedition to Mengzi, he alleged, was neither to discuss military reorganization nor to build roads, but to prepare the way for a revolution in Yunnan which could provoke French interference followed by the conquest and occupation of the province.[77]

Even if these claims were exaggerated, Pennequin's trip to Mengzi clearly violated Hanotaux's stipulation that contacts between French officers and Chinese officials were only to be to discuss the policing of the frontier, as specified in the convention, or for other unimportant or urgent business, in which cases the Consuls were to be informed.[78] The violation was deliberate. Pennequin's exploits over Christmas 1897 were the expressions of Doumer's policy of territorial expansion at China's expense, the beginnings of Indo-Chinese military imperialism across the frontier in China itself. The next agent of this policy was to be the Guillemoto mission.

Although it was intended only to investigate the options available to the French for extending railways into southern China, the Guillemoto mission was quickly transformed into an effective tool in Doumer's expansionist plans. Although its members arrived in Saigon early in December 1897, they did not begin work in Yunnan for nearly two months because Doumer sent them as tourists to Cambodia in order to allow Pennequin to operate unfettered by their presence. By the time Guillemoto arrived in Mengzi on 1 February 1898 his mission numbered 12 of whom three were officers and two non-commissioned officers who had been in Yunnan with Pennequin. The Chinese authorities, still smarting from the memory of that unwelcome intrusion, were hostile to the mission and it was nearly three weeks before its members could begin work on surveying the line. Once begun, its work was completed remarkably quickly. Guillemoto had been unable to survey a route between Longzhou and Bose due to disturbances the area, and rejected Rocher's proposal for a railway following the valley of the Song Gam on the grounds that it was too indirect, a criticism which applied with even more force to the Bose route. He found that there were two perfectly feasible routes between Laocai and Mengzi, one following the valley of the Xinjian, the other the valley of the Namti. On both routes a metre-gauge railway could be constructed with grades no steeper than one in forty. In less than a month Guillemoto destroyed the myth of the inaccessibility of the Yunnan plateau by rail from the Hong Ha at Laocai.[79]

From Mengzi Guillemoto's party advanced across the plateau to Kunming, now without the military personnel whose presence was so distasteful to the Chinese. They arrived at the provincial capital on 10 May 1898. Despite the departure of the officers and the fact that the railway had now been conceded to France, their reception once again was hostile.[80] This hostility evaporated when its instigator, the provincial treasurer, Tang Zhoumin, was transferred following allegations of financial maladministration, but it was nevertheless a harbinger of an unfortunate future.[81] Once the political atmosphere had improved, Guillemoto, reflecting the wishes of Doumer, suggested to Dejean de la Bâtie that the opportunity be taken to appoint a French financial inspector for Yunnan, who could help modernize the province's finances. Dejean de la Bâtie agreed that such a reform would be desirable, but considered that as it would constitute a sort of Protectorate, it could not be implemented without international consent.[82]

Whatever the vaguaries of the political situation, the engineering and economic prospects were bright. The railway could be easily constructed across the plateau to Kunming and the mining engineer, André Leclère, who was accompanying Guillemoto, had found interesting mineral deposits which would provide both traffic for the railway and fuel for its locomotives. Writing on the basis of Leclère's work, Dejean de la Bâtie reported:

> The mineral wealth of the country is considerable. Coal is almost everywhere. Mengzi itself is in the centre of an immense basin of high-quality coal. Iron and copper ores are equally abundant. We know that the tin mines of Gejiu export 3,000 tonnes of metal per year. Finally there are also several deposits of silver ore.[83]

Just as the old myth of the riches of Yunnan was dying, another, with more substance, was coming into being. Instead of gold the more prosaic minerals of coal, iron, copper, and tin were seen as the basis of potential wealth. The main difficulties for their exploitation were legislative. Apart from copper, which was an imperial monopoly, minerals belonged to whoever mined them. Dejean de la Bâtie did not consider this an insurmountable problem and felt that it would be relatively easy to convince either the *Zongli Yamen* or the provincial authorities to legislate to allow mining concessions to be granted, as they would benefit from the taxation of the concessionaires.[84]

While the news of the mineral wealth of the area around Mengzi improved the railway's economic prospects, it also fuelled the expansionist passions already aflame in Indo-China. For, in the eyes of Doumer and his circle, this wealth could be used to finance the ambitious programme of infrastructure development, notably railway construction, he intended to undertake. The reports of Guillemoto and Leclère were presented to Doumer in August 1898. Soon after, Guillemoto was rewarded with his appointment as head of the Indo-Chinese Department of Public Works and Doumer sailed for France intending to raise a loan of 250 million francs. Guillemoto was a divided man. Naturally he was flattered by his appointment and enthusiastic at the prospect of building 2,000 kilometres of railway throughout the Protectorate and into Yunnan. However, he also realized the difficulties that could occur in dealing with Chinese officials and was fearful that Doumer's direct approach and scarcely disguised expansionism could have unfortunate results. He was especially worried by Doumer's plan to visit Kunming himself and take control of France's diplomatic representation in Yunnan after his return from Paris. Guillemoto confided these fears in a letter to Dejean de la Bâtie. After telling the Consul of Doumer's fixed desire to build the Yunnan railway before any others, he wrote:

The Governor-General is talking about nothing less than going to Yunnan himself after his return from France in January or February 1899. He will carry out a veritable raid, accompanied by General Bichot, Colonel de Beylié, Commander Ecorsse, various government employees, your servant, and two escort parties, one French and one Vietnamese. This will be a real tornado for peaceful Yunnan, and you will have much to do to ensure that the trip makes Kunming without obstruction.

M. Doumer wants to build 2,000 kilometres of railway and borrow 250 million francs. Indo-China will pay for the railway to Kunming. He wants to ask France for nothing. He dreams of paying for the army, the navy ... and even for the consulates in China. So everything is rosy in Indo-China at the moment ... We will even have a French concession in Mengzi soon.[85]

Dejean de la Bâtie was horrified to hear of Doumer's plans and doubted that Guillemoto would have the force, or ultimately even the will to resist them. On 14 October he wrote a lengthy confidential dispatch outlining his fears to Delcassé. The dispatch was to be influential and is especially significant as at exactly the same time another colonial adventure had brought Britain and France into conflict over the upper Nile. Although he acknowledged that it would be quite legal for Indo-China rather than the French Republic to build the Laocai–Kunming railway, Dejean de la Bâtie believed that such a substitution would be politically dangerous in a number of ways. Firstly, until recently Annam had been a tributary state of China and the Chinese tended to look down on Vietnam and things Vietnamese. They would therefore find it humiliating for the government of Indo-China to undertake public works in Yunnan and might well be hostile. Secondly, a private company, unlike the Indo-Chinese government, could distribute shares to Chinese officials, thereby ensuring their direct interest in the railway's success. Thirdly, the calibre of many French officials in Indo-China was low and the introduction of this element into Yunnan would give the Chinese a bad impression of the French and ample legitimate causes for complaint.[86] The tendency of these officials to use the cane on any Asian who in any way offended them would inevitably lead to conflicts with the Chinese working on the railway. Dejean de la Bâtie suggested that the consequences of such conflicts could be grave:

It would be impossible thereafter to continue effective work on the line. The people would rise up against us. They would willingly follow mandarins and hostile gentry who would retain all their influence over them. In the end we would very quickly be driven into a most embarrassing situation, forced, to avoid a defeat, into a military expedition, into the occupation of the province by our troops.

Besides I am well aware that in Tonkin they very much want to conquer Yunnan. Many men talk about it openly . . .[87]

A rational imperialist of the modern type, whose first priority was profit rather than national grandeur, Dejean de la Bâtie believed that only an 'economic conquest' of Yunnan was desirable, and that if conquered militarily it would become another costly and burdensome colony. He believed that Doumer had been seduced into this risky plan of conquest by Pennequin, whose great reputation in Hanoi as a China expert was, in the Consul's opinion, exceeded only by his absolute ignorance of international affairs and the imbroglio which such a conquest would create. Doumer and the 'military party' were well aware that the Ministry of Foreign Affairs was opposed to such an adventure and hoped to reduce its influence in Yunnan. They wished to replace consuls with commissioners of the Indo-Chinese government who would be responsible to Hanoi rather than to the Quai d'Orsay. Dejean de la Bâtie described the attitude of the Indo-Chinese military clique as follows:

In the eyes of many of our compatriots in Tonkin, especially the military who know us only slightly, most of the agents of the Department [of Foreign Affairs], especially those who began their career at the Quai D'Orsay, appear to be snobs, infatuated with protocol, tainted with anglomania, frantic partisans of the status quo, showing I don't know what sort of aristocratic disdain towards both men and things from Tonkin.[88]

The allegations of Dejean de la Bâtie were corroborated in the pages of the *Quinzaine coloniale* of 25 August 1898. An anonymous article, which, according to Dejean de la Bâtie was obviously written by a military officer based in Tonkin, suggested that the only way to ensure the security of the French possessions in Indo-China was to extend their frontier to the Xi Jiang and the Yangzi, thus incorporating most of Guangdong, Guangxi, and Yunnan. Rather than conquer this vast area by force of arms, the author suggested that France ensure the nomination of two viceroys who would protect her interests. One, Marshal Su, would be charged with Guangxi, Guangdong, and the island of Hainan; the other, a Muslim, would be based at Kunming and protected by a guard of Algerian troops. In this way a *de facto* Protectorate could be established:

> Each of these Viceroys would be guided and under the surveillance of a French Consul-General responsible to the Governor-General of Indo-China ... who could impose on our protégés our railways of penetration, our colonists and their rich plantations, our mining engineers, our traders, and finally a few French instructors, ordered to organize the military area destined to make the new frontier along the Xi Jiang and the Yangzi invulnerable.[89]

Delcassé was alarmed both by the article and Dejean de la Bâtie's dispatch and raised the matter both in cabinet and with President Félix Faure. Faure spoke about the matter to Doumer, who was then in France in search of his loan, and to the military commander in Indo-China, General Borgnis-Desbordes. There is no record of what was said, for, as a marginal note on Dejean de la Bâtie's dispatch stated, 'it is rather too delicate to write about'.[90] Delcassé's reply to the Consul at Mengzi is no more revealing: Dejean de la Bâtie was thanked and told that the matter had been raised with the President and Governor-General.[91] There can be little doubt, however, that Delcassé, like Hanotaux before him, sought to prevent the interference of Indo-Chinese officials, whether civil or military, in Chinese affairs as much as possible. Additionally, the memory of Fashoda was very fresh indeed in December 1898. Delcassé was well aware that Doumer's plans violated the Anglo-French agreement of 16 January 1896 and that if they were implemented Britain would have just cause to retaliate. Moreover, Doumer's ambitions were contrary to France's established policy of supporting the territorial integrity of the Chinese empire. In his general instructions when Stephen Pichon was appointed Minister in Beijing in March 1898, Hanotaux had emphasized this policy and stated: 'We can only wish to maintain the integrity of China.' He had further pointed out that if the empire was partitioned, France would be doubly disadvantaged; firstly, because her interests were dispersed throughout China, frequently in areas to which she could not hope to make any successful territorial claim, and secondly, because the civil disorder which would accompany the dismemberment of China would have its echo in Indo-China. More specifically, Hanotaux had affirmed that the retention of the political status quo was the best means of ensuring the development of French commerce in the provinces bordering on Tonkin and that the policy of obtaining declarations of inalienability was aimed at ensuring this continuity.[92] Delcassé's sentiments on these matters were no different from those of Hanotaux, and it is reasonable to assume that Doumer was presented with an outline of French policy similar to that contained in Pichon's instructions.

Whatever was said, however, did not deflect the Governor-General from his ambitions. Immediately after his arrival in Saigon early in March 1899, Doumer was reproached for sending another group of military officers into Yunnan. He replied, 'very surprised', in an extravagantly lengthy telegram that, following discussions in Paris with Delcassé, Bompard, and Faure,

we were all absolutely agreed that I would act locally and most effectively in the frontier province. The Government must have confidence in the Governor-General of Indo-China and give him freedom to act ... You don't once again paralyze our action while others are marching and overtaking us. I was given a formal promise in Paris that it would no longer be like that, especially so far as Yunnan was concerned. M. Delcassé and M. Bompard have even assured me that Consuls on the frontier would be invited to get along with me and follow my instructions.[93]

This last claim is almost certainly exaggerated, as the Ministry of Foreign Affairs was not likely to have surrendered control over its agents and, by implication, over French policy towards China to the government of Indo-China. Doumer concluded his defence by asserting that he needed and expected liberty to act as he judged best in advancing French aspirations in East Asia.[94]

Doumer was assured that Paris did have confidence in him but was warned that the diplomatic situation was dangerous and asked to be prudent. He was also asked to reveal his aims in Yunnan and why he was using military officers to advance them.[95] His reply reveals that very different lessons were learned from Fashoda at the Quai d'Orsay and in Saigon. Doumer wished to avoid the repetition of such future conflict by being first on the ground with the largest force. He wired:

Our difficulties in the world come nearly always because we act only sporadically and then too late. Here, despite so much lost time, we can still arrive first and avoid the competition to which our lack of activity would give birth.[96]

This attitude overlooked the fact that there was already an Anglo-French agreement on Yunnan to protect the interests of both nations and that unlike in the Sudan there was in Yunnan a legal, universally recognized government whose authority no European power, least of all either Britain or France, seriously wished to destroy. Doumer's ambitions existed before the Fashoda crisis, but appear to have been encouraged rather than weakened by the French humiliation there. He defined French aims in Yunnan as the creation of something approaching a Protectorate:

I reckon that we are going to learn to know Yunnan and its resources, establish relations with its people and mandarins, attach them to us, quietly extend our sphere of influence, and establish a presence in the country thanks to the facilities which the survey and construction of the railway will give us.[97]

He intended to implement this programme of penetration by visiting Kunming himself, speaking to the Viceroy and establishing a Sino-French commission to resolve all issues concerning the railway. Military officers were useful agents of expansion because they were 'devoted missionaries, competent and cheap'. Their presence should not cause alarm because they did not carry any more arms than diplomats or engineers and were not accompanied by troops.[98]

Despite whatever was decided in Paris at the end of 1898, control of French policy in southern China was clearly slipping out of the hands of Delcassé and the officers of his department. The energetic Governor-General who was appropriating this control was bent on a policy of expansion which could provoke popular uprisings in Yunnan and conflict with both China and Britain. If such conflicts were to occur, the French position would be legally and morally wrong and politically untenable.

8

Doumer's Ambitions versus Mattress Diplomacy in Yunnan, 1899

IN one respect Doumer had suffered a setback through having his independence of action circumscribed during his visit to Paris late in 1898. In another respect, however, he had won a great victory. The main purpose of that visit had been to obtain the backing of parliament, finance for the proposed loan of two hundred million francs, and funds for the Yunnan railway. In all of this he was successful. It was relatively easy to persuade the banks to co-operate with his plans. Even though the interest offered by the Indo-China loan was much lower than that offered by the Beijing–Hankou loan for example, the security of a government guarantee which Doumer was able to promise meant that the prolonged negotiations on financial arrangements which had characterized the Franco-Belgian project were on this occasion totally avoided. Typically, and not perhaps unreasonably in view of the collapse of various South American speculations over the previous decade, including the Panama Canal project, the bankers preferred security to a high return. On 11 November 1989 Doumer met representatives of the Société générale, the Banque de Paris et des Pays-Bas, the Comptoir national d'escompte, the Crédit lyonnais, and the Crédit industriel et commercial. They agreed to the loans, provided the necessary guarantees were inserted in the legislation.[1]

The Bill was presented to the Chamber of Deputies on 25 November and debated on 15 December. It was introduced by Jean de Lanessan, a former Resident-General in Indo-China and a future Minister of the Marine. De Lanessan was supported in the Chamber by Doumer who, although not at that time a deputy, had the right to speak as *commissaire du gouvernement*. The Minister for Colonies, Antoine Guillain, played a minor role in the debate and Delcassé did not participate at all. Article 1 authorized the Indo-Chinese government to borrow 200 million francs over 75 years. The interest rate was not to exceed 3.25 per cent and repayment was guaranteed by the colonial government. These funds were to be used exclusively for the construction of five railways; including that from Haiphong to Hanoi and on to Laocai, which were authorized in Article 2 of the Bill. These two articles were passed on the voices and without lively debate.[2] Article 3, which concerned the Yunnan railway, was more controversial, largely because of its generosity to the company (as yet unformed) which would take up the concession. It read:

The government-general of Indo-China is authorized to guarantee interest to the company which will be the concessionaire of the railway from Laocai to Kunming and its extensions, except that the annual sum of payments may not exceed 3 million francs nor their duration 75 years.

The deposit of the sums which the government-general of Indo-China may need to make by virtue of the preceding paragraph to the concessionary company will be guaranteed by the French Republic.

The clauses and conditions of the future convention between the Governor-General of Indo-China and the concessionary company will be approved by a decree of the *Conseil d'Etat*.[3]

The *commission du budget* of the Chamber, headed by Camille Pelletan, unanimously recommended the suppression of the article entirely, not so much because all its members opposed the railway, but because all felt that an important matter such as this should be discussed separately from the Indo-Chinese loan. A future president of the Republic and enthusiastic opponent of all colonial expenditure, Gaston Doumergue, was the Bill's most vigorous opponent. His case was based partly on the traditional right-wing anti-colonial argument that money spent on such projects was wasted, especially when France's defence needs were so pressing, and partly on the grounds that not enough was known about the railway's engineering difficulties or cost. On this second point Guillain admitted that the exact cost was not yet known, but pointed out that Guillemoto had estimated it at a maximum of 70 million francs. Since the annual interest on this sum was about two and a half million, the three million provided in the Bill for its service should be ample.[4] It was a rash assurance, and both Guillemoto and Guillain, who was ironically also a civil engineer by training, were subsequently proved wrong. Many deputies were concerned that the third paragraph of Article 3 would allow cabinet rather than parliament to approve an increase in the sum to be guaranteed. The government conceded the point and deleted this paragraph.

On the issue of defence, the government, or more precisely its *commissaire*, Doumer, was more sure-footed. After all, he knew more about the military's role in the railway than anyone else. Early in the debate, when it was claimed that the money would be better spent on cannon and warships, Doumer interjected: 'The railway is also a military instrument.'[5] He later sought to persuade the Right to support the railway by appealing to history, to past sacrifices, and to nationalist sentiment:

It is now a matter of opening Yunnan to our influence, that is to say what was the original idea behind the conquest of Tonkin. And we have just now asked you if, after having given French blood and money without counting the cost to achieve this conquest ... you are once more going to stop and show that you are conquerors but that you are not colonizers, knowing how to profit from the advantages of your conquests.[6]

To this emotional appeal to the events of 1883–5, Doumer added another to the events of the previous few months. The evacuation of Fashoda had occurred only six weeks before, so Doumer's assertion that the railway was necessary to forestall a British move into Yunnan touched a particularly sensitive nerve. In revealing his concern about British competition, Doumer made an uncharacteristic reference to the supposed economic benefits of the railway:

Don't you know who they are who are marching with determination (Applause) and who, right now, are spending millions and millions to arrive there first? We now have the advantage of our position: it happens that we possess the shortest route to reach this commercial clientele which we must want to offer to French industrialists and traders. (Fresh applause) We have everything needed to succeed; we lack only one thing; that is to prove that we have as much initiative and determination as our rivals. (Fresh applause)[7]

Eventually Doumer won the day, but Article 3 was only passed by the relatively slim margin of 298 votes to 243.

Article 4 specified that all material needed to build and operate the railways either in Indo-China or Yunnan not found locally had to be imported from France. A successful amendment paid homage to the protectionist spirit then about: the Deputies insisted that this material be exported in French ships.[8] The Bill as a whole was passed by the Deputies on the voices and did not encounter any great difficulties in the Senate. President Félix Faure signed it into law on Christmas Day and on 27 December a meeting of bankers at the offices of the Banque de l'Indo-Chine resolved to form a consortium and send a survey party of its own to Yunnan to check on the work and cost estimates done by Guillemoto. Headed by a senior civil engineer, Léonce Guibert, the consortium's mission left France in March 1899 and arrived in Yunnan two months later.[9] Guibert discovered that he was far from alone. Since Doumer's return from France no less than half a dozen separate missions, mostly dominated by the military, had been sent from Indo-China into Yunnan. Early in 1899 Guillemoto's deputy, Wiart, had led a mission from Kunming to Yibin on the Yangzi to assess the potential for a railway along that route. Wiart ultimately returned to Hanoi via the Yangzi and Shanghai. Simultaneously Captain Bellot of the marine artillery led a group of two officers and two sergeants to conduct surveys between Mengzi and Laocai. Early in April Commander Bauzon and two other officers went to conduct topographical surveys around Mengzi. Soon after Captain Duprat and two lieutenants were sent to Kunming to build houses for Indo-Chinese officials who were soon to be stationed there and to make plans for the future railway terminus. Also in April Captain Auriac arrived in Kunming to establish his remount service for the cavalry, Dr Jeanselme came to study various diseases in the province, and Guillemoto and Wiart returned to supervise the survey work of the officers.[10] The presence of all these French missions seemed to be preventing rather than assisting work on the railway. Pichon bitterly compared the ineffectiveness of French operations in Yunnan, where the right to construct a railway had been conceded by the Imperial Government, with the achievement of the one English engineer, Kinder, in building the railways of Zhili without consent from Beijing and in the face of considerable popular and official opposition.[11] Moreover, Dejean de la Bâtie reported that a number of French officers had behaved badly towards the local populations, committing such indiscretions as shooting porters who did not understand orders. Although compensation had been given, such incidents had led the *daotai* in Mengzi to observe poetically that the French seemed to regard a Chinese life as of no greater value than a blade of grass, and

Dejean de la Bâtie was fearful of the effects such incidents would have on attitudes to the French in general and to the proposed railway in particular.[12]

The most important of these excursions into Yunnan, however, was that of the Governor-General himself. Doumer had signalled his intention of visiting Kunming to Pichon as early as January 1899. Pichon naturally contacted both the *Zongli Yamen* and Delcassé so that the protocol for the visit and topics to be discussed could be resolved.[13] Delcassé, who had been so anxious about Doumer's activities the previous December that he had arranged a meeting with President Faure to discuss them, was concerned about the dangers of such a visit and raised the matter in cabinet. Delcassé wanted the visit called off altogether, but if it was to go ahead insisted that certain conditions be met. He did not want more than a few military officers to accompany Doumer, and did not want either the commander-in-chief at Hanoi or the commanders of the military territories on the frontier to be amongst them so that the visit would not in any way appear to be a military demonstration. Delcassé also asked that Doumer be extremely circumspect in the topics he discussed with the Viceroy in Kunming. He was to avoid matters of general policy (which in any case could only be resolved satisfactorily in Beijing) so that there would be no possibility of Doumer and Pichon contradicting each other. Delcassé asserted that even though he was to confine his discussions to matters affecting relations between Indo-China and Yunnan, these would be 'numerous, interesting and important'. Such matters, in Delcassé's mind, would include the railway and the creation of a Sino-French commission to supervise the repossession of property needed for its construction and to organize its labour force.[14]

Delcassé's concern that Doumer's behaviour be correct and not be interpreted as hostile to China was prompted not only by a fear of any Chinese reaction, but also by an awareness of rival British ambitions in the region. Ironically, it was a desire to steal a march on the British, who were still building their railway towards Kunlong Ferry in upper Burma, that impelled Doumer to an active policy. Delcassé, however, believed that French interests were best protected from British pretensions by Article 4 of the London agreement of 15 January 1896. Both in cabinet and in a letter to Guillain, Delcassé was insistent on this point. As the agreement bound France and Britain to share all privileges in both Yunnan and Sichuan, that is to say in a good part of the Yangzi valley, it protected French interests better than any military or engineering contest on the ground could. Delcassé considered that Doumer's independent attitude, as expressed in his aggressive telegrams of early March, was dangerous, as Doumer's precipitate actions could give the British an excuse to violate the agreement. Delcassé insisted that Doumer be warned of:

the inconvenience resulting from any step which may be poorly understood or misinterpreted and would risk not only offending the Chinese administration ... but also offering British agents a pretext they would be truly happy to exploit.[16]

Guillain warned Doumer in these terms by telegram.[17] Clearly Doumer and Delcassé had learned very different lessons from Fashoda.

The warnings were futile and the trip was a disaster. Doumer left Laocai on 29 May and made an exceptionally fast trip on horseback arriving in Kunming on 7 June 1899 and leaving two days later. He was completely ignorant of Chinese protocol, which would have required so important a personage to travel by sedan chair, and the Viceroy, Songfan, offered his party neither an official welcome nor hospitality of any kind. Indeed, had not Captain Duprat insisted that a local banker provide his summer house, Doumer would have been obliged to stay in some inn.[18] His discussions with the Viceroy were unsuccessful, even though Songfan was a self-strengthener of the Tongzhi generation, and as such inclined to compromise with foreigners on matters like railway construction. Doumer wanted the provincial government to buy 18 hectares of land almost immediately for the site of the railway station. Although it was the responsibility of the Chinese to provide the land needed for the railway, Songfan refused on the grounds that, as the railway would be constructed from the frontier towards Kunming, it seemed premature to establish a station in the capital. In a veiled threat he stated that such purchases would arouse the resentment of both the notables and the people.[19]

If the vice-regal threat to provoke disturbances had been subtle, the same cannot be said of Doumer's negotiating style. It was not until two years later that the translator on that occasion, the elderly apostolic pro-vicar, Father Maire, told a French Consul what had been said. When asked to obtain the land for the station,

Songfan had told Doumer that as the surveys for the line were not yet complete, it was too early to discuss the construction of a station. In any case the area requested was too great and the matter was really one for discussion, not in Kunming but in Beijing between the *Zongli Yamen* and the French Minister. Doumer told the unfortunate pro-vicar to tell the Viceroy that he regarded his response as a *casus belli*. Father Maire translated Doumer's aggressive assertion reluctantly and apologetically. Songfan replied that, if that were so, he would give Doumer his seals, install him in his *Yamen* and leave him to govern the province. He then rose and swept out of the room.[20] It was quite a performance.

The very next day new weapons were ordered for the Yunnan army and the militia was called out.[21] When Doumer left Kunming a group of mandarins, including the provincial military commander, accompanied him some distance out of the city pretending to honour their visitor. In fact, they left him to attend a secret meeting at an isolated pagoda to discuss means of provoking popular hostility against the railway.[22] That hostility was not long in coming. On 16 June a horse being used in survey work only 27 kilometres into Yunnan from the Tonkin frontier fell into a rice paddy. The populace rioted in retaliation and the survey party remained barricaded in a pagoda it was using for a week.[23] Trivial in itself, the incident was the first of an outbreak of such attacks.[24] They culminated in an attack on the customs house and French consulate at Mengzi on the night of 21 June 1899.

The attackers were not residents of Mengzi, but a group of tin miners from Gejiu who feared, not without good cause, that the French intended to expropriate their mines and operate them on European instead of traditional lines. French and Chinese élites agreed that the miners were a dangerously rough bunch, and the immediate cause of the outbreak had been the arrest and subsequent violent escape of one of their leaders. Although their quarrel was therefore primarily with the Chinese authorities, it was against the Europeans in Mengzi that they directed their attack. After burning the customs house the leader of the band made it known that this was an error and that their hostility was directed against the French and not the American customs officer. For a week after the attack the countryside around Mengzi was dangerous, at least for Europeans and their agents. The Chinese caravans continued their work, but survey parties did not and a French Indian from Pondichéri and his Chinese interpreter were murdered and robbed on 25 June while carrying 8,000 piastres for the Guibert mission. Regular Chinese troops did not arrive in sufficient numbers to restore order until 29 June. For that week the Europeans and Vietnamese in Mengzi remained barricaded in the town.[25] It was not until 3 July that 17 Europeans and 13 Vietnamese were evacuated from Mengzi and escorted to the frontier by Chinese troops.

French reaction to the raid was confused and hampered by the increasing bitterness in relations between Doumer and officials of the Foreign Ministry. Moreover, at the time of the raid France was without a government. When a new one was formed there was a new Minister for Colonies. Simultaneously Dejean de la Bâtie was on the road between Kunming and Mengzi and did not reach his posting until 25 June. His deputy, Reygondans, sent a message to Tonkin (the telegraph line was cut) requesting the dispatch of European troops to protect the consulate. Doumer was delighted to receive the request and less than honestly wired his new Minister in Paris, Albert Decrais, himself a product of the Quai d'Orsay and former ambassador in Italy, Vienna, and London, that 'at the request [of the] Consul' he was sending troops to Mengzi.[25] Doumer ordered two companies of the Foreign Legion to the frontier and informed Decrais that he would take 'all responsibility' for the operation.[27] It was a responsibility Decrais would not give him. He replied that the government wished to avoid if possible French forces entering China, but that if it was judged absolutely necessary that they do so to protect French lives, then they were only to go to Mengzi and immediately evacuate Europeans to Tonkin.[28] It was not the *carte blanche* Doumer had wanted. Opposition from the Quai d'Orsay and its agents was strong. As soon as he returned to Mengzi Dejean de la Bâtie wired that it was 'urgent' to put a halt to Indo-Chinese political and military intrigues in the province.[29] Pichon considered that as Chinese regular troops were moving into the area and the *Zongli Yamen* had promised to protect the Europeans there, French troops would not be admitted without a struggle which they could well lose or would win only after bloody encounters which would be destructive of the French position in China overall. He blamed the incident on the behaviour of the Indo-Chinese government which had created the feeling amongst both the people and mandarins in Yunnan that the French intended to seize the country. Under these circumstances it was not surprising that a simple criminal act could develop into a virtual anti-foreign uprising.[30] He also argued that even if the Chinese did allow French troops into Yunnan, Article 4 of the London agreement meant that the British would be entitled to send a detachment to the same region, something that was to be avoided if at all possible.[31]

Doumer halted the legionnaires at Laocai, but ordered them to be ready to march into China at short notice. Dejean de la Bâtie sent an urgent telegraphic message by a devious route urging his colleagues to protest with their last energy against the dispatch of troops, describing such a move as 'an irreparable error' and claiming that the Chinese authorities in Mengzi at least were doing everything possible to protect the Europeans there.[32] From Hanoi Doumer attempted to destroy Dejean de la Bâtie's credibility, claiming that the Consul knew neither the country (as he was not a good rider) nor its people (as he did not speak Chinese), and asserted that those who knew Yunnan, such as himself and his officers, were certain that troops had to be sent in.[33] Despite streams of advice from Paris not to do so, he claimed that he did not consider himself prohibited from ordering an invasion if French nationals were in danger, said of Dejean de la Bâtie's assertion that it would be dangerous to send in troops that 'it passes all the bounds of probability', and expressed the opinion that there was nothing to fear from the English. He cited the opinion of the normally francophobe Hong Kong *Daily Press* that there would be no opposition to the French frontier being extended into Yunnan to include Mengzi.[34] Doumer even sent a telegram to Guillemoto in Kunming asking him to forward gubernatorial salutations and an ultimatum threatening invasion to the Viceroy. Guillemoto discreetly ignored the request.[35]

Dejean de la Bâtie reported the incident to Delcassé who, furious, insisted that Decrais inform Doumer that it was the task of Consuls, not the Indo-Chinese government, to represent France's political interests in China.[36] Doumer replied in a long self-justifying telegram in which he took the opportunity to abuse the Consul for his inactivity, accuse the Quai d'Orsay of 'a lack of confidence and hostility towards my proposals', and defend his right to 'a sufficient liberty' of action in Yunnan.[37]

These were the words and actions of a threatened and out-manoeuvred man. Largely on the basis of Dejean de la Bâtie's reports, Decrais and Delcassé decided that the situation had to be brought to an end. Delcassé drafted in his own hand a telegram to Doumer pointing out the dangers of his actions, reminding him that the government, not himself, was responsible to the nation and parliament for his actions, asserting that Article 4 of the London agreement protected French interests in both Yunnan and Sichuan, and ordering him not to take any action which would violate that agreement.[38] Although he continued to complain, that telegram put an end to Doumer's plans for an invasion for the time being at least. Doumer had lost credibility in Paris and was lucky to survive. So serious was the situation that Delcassé informed Paul Cambon in London of the dangers of Doumer's impetuosity and authorized him to use this confidential information to reassure the British government that, whatever aspirations existed in Hanoi, Paris had no intention of seizing Yunnan.[39]

Despite the riots, murders, and suspicions it had provoked and the fact that it almost brought war in its wake, Doumer himself seemed thoroughly pleased with his visit to Kunming. He reported that it had been undertaken 'in the best possible conditions', and even claimed that Songfan had greeted him with 'much cordiality'.[40] Even if he had been unsuccessful in his attempt to acquire land in the capital, at least while there he had created a French commission to supervise the railway's construction. Considering that its role was intended to be technical, rather amazingly only one of its nine members was an engineer. Headed by a senior Indo-Chinese administrator from Saigon, M. Masse, its vice-presidents were Commander Bauzon and Wiart, the sole engineer. Other members included three army captains, Bourguignon, Duprat, and Lacoste. Its secretary was a *commissaire* of the Laos government called Ganesco, described as an adventurer by Angoulvant, formerly a Vice-consul at Hekou and then a senior official in the Ministry of Colonies in Paris. Ganesco was to be assisted by another Indo-Chinese official named Gérard and a Chinese interpreter, Liu Wenquan.[41] Dejean de la Bâtie was highly suspicious of the commission and its role. At the height of the crisis in Mengzi he wired Delcassé: 'I insist that the appointment of M. Masse to head a secret political mission not be ratified.'[42]

Reaction in Paris to the establishment of a French railway commission dominated by military and official rather than technical personnel was to be far from enthusiastic. In Kunming it was hostile. The commission obtained a 99-year lease on a large pagoda on the outskirts of the town. The French closed down its religious functions and expelled the stall holders from the market which was held in its grounds. As well as the railway commission, Captain Auriac moved in with 35 horses to establish his remount service. In addition to disrupting the life of the area in this way, Ganesco inflamed public opinion by constantly entertaining prostitutes in the pagoda. Not surprisingly, these circumstances coupled with the hostility of the senior Chinese officials in Kunming led to a riot. On 16 July 1899, a little over a month after Doumer's visit, a large group of local inhabitants, led by and inspired by the gentry, attacked the pagoda and expelled the French. They retreated to another pagoda, being attacked *en route* with stones. At this stage the mandarins and police arrived. Not wishing to see the demonstration get out of hand and having had their point made, they restored order and apologized to the French. The crowd took vengeance in assaulting the pagoda's caretaker and massacring a monkey and some birds.[43] While this riot was a less serious affair than the events at Mengzi, it was a discouraging beginning for Doumer's commission, all the more so as the riot was held with the connivance of the Viceroy.[44]

In both Paris and Beijing, French officials expressed dissatisfaction with Doumer's unilateral decision to establish the commission and more specifically with its personnel. Pichon condemned the commission as a political rather than a commercial or technical body and suggested that France's problems in Yunnan arose from exactly such a confusion of commercial and political aims. He commented bitterly on the effects of Doumer's operations in Yunnan:

a year has not yet passed, the surveys are not finished, we have already sent at least 50 engineers, officers, NCO's drivers, draughtsmen, bookkeepers, plus their escorts, and we have alienated the people to the point that a criminal fleeing from justice has been able to exploit this animosity to stir up a revolt in one of the province's principal towns. From this I conclude that if we do not want to stir Yunnan to the point where the success of our projects is either seriously compromised or dependent on a military expedition which would be a true conquest, then we have to change our ways.

He argued that the railway must be more 'a private industrial business affair' and have less of the character of a take-over of the province. To ensure this there must be no more military missions in Yunnan, and any Indo-Chinese officials there should only act on the advice of the Legation in Beijing. Pichon believed that the aim of French policy in China was 'the creation and exploitation of economic interests' which would themselves bolster France's political interests. He feared, though, that the railway:

will furnish ardent colonialists with hopes of expeditions and conquests. Was that Parliament's aim in voting the 80 million franc guarantee? Is not general opinion, on the contrary, that we have conquered enough territory and appointed enough functionaries? Is not the Yunnan railway's commercial aim more important than its political interest? . . .

If the political aim is to surpass the commercial aim, it must be well understood that the guarantee on the 80 million loan is only the prelude to considerable expenses, not to mention the diplomatic problems . . .[45]

Pichon's sober analysis was accurate in its awareness of the consequences of prolonged political intervention in Yunnan: it would almost inevitably lead to conquest of the province at great financial, human, and diplomatic cost to France. After conquest would come the problems of security and administration of a Chinese province. For Pichon that was more than a responsible government would desire. There was little disagreement in Paris. whatever opinion in Saigon and Hanoi might have been.

So far as the immediate problem of the Masse commission was concerned. Pichon believed that the best approach was to remove all officials from it and reconstitute it as an arm of the consortium whose job it was, after all, to build the railway. Such political representation as was needed in Kunming could best be effected, he felt, by the appointment of a Consul-General there. Part of the Consul-General's role would be to act as an intermediary between the Chinese and the railway commission.[46] Delcassé agreed with his Minister in Beijing and told Decrais as much.[47] By this stage Decrais and his senior officials in the Ministry for Colonies were far from sympathetic to Doumer's behaviour and were inclined to similar views to those current in the Quai d'Orsay. Angoulvant, Decrais's *sous-chef* and a former Vice-consul in Yunnan, complained that:

We are very badly informed, as much at the Pavillon de Flore [where the Ministry for Colonies was housed] as at the Quai d'Orsay, on this commission's composition. M. Doumer is always promising reports which he never sends.

Angoulvant had no illusions about Doumer's intentions or those of the Indo-Chinese officials, who, he claimed, were 'inclined to act in the most arbitrary fashion and imbued with the idea that they were preparing for the conquest of Yunnan or even of Sichuan'. He considered that the best way to halt Doumer's 'policy of encroachment and annexation' was to replace Masse with an official of the Department of Foreign Affairs, preferably with the title of Consul-General.[48] While Angoulvant's aims accorded with those of Pichon and Delcassé, his advice on securing them was different. Decrais shared Angoulvant's views and was 'very much opposed to the interference of M. Doumer and his functionaries in China and ... was not afraid to explain that in the clearest manner to the Governor-General'.[49]

The problem of what Dejean de la Bâtie had called 'the dual representation which Indo-China is striving to achieve' in Yunnan was brought to a head by the appointment of Masse, which was so overtly political that it could no longer be ignored.[50] Delcassé decided to accept Pichon's advice. The Consul-General would be appointed but he would not head the commission, unless the latter refused to comply with any agreements it might reach with the Chinese. In this way Delcassé hoped to ensure that the commission assumed both the appearance and the reality of an entirely commercial body and that political control remained firmly in the hands of his department.[51] Decrais accepted the arrangement willingly and informed Doumer.[52]

The Governor-General was far from pleased at this attempt to interfere in what he regarded as his Chinese bailiwick. He objected that there were no political issues to be discussed at Kunming, only railway matters, and that the Consul-General would have nothing to do. Without conscious irony he observed, 'there is nothing as dangerous as an unoccupied man'. If the appointment had to be made, Doumer was adamant that 'he must be placed directly under my orders, like the engineers and officers now operating in Yunnan'. Graciously he was willing to allow the Consul-General to correspond with the Beijing legation and with the Quai d'Orsay, although all such correspondence would have to pass through Doumer's office.[53]

Doumer was clearly anxious to avoid the continuation of surveillance of his activities by the agents of the Department of Foreign Affairs. More than anything else it had been the reports from Dejean de la Bâtie in Mengzi that had alerted Paris to what was going on in Yunnan. As a result of his accurate reports, Dejean de la Bâtie earned the hostility, disdain, and abuse of Doumer and his clique. As early as March 1899 while visiting Hanoi the Consul was described by Doumer's *chef du bureau militaire*, Commander Lassalle, as a 'bad Frenchman'.[54] The consequence of that insult was that Delcassé insisted that Doumer reprimand Lassalle.[55] By November Doumer was referring to 'the ill-will and clumsiness of the consul in Mengzi, who is easily the most dangerous adversary of French influence in Yunnan'.[56] In view of the extent to which relations had deteriorated, Decrais and Delcassé agreed to recall Dejean de la Bâtie, who had, in any case, requested leave. Doumer's satisfaction was rather restrained by the award of the cross of a *chevalier de la Légion d'honneur* which the Consul received.[57]

Although Doumer would no longer have Dejean de la Bâtie with whom to contend, the influence of the Quai d'Orsay was far from finished. On 29 October Auguste François, the victor in the fight against Fives-Lille's Longzhou railway, had arrived in Kunming overland from Guiyang. He had the title of *Consul en mission* and the task of supervising relations between the French railway builders and Chinese in the city.[58] Ultimately, François was appointed Consul-General, but not on the terms which Doumer had suggested. As there was no treaty provision for the creation of a French consulate at Kunming, and since such a provision, if obtained, would also give the British the right to a consulate there, François was styled a *Consul général honoraire*. He was to deal only with matters arising from the construction of the railway and his appointment did not imply any French right to a permanent consulate.[59] François had the right to correspond directly with Doumer, but the latter could not give him orders, only 'opinions and advice'.[60] Apart from the replacement of Dejean de la Bâtie with Saison at Mengzi, Doumer gained nothing from Paris as a result of his activities in mid-1899. In terms of diminished credibility and increased surveillance by agents of the Quai d'Orsay, he lost a great deal. The decision not to allow Doumer to issue orders to François was taken in cabinet on 28 November on Delcassé's advice.[61] Thus, the entire ministry was by then aware of the problems Doumer was creating for the cohesion of French policies in Yunnan.

There remained the problem of the commission headed by Masse in Kunming. Dejean de la Bâtie considered that Doumer had created it in order to establish the machinery of the Indo-Chinese government in Yunnan, that Masse was intended to be Indo-China's *commissaire général* in the province, and that the parties of officers and officials sent into Yunnan under the pretext of surveying the railway were in fact regarded as the core of the future French civil and military administration of the province.[62] François's impressions were similar. Masse had obtained the lease of three properties in Kunming. None of these was near the site of the proposed railway station and on one Captain Auriac had established his remount service, on the second a base for military carrier pigeons was under construction, and on the third a secret local militia was being trained and armed.[63] Commander Bauzon, Captain Bourguignon, and Captain Lacoste were in charge of this extraordinary militia and were recruiting Muslims and members of minority ethnic groups who might be hostile to the Chinese. Under Doumer's orders, Lacoste was importing arms and ammunition from Tonkin, a measure which had provoked a harsh response from the English customs official at Mengzi. Lacoste was worried about the activities he was being ordered to undertake and so requested his recall to France.[64] Many of the officers in the area around Laocai, who knew the difficulties of maintaining French control even in upper Tonkin, were also very fearful of a campaign to conquer Yunnan and considered Doumer's plans dangerous.[65] No such doubts, however, plagued Bourguignon, who told François formally that Doumer had ordered him to undertake the conquest of Yunnan in seven or eight months. What Bourguignon did not realize was that his Chinese translator was an agent of the Yunnan chief of police and the son of Liu Yongfu, the leader of the Black Flags who had resisted so strongly the French conquest of Tonkin 15 years before. In François's opinion the military preparations of Masse's deputies were anything but secret so far as the senior provincial administration was concerned.[66]

The pretensions of Indo-Chinese officials in Yunnan were not only military. Masse embarrassed François and the Viceroy by using grandiose titles for himself which Doumer had invented and which were not accredited with the Chinese.[67] He further told Pichon that he had been appointed with a political role and wished to correspond directly with the Beijing legation.[68] Under Doumer's instructions, the commission subsidized Catholic missions so that they could open three French schools in the province and

Guillemoto proposed to extend the Indo-Chinese postal and banking system into the province.[69] All these activities are clear evidence of the tendency in Tonkin to regard Yunnan as an annex of Indo-China, soon to be absorbed into the union. Such was not the aim of French policy makers in Paris. The *Direction politique* at the Quai d'Orsay observed darkly that 'the interference of Indo-China agents in Yunnan has already gravely compromised our interests in this region'.[70] Delcassé pointed out to Decrais the dangers of Masse ascribing a political character to his activities and his 'far too numerous military entourage'. He wanted the Indo-Chinese government to cease all political activity in Yunnan, to keep the number of engineers working in the province to a minimum and, as far as possible, to eliminate the military element.[71] Decrais agreed. He had already reminded Doumer that responsibility for French relations with China lay in the offices on the Quai d'Orsay, not in a tropical gubernatorial palace. The policy to which Paris was committed was one of peaceful economic penetration:

The Government does not want an armed conquest of Yunnan; it wants the region progressively placed into Indo-China's economic orbit and our influence made predominant there. That result can only be achieved through the activities of the government general of Indo-China; but to be effective, this activity must not be taken in isolation. It would lead to a certain defeat if it is not closely planned with our diplomatic activity.[72]

Doumer insisted that his civil and military officers in Yunnan were behaving well:

in undertaking with diligence and energy their special task, gaining the sympathy of the people, the confidence and friendship of the mandarins, in a word making the name of France loved in this part of China.[73]

However well loved the name of France was in Yunnan, Delcassé and Decrais had decided that the political activities of Indo-China officials in Yunnan had to end. The appointment of François provided Decrais with the opportunity to instruct Doumer to terminate Masse's appointment without appearing to be too imperious. On 15 December 1899 Doumer was ordered to replace Masse with an engineer and the military members of the commission with technical personnel.[74] Doumer's reply verged on insubordination. He claimed that Masse was well regarded by the

Chinese and that François had come to accept his role. If Decrais insisted absolutely on his replacement, he requested a delay of a few weeks.[75] For once Doumer was on sound territory in his support of Masse. Although he had a pretentious love of titles and an excessive concern for his own dignity, Masse had worked hard and effectively to gain the respect of the Chinese officials and of François. On 9 October he had presided over the first meeting of the mixed railway commission, consisting of four Chinese and three French members. The minutes reveal that its proceedings were business-like and devoid of recriminations which could have been expected. Both sides seemed to be co-operative, even amicable. The main issues concerned the sums to be paid as wages to Chinese labourers and as rent for various properties.[76] In December Songfan sent Masse a highly complimentary address, a translation of which Doumer did not blush to forward to Paris.[77] François, despite his concern at the activities of some of Masse's deputies, praised his work and contrasted it strongly with that of Guillemoto, whom he considered to be incompetent.[78]

Doumer's tactic of delaying any replacement of Masse was effective. A little over a month after the order was sent to Doumer, Pichon suggested that he go to Hanoi himself to meet the Governor-General and François to resolve the procedures for conducting French policy in Yunnan in general and the question of Masse's political role in particular.[79] In Paris Delcassé and Decrais agreed that such a meeting would be valuable and that a decision on Masse's future could be left to the three most senior men on the spot.[80] The meetings took place during the first week in March 1900. François made the two-week journey down from Kunming while Pichon sailed from Tianjin in a French naval vessel. Camille Sainson, the new Consul at Mengzi, also participated.[81] All those present seemed pleased with the results of the meeting. Doumer was given satisfaction on a number of points. Masse would remain at the head of the railway commission. Pichon conceded this point because François had influence over Masse and did not wish to risk having to deal with a possibly more difficult replacement. The Indo-Chinese government, it was agreed, would establish a postal service in Yunnan, subsidize French education, and provide François with funds to operate an intelligence service in the province. On the other hand, Doumer conceded that only consular officials could deal with political matters and discuss official business with mandarins, except for contacts provided for in the frontier police regulations, and François agreed to resolve as many issues as possible without reference to the Beijing legation.[82]

Pichon felt that his visit had enabled him to correct a number of illusions held in Hanoi about conditions in China. He found that French opinion in Indo-China inaccurately regarded the Chinese empire as on the point of collapse. It was widely believed there, not least by the Governor-General, that: 'It was only due to chimerical fears and a pure lack of boldness in the face of England that we did not march with deliberation into Yunnan last summer.' He was told, moreover, that:

To add new territories to our Far Eastern possessions we would only have to follow the example of Russia to the north of the Great Wall and not let ourselves be intimidated by vain threats.

It was against such widely held opinions that Pichon laboured in Hanoi. He attempted to convince Doumer that the realities of international politics in China meant that it was not in France's interests to provoke the dismemberment of the empire, and that any territorial issue would involve the European powers.[83] Pichon was pleased with Doumer's response to his arguments and felt that he had destroyed the Governor-General's illusions of easy territorial aggrandisement:

In sum, provided that the sentiments expressed to me persist and that the arrangements on which we agreed are executed, I judge that my trip will have had fortunate consequences. It will have dissipated misunderstandings, rectified annoying ideas and, I hope at least, thrown light on errors which it would be alarming to allow to continue.[84]

Delcassé was less sure and indicated his scepticism with a wry annotation 'Voila!' beside the conditions Pichon had specified for the future success of the agreement.

François was even more delighted than Pichon. He had good reason to be as Doumer had in effect given him supervisory control over the human and financial resources which the Indo-Chinese government was deploying in Yunnan. He modestly ascribed this to the 'very fortunate influence' of Pichon at Hanoi, although his own success in cultivating good relations with Masse was also largely responsible. François convinced Doumer that the structure of his railway commission was badly flawed: Masse's political role had led him to get involved in negotiations which had nothing to do with the railway and military officers had refused to take orders from either Masse or the engineer, Wiart, claiming that they owed obedience only to their superior officers. François thought that he had convinced Doumer that military officers caused more problems than they were worth and should be removed from the commission, and, indeed, from topographic work in Yunnan generally.[85] Doumer conceded ground on this matter and on the whole seemed to be able to work well with François, whose 'devoted assistance' he claimed to have received.[86]

The repudiation of Doumer's political and military ambitions in Yunnan was not just the result of difficulties in relation between two distinct French bureaucracies or of confusion as to the nature of French representation in the province. During 1899 French policy makers, like those of other imperialist powers, had been forced into making a decision on their attitude to the future of the Chinese empire. Events beyond French control meant that it was no longer possible both to attempt to establish an area of predominant French economic and political interest in one part of China and to seek simultaneously to maintain a serious French presence throughout most of the empire. To use the terminology of the age, a choice had to be made between the policy of the open door, which implied the more or less equal access of all imperialist powers to all parts of the empire; and the policy of spheres of influence, which anticipated the ultimate partition of China. Doumer's plans were an aggressive expression of the policy of spheres of influence, but by the end of 1899 France had rallied to the policy of the open door. This is, perhaps, surprising. After all, her Russian ally was clearly carving out a sphere of influence, indeed claiming, it would seem, an area for future annexation in Manchuria and northern China. This was being done with French money. Moreover, apart from Russia, France was the Western power with the longest and most sensitive frontier with China and the policy of Paul Doumer in Yunnan is ample evidence that many French citizens wished to emulate in southern China the successes of their Russian counterparts in the north. Yet, ironically, it was the very success of Russia in laying claim to her sphere that prompted serious debate in the Quai d'Orsay leading to repudiation by France of the Russian approach.

The issue of Western powers laying claim to spheres of influence had first been raised by Germany in the wake of the seizure of Qingdao. German pretensions to a monopoly of concessions in Shandong were in a sense no more significant than, for example, the very tenuous French mining privileges in southern China that Gérard had extracted from the *Zongli Yamen* in 1895 and 1897. Nevertheless, there began an unedifying period during which various Western powers sought, with varying degrees of enthusiasm and success, to claim similar spheres of influence elsewhere in China. The problem for the powers with the notion of spheres of influence was that most of them had commercial and political interests throughout the empire, not concentrated in one or a number of provinces. Probably the power whose interests were most widely diffused was Britain, while the power whose interests were most nearly concentrated was Russia. It was an agreement to resolve the differences between the diverse interests of these two powers that brought the issue of the relative merits of spheres of influence and the open door to a head. The immediate issue between the two powers was the northern railway from Shanhaiguan to Niuzhuang in Manchuria. The railway was being built by British engineers with British money, although, of course, its ownership remained, nominally at least, Chinese. The area it served, however, was one where Russian political and strategic interests were involved. During early 1899 Salisbury and Muraviev began discussions to resolve the dispute. These culminated in the signing of an Anglo-Russian agreement on 28 April 1899. Britain effectively agreed to play the game by Russian rules. The two powers recognized each other's spheres of influence in Manchuria and the Yangzi valley. Britain promised that the northern railway would remain in Chinese control while Russia promised to confine her interests to the north. Russia also agreed to refrain from supporting any claims by third powers for railway or other concessions.[87] The third power whose efforts in China the Russian legation in Beijing had hitherto supported was France. If Britain was giving away her commercial principles and possible future control of the northern railway, then Russia was giving away France.

French policy in China had hitherto relied extensively on the Russian alliance. Although in some respects the alliance had seemed to work more effectively in Russian than in French interests, France had enjoyed consistent support from the Russian legation in her quest for concessions throughout China. That situation had now come to an end and within the Quai d'Orsay there began a search for an alternative. Bompard commented bitterly on the Anglo-Russian agreement:

Russia, with whom we have been formally linked in foreign policy, has thus now solemnly contracted with England to leave us isolated in the face of our rival, in a way and in a country precisely where her assistance would be especially useful.[88]

Bompard proposed that France break out of her isolation in Chinese affairs by seeking to link her policy with a power with similar ambitions and interest. This power seemed to be Germany. Bompard considered that Germany's need for markets for her industrial production was even greater than that of France, and that Shandong would be insufficient. German diplomats had indicated that they did not recognize any British monopoly on railway construction in the Yangzi valley and Bompard therefore felt that it was desirable to open an exchange of views with the German government on Chinese industrial questions generally and railways in particular, especially those in the Yangzi valley.[89]

Paul Cambon's views were similar, although he was not as confident of Germany's will to maintain an open door in China as was Bompard. Certainly he acknowledged that Germany's interests lay in the open door: Salisbury had confessed to Cambon that he feared German economic rivalry throughout China. Cambon concluded that Britain would try to encourage the Germans, against their economic interests, to retreat into their sphere of influence in Shandong. Cambon acknowledged the possibility of British success here. Although Germany would prefer to have access to all of China, Shandong was a wealthy and populous enough prize if British opposition made her operations elsewhere difficult. France, however, did not have this option which was available to Germany. There was no wealthy potential sphere of influence which would compensate France for her loss of access to the rest of China. Cambon examined the provinces bordering on Tonkin: Guangdong was already under the economic influence of the British in Hong Kong; Yunnan would have to be divided between French and British spheres; and Guangxi was picturesque, but poor and mountainous. The pursuit of a sphere of influence held no interest for France, argued Cambon:

For us the conquest of our sphere of influence would lead to serious difficulties without economic compensation. In addition, just as we have sustained the policy of spheres of influence in Africa, with Germany's assistance and against the encroachments of England, so in China we must force England to remain faithful to the principle of free commerce which alone will allow us to challenge the monopoly which she now enjoys.[90]

The problem was to find other powers to support France in this approach. Her only ally, Russia, was committed to her sphere and hence of no use. Germany, the USA, and Japan seemed to Cambon to have similar interests to France. Cambon considered that the situation in the Yangzi was similar to that in Egypt in 1882. In order to prevent the successful establishment of a *de facto* British Protectorate in China, a coalition of other powers would have to be created. However, Cambon felt that in the past Germany had benefited greatly from Anglo-French rivalry and that on this occasion it would be advantageous to talk the Germans into making the running. He acknowledged that this would be difficult and that the negotiations would have to be conducted with great prudence as a powerful element in Berlin, including Foreign Secretary Bülow himself, favoured an agreement with Britain which might involve acknowledging spheres of influence. Cambon, therefore, agreed with Bompard that a *démarche* should be made in Berlin. He felt that it had to be made with great delicacy and that if it did not meet with the immediate approbation of William II, the matter should be dropped. In that case,

we must come back to Lord Salisbury and seek the basis of a common policy in China with him. There our principles are still the same and for England the renunciation of the policy of spheres of influence and adhesion to that of the Open Door would have the advantage of being a return to British traditions.

Whether France was able to work with Germany or with Britain, the matter was urgent. The British mercantile class had not yet resolved to make a second Egypt of the Yangzi valley, and Cambon considered that this class would decide what would be done. An arrangement had to be reached, however, before British imperialistic passions were aroused and the maintenance of the open door became impossible.[91]

Delcassé took up the suggestion of the head of his Direction commerciale and ambassador in London, forwarded a copy of Cambon's dispatch to Noailles, the Ambassador in Berlin, and set about trying to extract Germany's views.[92] On 19 June 1899 Noailles contacted Bülow and discussed unofficially the question of Chinese investments and commerce. Bülow indicated that he believed in the necessity of preserving open competition and seemed willing to hear more about French views on the matter when told they were similar. Noailles felt that Bülow did not much like the Anglo-Russian agreement, but it was impossible to seduce the Chancellor into defining German attitudes more clearly.[93] Delcassé had no more success in extracting anything more specific from the German embassy in Paris and by mid-June had concluded that the Germans were not sufficiently forthcoming to permit an exchange of views to take place.[94] By August Noailles considered that Germany was drifting more towards concentrating her efforts in Shandong, although she was not as yet totally committed to the policy of spheres of influence.[95] The Germans, with a rich sphere of their own to exploit but possessed by a desire to play a large role throughout China, had not made up their minds.

Nor, in a definitive way, had the French. For although Delcassé's instructions to Noailles indicated that he favoured the approach of Bompard and Paul Cambon, and advice he was receiving was far from unanimous. Stephen Pichon in Beijing believed that it was foolish for France not to claim a sphere of influence while other powers were doing so. He feared that most of China would be partitioned between Britain and Russia. Germany would succeed in claiming Shandong, but France would lose her commercial and industrial advantages in the south. If France were excluded from other powers' spheres her religious Protectorate of Catholic missions would perforce lapse, as the Chinese only acknowledged it because France could make a military or naval demonstration if necessary.[96] Pichon considered that France's position in China was vulnerable and from the time of his arrival had favoured supporting French policy with naval and military might. The previous year he had fought successfully against a proposal to reduce French naval strength in East Asian waters and, at Pichon's request, Delcassé had even brought the matter to cabinet. As a result the French fleet was retained and its effectiveness improved by the redeployment of marines from occupation duties in Guangzhouwan to their ships in October 1898.[97] At the same time Pichon had sought, again successfully, to ensure a strong French naval presence in the Bohai Sea and had even suggested that an occupation of Beihai might be a useful demonstration to Britain and China of French determination to preserve her position.[98] All this bellicosity on Pichon's part was not due to any specific crisis, but only to the less pliant attitude of Chinese officials following the return to power of the Dowager Empress. As an advocate of a strong policy in China, Pichon was disappointed that France had not acquired a naval base in northern China when she had obtained the lease of Guangzhouwan. The withdrawal of Russian diplomatic support in the wake of the Anglo-Russian agreement had worsened France's already weak position. Pichon's proposed solution to this perceived weakness was to cut French losses in the north and claim a sphere of influence in the south. He therefore suggested that negotiations be opened with the British government with a view to defining a French sphere of influence with British consent. This sphere, he felt, should include that part of Guangdong west of the Bei Jiang and south of the Xi Jiang and all of Guangxi. The British should also agree to share equal rights with France in Yunnan and Sichuan (they already had done so, of course), recognize French mining privileges in Yunnan and Guizhou, agree to maintain the neutral status of Guangzhou and respect France's religious Protectorate, even within the British sphere of influence.[99] In fact for all his earlier bellicosity, Pichon's proposed French sphere was quite modest and he concluded by recognizing this:

We would thus have a considerable region for our activities such as investment of our capital and commercial and industrial penetration by Indo-China, and we would be free there to make any efforts in proportion to our resources and to their potential results. It seems that this would be enough for a nation which seeks to use its colonies and which must be little tempted to add further to its vast possessions.[100]

When asked to comment on Bompard's and Cambon's suggestion that France attempt to enlist the assistance of Germany in persuading the British to revert to a policy of free and equal competition between all powers in China, Pichon replied that it was too late. Although he personally would prefer to see the maintenance of the open door, and agreed that the Germans, Japanese, and Americans no doubt held similar sentiments, it was impossible to prevent Britain claiming her sphere in the Yangzi valley. All powers apart from France and the USA, including Germany in Shandong and Japan in Fujian, had recognized this. The Americans, he felt, would have emulated the other powers in seizing a naval base. and claiming a sphere of influence had they not been preoccupied with Cuba and the Philippines. Pichon considered it to be impossible to deflect Britain from her 'dream of making another Egypt out of this immense country drained by the Yangzi and its tributaries' because Britain's policy reversal had been prompted by the behaviour of only one power, Russia. If it were not for the Russian seizure of Port Arthur and Dalian, the 'progressive annexation' of Manchuria, and other clear evidence of Russia's intentions to claim northern China, Britain would not have claimed the Yangzi region in contradiction of her past policy. In view of Russia's determination to claim the north, nothing other powers could do would change British plans. In short, Pichon considered that, attractive and desirable as Cambon's proposal might be, 'it now no longer seems practicable to me'.[101]

Bompard and Cambon were unimpressed, both with Pichon's proposed French sphere of influence and with his argument that it was unrealistic to expect to be able to maintain an open door policy. Cambon considered that Pichon's proposed sphere would give France nothing more than she had already under the terms of the 1896 Anglo-French convention apart from preponderance in Guangxi, and that was hardly worth having. Thus, France would gain no real advantage from claiming a sphere of influence. Cambon also disputed Pichon's assertion that the British were irrevocably committed to claiming their sphere on the Yangzi. Although he acknowledged that imperialist feeling in the country was running high, there remained doubts as to the wisdom of pressing this claim. France, Cambon believed, could exploit these doubts, especially as British and French interests in China were not, in his opinion, so much competitive as complementary:

If in fact the British government and public opinion are now strongly seduced by the allure of ideas of imperialism and conquest, nevertheless they consider the heavy responsibilities which would result for England from grabbing the most populous half of China with a certain anxiety. If the English were assured that their trade and enterprises in the Middle Kingdom would not be troubled, they would willingly renounce its dismemberment. We could profit from these dispositions and strive to develop them.

Our interests in China, which are considerable even in the Yangzi basin, are of two sorts: moral, those of our Catholic missions; and material, those of our silk exporters and our financiers involved in railways and mines. Neither our missionaries, nor our silk buyers from Lyon, nor our financiers, in any way threaten England. Their true rivals are the Germans or the Americans. Why then should our action necessarily be in opposition to the British Cabinet?[102]

Cambon's suggestion that, as Britain's major economic interest in China was exporting manufactured goods whereas France's was in importing raw silk and exporting capital, Britain and France were not true rivals, was revolutionary. It was almost a truism, shared even by a perceptive economic analyst like Bompard, that Britain was France's great rival in China. Cambon even argued that, given the lack of enthusiasm that British investors had shown for Chinese loans, London financiers would probably have to win the support of the Paris market in their search for capital to implement British concessions in China. 'Almost out of necessity', there would arise 'a collaboration of the nationals of the two countries, at least in business matters'.[103]

The theme of potential Anglo-French economic collaboration in China was one to which Cambon was later to return. In the meantime, within a fortnight his views about British reluctance to conquer the Yangzi valley received corroboration. An article in the influential *Edinburgh Review* expressed grave doubts that Britain, whose interests were widely distributed through China, would gain anything from a dismemberment of the empire.[104]

Soon after Bompard made almost the same observation about French interests. Asked to comment on Pichon's views, the head of the *Direction commerciale* observed that in Russia's case the political and strategic importance of Manchuria and relative insignificance of her economic interests elsewhere in China, drove her almost inevitably to a policy of spheres of influence. However, French interests were very different:

France's position is quite the opposite. Her economic and political interests are not localized in China's southern provinces, but, on the contrary, are diffused throughout the different regions of the Middle Kingdom.[105]

We went on to list a number of French enterprises in central and northern China, emphasizing the crucial significance of the Beijing–Hankou railway. If France were to recognize the Yangzi valley as a British sphere of influence she could neither give official protection to her investments already under way nor solicit any new concessions such as the Hankou to Guangzhou railway outside of her own sphere. In return all France would receive was the right to preponderance, which he considered would soon lead to conquest, in the provinces adjacent to Tonkin. Bompard considered that to be a very poor bargain indeed.[106]

Although nominally responsible for evolving policy on such matters, the contribution of the *Direction politique* at the Quai d'Orsay to the debate on the relative merits of spheres of influence and the open door was not very distinguished. In a rather brief note it sought to reconcile the irreconcilable views of Cambon and Pichon. As a result the hybrid policy it advocated was neither particularly convincing nor did it have the slightest chance of being accepted by any other power, except possibly Russia, whose interests were not involved. The policy suggested that Britain and France agree to establish their spheres of influence in southern China as proposed by Pichon. Sichuan would be discussed later, but in the rest of the Yangzi valley France would claim her rights to freedom of trade and equality of treatment. In doing so she would co-operate with other powers against the 'exclusivist tendencies of England'. Nothing came of this rather hypocritical proposal, which would have established a French sphere in southern China but maintained the open door in the potential British sphere.[107]

By then the American Secretary of State, John Hay, had declared that the United States favoured the maintenance of the open door. He was circularizing a note calling on the powers to continue to recognize the equality of all nations in China as provided for in the most favoured nation clauses in various treaties even within their spheres of influence. Although Hay's note was prompted mainly by the concern of some influential South Carolina cotton mill owners that their products would be excluded from a China divided into spheres dominated by European powers, ultimately it was to have great influence.[108] The reaction of the powers was at first dilatory. Apart from Russia, all recognized that their interests were best served by preserving an open door. However, Russia's adherence to her sphere of influence was tending to generalize the policy only she really favoured.

Despite the American note, Pichon's position against the open door hardened during October. He even questioned if, ultimately, the open door was in French interests, as other powers seemed so much better able to compete in China. This view contradicts his opinions expressed early in September, when he appeared to support elements at least of the open door. These rapid changes in Pichon's advice suggest that his strength of expression was not equalled by a comparable strength of conviction. His renewed support for a French sphere of influence led him to overstate the weakness of the French position elsewhere in China:

It is not enough for the door to be open; it is even more essential that we are able to go through it with our resources, our propaganda, and our products. We have only a few French trading houses in China, no merchant marine, an inadequate naval presence, no serious consular network, no businessmen of capital and initiative ...[109]

After questioning in this way whether the open door really was in France's interests, he pointed out that the spheres already existed, that they were tending to increase and that England was determined to take hers on the Yangzi. As a result, he concluded:

If we keep our present intermediate position, we will have neither the advantages of the open door nor those of the sphere of influence and the exploitation of China will take place without us.[110]

Pichon's views, however, found little support in the Quai d'Orsay. Ultimately, it was probably the paucity of the sphere Pichon was able to offer rather than Hay's note that decided the issue. In a marginal note on this dispatch Delcassé observed that the only portion of the proposed French sphere where British interests were not legitimately involved was Guangxi, and it was 'detestable'.[111]

At the same time as Hay was circularizing the open door note, negotiations were under way that were to result in the establishment of the Yunnan Syndicate. These negotiations inspired Cambon to suggest, using an extraordinary metaphor, that it might be possible to hold the door open and prevent the creation of a British sphere by creating a 'mattress' to be shared by both powers:

The establishment of an English Protectorate, more or less disguised, ... would be a real peril for French interests already existing on the Yangzi and if it is recognized that we cannot seek compensation by territorial expansion to the north of Tonkin, we must try to safeguard our enterprises and prevent an English seizure of the heart of China. The best means is to create a sort of protective mattress by mixing as much as possible French and English interests, encouraging agreements between the two countries' financiers, pursuing the creation of international syndicates to build and operate factories, railways, etc., in a word, forcing the British government to injure British interests if it wants to touch French interests.[112]

Cambon asserted that such a policy of 'industrial and commercial *entente*' was well regarded in the City and by many members of the government. The involvement of Herbert and Bourke in the Anglo-French Syndicate was clear evidence of this.[113] The fundamental and reassuring belief behind Cambon's mattress doctrine was that international finance could be used to promote better relations between the powers. Certainly Cambon put a great deal of effort into his mattress diplomacy in London in late 1899 and early 1900.

Whatever the impact of Cambon's mattress doctrine on Delcassé's thoughts may have been, the Minister certainly rallied to the policy of the open door. Hay's note was no doubt influential, but Delcassé chose to announce the decision, not in a reply to Washington but in a speech to the Chamber of Deputies on 24 November 1899, albeit in the presence of General Porter, the American Ambassador. He used the opportunity not merely to announce French adherence to the open door but also to rebuff those who, like Doumer, longed for a conquest of Yunnan. He began by answering those expansionist critics of his failure to take a naval base on the Bohai Sea. He claimed that France's possession of Indo-China and a good series of coaling stations between Marseilles and the South China Sea meant that, unlike Germany, which lacked these advantages, she had no need for such a base. He then went on to examine France's prospective sphere of influence, pointing out that as Britain had legitimate rights in Yunnan, Sichuan, and Guangdong, France could only claim a monopoly in Guangxi, which had few resources but plenty of bandits. He claimed, 'such is, in total, the region which, if the policy of zones must prevail, could constitute the French zone', and asserted, in effect, that it was not worth conquering, whatever the sentiments expressed in the nationalist press. Delcassé argued that it was essential for the security of Indo-China that no other power be permitted to establish itself on the Tonkin frontier. As China had promised not to alienate any of this area, it only remained to ensure that China honoured these promises.

From political and strategic considerations Delcassé turned to French economic interests in China and argued that their diffusion throughout the empire impelled France towards support for the open door policy. He asserted:

These interests, gentlemen, are not small... The Chamber will learn with pleasure that in this task of the economic modernization of China which the great powers seem to have given themselves, France is far, very far, from being backward. Of the 10,000 kilometres of railways conceded, France has obtained 2,000, of which more than a half is already under construction. Moreover we have obtained for French individuals or companies a large number of industrial concessions ... [which] are in other regions than that which, if the policy of zones carries the day, could constitute the French zone. Does it not follow from this that we must be wary of the serious encroachments [on China's territorial integrity] and rally to maintain China open to the free competition of minds and capital of the entire world.[114]

The following month, when asked by General Porter for a reply to Hay's note, Delcassé referred the American Ambassador to this speech and indicated his support for the principle of the open door, 'under the quite natural reserve that all the other interested powers affirm their willingness to do the same'.[115] Eventually even Russia agreed to allow equal access to the trade of all nations to the port of Dalian, thereby allowing Hay to claim that all powers had responded favourably to his note.[116]

Delcassé's speech had played a significant role both in ensuring the success of Hay's note and in reducing Anglo-French tension. It came at a time when the more disreputable elements of the press on both sides of the Channel were attacking, from one side the Royal Family, from the other everything French. Delcassé's speech on China was excellently received in London and the reaction of the press was unanimously favourable. The conservative *Morning Post* even went so far as to reproach Salisbury for not being as forthright in his support of the open door as had Delcassé. At the same time, British reverses in the Transvaal in mid-December and the subsequent loss of prestige suffered by the ardent imperialists, weakened British enthusiasm at the prospect of ruling half of China.[117] The adoption of the open door policy by Britain and France and the movement towards the creation of Cambon's protective mattress seemed to suggest that the early years of the new century would see better Anglo-French relations, both in China and elsewhere, than had the last years of the old.

9

Railways and Mining:
Steps towards Anglo-French
Co-operation in Yunnan, 1900–1901

FRENCH adhesion to the policy of the open door in China at the end of 1899 meant that Doumer's dream of absorbing Yunnan into Indo-China was now directly contrary to the principles which Delcassé had expounded in the Chamber of Deputies. The appointment of François to Kunming; his establishment of good relations with Doumer; and the meeting of Doumer, Pichon, and François in Hanoi in March 1900 suggested that the dispute between the authorities in Hanoi and the Quai d'Orsay had been resolved in favour of the latter. To a certain extent that was true: Doumer could no longer expect to receive any support for expansion into Yunnan from the Pavillon de Flore or in cabinet. Nevertheless, the fundamental causes of the dispute remained. A powerful element around the Governor-General continued to refused to accept the decision from Paris not to invade Yunnan as anything more than a postponement of their plans. Léonce Guibert, the engineer in charge of the consortium's survey mission, had been jovially informed in Hanoi that the military hoped his party came to some grief so that there would be an excuse to come and rescue him. More seriously, General Borgnis-Desbordes told Guibert what use he intended to make of the railway: 'Just now it is no longer a matter of immediate action, but when things are a little better prepared we will climb [into Yunnan] along your railway'. Such talk naturally horrified François.[1]

He would have been even more horrified had he seen a 'Notice sur le Yunnan' prepared by the *Etat-major des Troupes de l'Indo-Chine* at the end of 1899. This virtual blueprint for the conquest of Yunnan was prepared on the basis of information gathered by the military members of the various missions which had entered the province over the previous two years. It was a large document of ten chapters in which the geography, demography, administration, and economy of the province were examined in some detail. The document concluded with a discussion of the province's penetration and future conquest by France. After discussing the apparent rise in anti-French feeling during 1899, its anonymous author asked,

Must we therefore make the same efforts as were needed for the conquest of Tonkin for our commerce to penetrate Yunnan? Must we stop hoping that we can annex Yunnan peacefully, following a course of firm conduct and learning from the lessons of the past and above all from our knowledge of what the Chinese empire really is?

The question is worth studying. From our study of the history of Yunnan and of the peoples who live there, we have drawn one conclusion: *Yunnan is not a Chinese country but a Chinese colony*.[2]

The author went on to assert that the Yunnanese and their tribal leaders were discontented with the domination of Chinese officials. Moreover, the commerce of the province was dominated by Cantonese. The task, therefore, was to convince the Yunnanese that 'it is in their interest to exchange Chinese domination for ours'. The author did not consider this to be a particularly difficult task in view of the resentment the Yunnanese and Muslims felt towards the Chinese. The only serious opponents of a French take-over would be the mandarins. By acting with sympathy and compassion towards the people, and treating the mandarins with firmness, this military officer considered that Yunnan could be secured. He concluded:

If, despite our frankness and our firmness, we have to use force to have our rights respected [*sic*], it must be done without hesitation, and, once begun, completed. We came to Tonkin to open the Hong Ha route, we must not forget that, and when all the other European powers are creating their spheres of influence in China, the French should not stay behind, at the risk of losing even what they already have.[3]

Other than to provide further confirmation that expansionist ambitions were still current in Hanoi, this document probably had no effect on policy formation in Paris. It does, however, reveal that some army officers in Indo-China were doing more than talking about a conquest: they were carefully (if not always accurately) analysing conditions in the province and how best the conquest could be effected.

As it was largely the task of agents of the Foreign Ministry to prevent the execution of any such plans, it is ironic that it was soon to be François himself who solicited a French military intervention in the province. For, at the same time as the Boxers were converging on Beijing, anti-European agitation became widespread in Yunnan. In December 1899 Beijing had appointed a new Viceroy to Kunming. Ding Chenduo was a protégé of Yuan Shikai and a vigorous defender of Chinese sovereignty. Assisting him was Li Jingxi, the Western-educated nephew of Li Hongzhang, a decade later the last imperial Viceroy of Yunnan and Guizhou. Both Ding and Li were willing to exploit popular movements in the defence of Chinese interests against French encroachment, but unlike their superiors in Beijing in 1900, their attitudes were more those of modern naitonalists than of obscurantist reactionaries.[4] Inspired by Ding and Li, both officials and the populace were increasingly hostile to the French in Yunnan during the spring of 1900. While on his return journey from the meeting in Hanoi, François was told by Sainson, who had preceded him to Mengzi, that the miners were agitating once again. The *daotai* in the town had recalled the arms he had issued to its European population in the wake of the previous year's riot, and Sainson felt that his consulate was in need of protection.[5] François was well aware of Sainson's dislike of military intervention and so supported his request suggesting that 20 troops would be sufficient. They should travel in *mufti* so as not to excite resistance. In addition, François himself was importing 125 guns for distribution to Europeans.[6] François's request and action indicated that the situation was serious. An official of the Department of Foreign Affairs was asking Indo-China to send in the military presence which he and his colleagues had always sought to prevent. Delcassé and Decrais agreed to instruct Doumer to provide the troops requested.[7]

The troops were not sent as the situation deteriorated so rapidly that it became obvious that the 20 would merely be marching to their deaths. The immediate cause for this deterioration was François's gun shipment. After he arrived in Kunming on 10 May, his baggage, including the guns, was held at the *lijin* barrier at the South Gate. François and Beauvais went to the barrier, pistols in hand, to demand delivery of the goods. There followed a week of acrimonious correspondence between François and Ding over the issue. François by then considered Ding to be violently xenophobic. The Chinese accused the French of importing arms to start a war in the province and this allegation was backed up with an organized demonstration and placards accusing François of importing 6,000 rifles. During this week missionaries, who had been untouched during the war of 1883–5, were assaulted by mobs who were left unpunished. François gathered all the Europeans in the building he was using as his base and they prepared to defend themselves. The building was besieged by demonstrators for several days, but there was no serious violence. By 19 May the city was calm.[8] The incident was followed by recriminations at a senior level both in Paris and Beijing, where they added to Pichon's increasingly difficult task.[9]

François believed that the most appropriate response to the officially inspired riot in Kunming would be to send a larger uniformed detachment to Mengzi than was planned. Pichon and Delcassé disagreed, Pichon on the ground that a military presence would not aid François's negotiations and could be dangerous.[10] Delcassé was unwilling to send in troops because of the delicacy of France's position in central Africa and because the deteriorating situation in northern China commended a policy of restraint.[11] Doumer was ordered to cancel the dispatch of troops into Yunnan.[12]

The respite from anti-foreign agitation in Kunming was brief. A fortnight after the riots were quelled another violent outbreak occurred. This time it was not confined to the capital. From Kunming François wired, 'Situation violent, of terminal gravity', telling Delcassé that the Viceroy had himself suggested that the Europeans return to Tonkin and that he was proposing to take this advice. Delcassé, realizing that the international crisis in northern China was being matched by one in the south, had copies of this alarming telegram sent to all the major capitals.[13] In Mengzi Sainson described the situation as 'very serious at the moment. The gentry on one hand, the miners at Gejiu on the other, speak openly of killing the Europeans and burning their houses'. Moreover fresh Chinese troops were being sent from Kunming to the frontier.[14] Under the pressing circumstances François decided to evacuate all Europeans from the province. He gathered all those in the capital, mostly French and some English, on 9 June and departed the following morning. They marched into a trap, were attacked by a large group of Chinese who seized their luggage and pillaged it, and retreated to François's house which once again was besieged. The French and English missions and schools as well as all the railway and other Indo-Chinese installations were burned. François lost all his papers, archives, and code books, indeed everything except for the clothes in which he stood. His telegram to Paris informing Delcassé of the débâcle was in Spanish, a language which he hoped none of the Chinese operators would understand. In it he requested energetic action so that they could leave safely.[15] He specifically requested, however, that Indo-China refrain from sending in troops until the evacuation was complete.[16]

Delcassé's response was energetic. On 18 June he summoned the Chinese Minister in Paris, Yu Geng, to the Quai d'Orsay and told him the Ding's life depended on that of François. Yu Geng promised to wire the threat to Kunming immediately.[17] Delcassé further told François to inform Ding that the only way to prevent an invasion by French troops already massed on the frontier was to ensure the safe conduct of the Europeans in Yunnan to the security of Tonkin.[18] The threats were unnecessary. Ding Chenduo had lost control of the riot on 10 June so completely that even his own *yamen* had been attacked. He had over-played the popular hand and knew it: popular anti-foreignism had drifted into rebellion. That he could not tolerate. He immediately moved to restore order and protect the Europeans, executing four rioters.[19] He then arranged for the safe departure of the Europeans on 24 June and gave them a strong guard of reliable troops. Finally he wired Delcassé blaming 'ignorant evil-doers' for the attack on 10 June and asserted that his strong measures had removed any need to send in French troops.[20]

François's party reached Mengzi on 2 July and Laocai on 5 July. There were no serious incidents *en route* and they were accompanied by General Su, an ardent francophile whose expenses had been subsidized for years by the Indo-Chinese government.[21] François felt, however, that Su did not trust his men and that the party was in constant danger. Su, in any case, had orders to oppose any French entry into the province. The Chinese preparations for war impressed François. Clearly they considered that an invasion from Tonkin was imminent. Almost 2,000 soldiers, half armed with modern rapid-fire rifles, were stationed at Mengzi, where there was also a battery of Krupp artillery manned by graduates from the Tianjin artillery school. Additionally, the passes between Mengzi and the frontier near Laocai were heavily guarded by

6,000 regulars. In contrast, French preparations were negligible. The French in Mengzi were riven with dissention. Captain Duprat was intriguing and attempted to persuade the evacuees from Kunming to make a stand at Mengzi. He converted Masse to this reckless position and the situation was only resolved by François issuing an order of expulsion against Duprat.[22]

The French were no better prepared on their own side of the frontier. Delcassé's threat to launch an immediate invasion of Yunnan had been empty and Doumer's desire to fight the Chinese on the plateau near Mengzi rather than in the valleys on the frontier was foolish. He may in a sense have been right when he impatiently cabled Decrais on 17 June 'in the present situation the dispatch of troops of Yunnan is indispensable'[23] but he was, in fact, in no position to send those troops. The forces available to Doumer were very thin in June 1900. He had been ordered to send his best French and Zouave (North African) troops to Dagu to participate in the international column preparing to relieve Beijing. Delcassé and Decrais had further ordered him to maintain a strong force on the Yunnan frontier ready to march if ordered to do so.[24] These were the type of instructions which the Governor-General loved to receive and he talked loudly about the invasion his forces were about to undertake. He supported the intrigues of Duprat and Masse in Mengzi and cabled François, urging him to make a stand there while a relief column was sent up from Laocai.[25] Unlike Ding, Doumer wanted to have a siege of the Legation Quarter of his very own in Yunnan. Doumer was absurdly optimistic and cabled Sainson that he was sure that the military mandarins and bandit chiefs would welcome a French invasion.[26] In fact, they were making energetic preparations to repel one and even General Su had said that he would reluctantly obey orders to do so. Moreover when François reached Yenbay, midway between Laocai and Hanoi, he observed that the hostile attitude of the Vietnamese was such that it was a question more of assuring the security of Tonkin than of invading Yunnan.[27]

François was horrified to discover not only that was Doumer ignorant of the deterioration of security in Tonkin itself, but also that the preparations of the invasion on which his life and those of most foreigners in Yunnan could have depended were derisory. At the frontier post of Long Po he found *one* French sergeant and 55 Vietnamese riflemen. At Laocai there were only 150 effective

legionnaires and they were inadequately armed and supplied —
over half of the 12,000 cartridges obtained from Hanoi were
empty. François commented bitterly:

> M. Doumer asserts that he has a expedition all ready to go. He had
> prepared not one man, nor one horse, nor rations nor amunition. He
> knows that he has no one to march in and that the effective military
> strength of Indo-China has been weakened by the despatch of its strongest
> elements to Tianjin.[28]

The military, who were normally anxious to undertake new
conquests, were as shocked by Doumer's aggressive plans as
François. On the whole, they ardently desired the conquest of
Yunnan, but General Borgnis-Desbordes declared that he would
not attempt the enterprise with anything less than 17 battalions
direct from France. Colonel de Beylié spoke of the folly of
invading with three companies as Doumer proposed and Colonel
Louvel claimed that the almost certain failure of a march into
Yunnan would lead to a widespread revolt of the Vietnamese.[29]
One officer estimated that there was only sufficient ammunition in
Tonkin for a campaign of *three hours* duration.[30] The Chinese
certainly considered the French position more at risk than their
own: in Mengzi the militantly modernizing *daotai* Liu Chunlin,
had set a highly relevant topic for official aspirants to discuss in the
annual examinations: 'the best means of chasing the French from
Vietnam and throwing them into the sea'.[31]

On the basis of François's reports of the relative strength of the
French and Chinese forces facing each other across the frontier,
Delcassé decided to evacuate all French personnel remaining in
the province and Doumer was ordered to do so 'without one
exception'.[32] Guillemoto, however, by now totally committed to
an aggressive policy, ordered his men to remain in Mengzi. Some
refused to stay, so Guillemoto threatened to send them to safer
positions in the most malarial regions of Indo-China and replace
them by 'true Frenchmen'. François was convinced that these true
Frenchmen would soon be dead Frenchmen: '[Guillemoto] knows
that his mission is devoted to massacre, but M. Doumer is looking
for a pretext for an expedition and he executes his will with
servility'.[33] The Indo-Chinese military officers and engineers who
had remained with Sainson in Mengzi were being given impractical
orders by Guillemoto which contradicted orders from Paris. When
Sainson received orders from Delcassé to complete the evacuation

of Yunnan by bringing all foreigners down from Mengzi,[34] he was unable at first to carry them out owing to the defiance of some officers and Wiart. After two attempts were aborted by French officers on 11 and 12 July, he eventually left with a party of 44 French, three English, and about a hundred Vietnamese refugees.[35] When they reached Laocai on 17 July, after a very difficult trip, the evacuation was complete.[36]

It was a serious blow to French aspirations in southern China and an event which François and Delcassé had delayed as long as possible and Doumer had attempted to prevent altogether. François had hoped to be able to maintain a presence at Mengzi in order to conciliate Doumer, but when he saw the military situation near the frontier he realized that it was not possible. Anxious to maintain good relations with the Governor-General, François blamed Guillemoto for encouraging plans for an invasion even though no preparatoins had been made, described the engineer's behaviour as 'a real danger', and asked Delcassé to ensure that he no longer interfered with developments in Yunnan.[37] As for Doumer, who was bitterly disappointed by the turn of events, François attempted to persuade him that all was not lost in Yunnan and that it was best to wait until the Boxer Rebellion was suppressed and serious negotiations about France's rights in the province could be opened with the government in Beijing.[38] Doumer listened and waited to avenge the Quai d'Orsay for its weakness.

Delcassé was unhappy. He had been willing to commit troops to protect European security in Yunnan, believing, probably rightly, that there would be none of the international complications in the summer of 1900 which he had so feared the previous year. After all, these troops would not have been intended to annex territory, any more than those around Beijing or in Shanghai. Only French weakness and a desire to see military operations confined to northern China prevented the invasion. He explained the dilemma to François:

The only measure worthy of France would have been to send to Mengzi a military force capable of breaking probable Chinese resistance on the way and of ensuring effective protection to our consular representation.

Now you signalled yourself the difficulties of a military action against the Chinese forces, whose strength along the frontier you ascertained. On the other hand ... the entry of our troops into Yunnan could provoke an uprising not only in this province but also in neighbouring provinces ...

That is a risk the government could not take just at the time when the efforts of all the powers and of France in particular were aimed at encouraging the peaceful attitudes shown by the central and southern Viceroys and to restrict the warlike movement to the north.[39]

The French diplomatic response took the form of a formal note of protest which Delcassé sent to Yu Geng, the Chinese Minister in Paris, and in which he asserted the principle of France's claim to compensation. This claim, however, would not be pressed until the re-establishment of a regular and legitimate government in Beijing.[40] Delcassé insisted on recognition of China's obligation to pay reparations and refused to allow any French officials to return to the province until such recognition was given.[41] François returned to Paris while Sainson waited in Tonkin.

Guillemoto prepared an estimate of the compensation due for the damage done. The costs of delays and damage to roads caused by the evacuation and their subsequent neglect were quite low, a total of 247,000 piastres or about 630,000 francs. To this had to be added the 200,000 piastres (500,000 francs) advanced to General Su. The largest cost, however, in Guillemoto's opinion, would be the higher interest rate which would be expected from a Chinese investment following the Boxer Rebellion. He considered that the rate would rise from three and a half to five per cent, which would ultimately increase the cost of an 80-million franc loan by 35 million francs. Guillemoto suggested that compensation for this cost be in the form of concessions of mining leases and of a Kunming to Chengdu railway.[42] François also prepared a report on the compensation to be claimed. He agreed with Guillemoto that much of the indemnity could well be paid in the form of concessions, although he also wanted land for various French institutions such as post offices and sanitoria for jaded residents of Indo-China seeking a cooler climate. He also favoured extension of the railway, not to Chengdu, but to either Yibin, or Guiyang, or both. He also considered that France should have a military presence in Yunnan. He suggested that French troops be permitted to protect the railway in the way that Russian troops could in Manchuria and believed that a large guard of about three hundred troops, either French or Algerian, should be stationed at Mengzi to ensure the security of the consulate and those working on the railway.[43]

François's suggestion for a military presence did not find much support from outside the ranks of the military themselves. Pichon considered that a force of the size proposed would be large enough to provoke resentment but too small to be militarily effective.[44] Sainson felt that his return to Mengzi should not only be peaceful but be seen to be so and thought that a consular guard of 30 Vietnamese, as had been proposed before the evacuation, would be adequate.[45] Li Hongzhang negotiated an agreement with Pichon for the return of the French to Yunnan which would protect the position and status both of his French friends and of the Yunnan government with which he was so closely connected. A prefect first class would apologize to the returning French officials and the Yunnan authorities promised to protect them.[46] Delcassé therefore decided to provide no guard, unless the guarantee of security proved to be worthless. The question of reparations was to be discussed in Beijing rather than in Kunming and the claim for damages in Yunnan submitted with the total for the Boxer Rebellion. This would remove the source of potential bitterness between the provincial authorities, who were never particularly affluent in Yunnan, and the returning French Consuls.[47] On 29 April 1901 François and Sainson returned to Mengzi and received the anticipated official reception and apologies. It was obvious that the Chinese were pleased that no troops came with them, but in view of the continued support for an invasion amongst some officers in Tonkin, François suggested that no further contact should be permitted between Chinese officials and the French military.[48]

In the wake of the evacuation there was some revaluation not only of the French political position in Yunnan but also of the financial viability and economic potential of the Haiphong to Kunming railway itself. François had researched the demography and economy of Yunnan during his stay in Kunming and on his return to Paris reported to Delcassé that the potential of the province was limited. Even before the tragedies of rebellion, plague, and depopulation which had ravaged the province in the mid-nineteenth century, it was neither populous nor rich. In 1899 its foreign imports via Mengzi totalled only 8,000 tonnes and François agreed with Dejean de la Bâtie that, even when the province regained its previous modest prosperity, this figure would only rise to 40,000 tonnes. As François observed, this was an average of only a little over a hundred tonnes per day, a dozen wagons, indeed 'a very modest load for a train'. Even though the daily train up to Kunming would not be particularly heavily laden,

François still believed that the construction of the railway was necessary, not for economic reasons, but 'for a strategic end and to establish our influence in Yunnan and to assure ourselves of a means of penetration towards the Blue River [Yangzi]'.[49]

In a subsequent and expanded report François repeated that the railway would probably run at a loss and concluded that:

from a commercial point of view, Yunnan by itself has no serious importance, and as a potential trade route it can only be considered as a passage leading to the Yangzi ... to link Tonkin to Sichuan and to bring this province's traffic to Haiphong.

He was, however, sceptical about this plan. He argued that the natural route into Sichuan was along the Yangzi, especially since one steamer had already made the difficult voyage through the gorges between Yichang and Chongqing. Improvements to navigation of the river at this point or construction of a parallel railway would be far more attractive than shipment to Haiphong along 1,800 kilometres of narrow gauge railway constructed at great expense largely through mountainous terrain. Moreover, François considered the assertion that the railway would enable temperate Yunnan to become the market garden of Tonkin was utterly ridiculous:

it would truly be a singular financial scheme that would be put together to build 500 kilometres of railway through such wild country, in order to provide fresh food for the tables of 3,000 or so Europeans in Hanoi and Haiphong. .

The only export trade of any consequence would be from the tin and copper mines in the province. There was also the possibility of developing the exploitation of coal deposits that had been discovered in both Yunnan and Guizhou. The possibility of these deposits sustaining French industry when those in Europe became exhausted was the only substantial economic benefit he saw in the railway. From a strategic point of view, however, he considered its construction to Mengzi highly desirable as it would be far better either to defend Tonkin or attack China from a point on the plateau than in the jungles of the valleys near the frontier.[50]

At the same time as François was expressing his doubts as to the economic value of the railway, the consortium received the report of Guibert, the engineer it had sent to Yunnan to verify Guillemoto's surveys and estimates of costs. His news was no better than that of François. Instead of requiring up to 70 million francs, Guibert estimated the railway's cost at 100 million.[51] According to François, the difference was explained by the haste with which Guillemoto prepared his report, without even surveying the entire route and using impressionistic estimates rather than working from detailed surveys. As well as increasing the estimate of the railway's cost, Guibert agreed with François that there was little prospect that it would be a profitable venture, although he was more optimistic about the potential of an extension into Sichuan. Such ardour for the project that the bankers in the consortium possessed, cooled in response. No doubt already extremely wary of any Chinese investments following the Boxer Rebellion, their unenthusiastic reaction was scarcely surprising. That of the Crédit lyonnais was the most extreme. Unless the government took over the railway's construction and the banks' role was confined to lending it the funds required, the Crédit lyonnais would withdraw from the scheme.[52] This proposal was unacceptable to the Quai d'Orsay, which continued to insist that the direct action of the Indo-Chinese government in Yunnan be replaced by that of a private company: 'it is important to retain the purely industrial character of the Yunnan railway'. The law of 25 December 1898 had provided for the railway's construction by private enterprise and, in the opinion of the officials at the Quai d'Orsay, the difficulties which arose over the increase in estimated costs were for the consortium and the Indo-Chinese government to resolve.[53]

In March 1901 Doumer returned to Paris and began negotiations with the bankers. He had to make considerable concessions to ensure their co-operation. He offered them the concession of the railway from Haiphong to Hanoi, which, it had been planned, would be owned and operated by the Indo-Chinese government. This railway would obviously be the most profitable in Tonkin, linking its capital and largest city to its major port. Additionally, he offered to raise the subsidy provided by Indo-China from eight to ten and finally to 12.5 million francs. Despite these concessions on Doumer's part, the Crédit lyonnais and, more surprisingly in view of its connections through the Gouin family with the Société des Batignolles which was to construct the railway, the Banque de Paris et Pays-Bas refused to remain in the consortium on these terms. Both, however, promised to purchase shares totalling one and a half million francs in the company when it was constituted.[54]

Thus, on 15 June 1901, Doumer, on behalf of the French and Indo-Chinese governments and representatives of the four banks remaining in the consortium signed a convention in Paris which handed the concession of the Haiphong–Kunming railway to the consortium. The banks concerned were the Banque de l'Indo-Chine (represented by Homberg and Simon), the Comptoir national d'escompte (Mercet and Rostand), the Société générale (Hély d'Oissel), and the Crédit industriel et commercial (represented by Desvaux).[55] Under the terms of this convention a company, ultimately known as the Compagnie française des Chemins de fer de l'Indo-Chine et du Yunnan, was established with a captial of 12.5 million francs. This company would operate the line from Haiphong to Laocai, which would be constructed by the Indo-Chinese government. The company would also both construct and operate the line between Laocai and Kunming. In return it would receive a subsidy of 12.5 million francs from the Indo-Chinese government and a guarantee of interest of three million francs per annum for 75 years. The guarantee would permit the company to raise 76 million francs in share capital, thus giving it a total of 101 million francs with which to build the Yunnan section of the railway. The contractors for this section were to be Gouin et Vitali.[56] A separate 'Cahier des charges' specified that the railway was to be built to metre gauge (Article 4) and opened to Mengzi within two years and Kunming within five years of the rails reaching Laocai (Article 3). The Indo-Chinese Department of Public Works was to have powers of control and surveillance during construction to ensure that the works conformed to the specifications in the 'Cahier' (Article 14). The concession was for a period of 66 years (Article 25), after which time the railway would become the property of the Indo-Chinese government (Article 26). The government had the right to purchase the railway after 15 years (Article 27) and could set maximum fares and freight rates (Article 32). All of the company's European employees were to be French (Article 51).[57]

The terms of the law of 25 December 1898 required parliamentary sanction for this change in the financial arrangements for the railway. The necessary bill was introduced into the Chamber of Deputies on 24 June 1901. The report in support of the bill was highly optimistic. It contained some familiar assertions of doubtful accuracy, notably in claiming that the railway would capture the trade with Sichuan, inevitably 'one of the most populous and beautiful [provinces] of the Chinese Empire'. Seeking to dispel any fears that the Boxer Rebellion may have created amongst deputies, the Chamber was assured that 'the population of Yunnan, which is moreover not of Chinese origin, has always been welcoming and good-natured in its relations with the French'. The report concluding with a discussion of the benefits that the railway would bring, not just to Indo-China but to metropolitan France herself:

If these works have opened a new era in Indo-China, spreading significant sums of money among the natives in the form of wages [!], leading to the growth of workshops which will make Tonkin well armed for its commercial and industrial attack on the immense neighbouring empire, then the metropole, whose nominal guarantee will in reality never be called upon, will not find them any less profitable. A new outlet for its capital, *more than 100 million francs worth of orders* (iron and steel, rolling stock, machinery, lime, cement, and explosives), which the law obliges the company to place in France, such will be the first results for the metropole, which will later benefit from the new markets which the railway will conquer.[58]

Doumer, acting as *Commissaire du gouvernement* as he had in 1898, returned to the theme of social imperialism when the bill was debated in the Chamber three days later:

There is no antipathy between clearly democratic domestic policies and a policy of national pride and energetic defence of French interests abroad ... I felt how essential it was for our country to extend its outlets to give work to all these workers. I recalled that Indo-China has given 50 or 60 million francs worth of order to the metallurgical industry, allowing it to avoid a crisis and unemployment. (Applause)[59]

The strongest opposition to the bill, as in 1898, came from Gaston Doumergue. As well as complaining about the increase in costs, he objected violently both to the company being granted the lucrative concession of the Haiphong to Laocai section and to the 'gift' of 12.5 million francs which Indo-China had to make in order to persuade the banks to participate. In Doumergue's opinion, it was 'an excellent affair' for the company at the expense of the Indo-China government.[60] On this occasion, however, Doumergue's objections did not produce the close vote which they had in 1898. The bill was approved by the large majority of 415 in favour and 103 against. After gaining easy senatorial approval, the law was enacted on 5 July 1901.[61]

The success which Doumer was able to achieve both in his negotiations with the bankers and on the floor of the Chamber of Deputies was probably in part due to an efficient press and publicity campaign which he had organized during his visit to France. There were in effect two campaigns, one quite reputable, the other thoroughly scurrilous. The reputable campaign was conducted in the columns of various journals sympathetic to the cause of French imperial ventures. These included the *Revue Géographique*, the *Dépêche Coloniale*, and the *Revue Française de l'Etranger et des Colonies*.[62] More importantly, Doumer used as a forum the Comité de l'Asie française, which had been founded by Eugène Etienne[63] as recently as February 1901, just one month before Doumer returned to France. Doumer addressed a banquet held under the auspices of the Comité de l'Asie française at the Hotel Continental on 2 April 1901. Amongst the 400 guests, described as 'the highest and most notable men of the diplomatic, colonial, industrial, and commercial worlds', were Etienne, who chaired proceedings, Decrais, de Lanessan, then Minister of the Marine but a predecessor of Doumer in Indo-China, Bompard from the Quai d'Orsay, Paul Beau, newly appointed as Pichon's successor in Beijing, Simon, head of the Banque de l'Indo-Chine, Ulysse Pila, the Lyon businessman and propagandist, and Guillemoto. Apart from his brave declaration that France 'is beginning to be a great Asian nation, as she is a great African nation', there was nothing particularly original or startling in Doumer's speech, but it was certainly a fine opportunity to influence an important group in congenial surroundings.[64]

The scurrilous campaign was waged in the nationalist press, which was always anxious to condemn the allegedly timid or cowardly attitudes of those responsible for French foreign policy. On the very day of Doumer's arrival in Marseilles, 10 March 1901, an article by Albert de Pouvourville appeared in a number of right wing French papers. It was subsequently reprinted in the *Avenir du Tonkin* in Hanoi. De Pouvourville claimed that a French battalion had been waiting on the frontier at the time of the evacuation of Yunnan, that a barracks had been constructed at Kunming in preparation to receive it and that the mandarins would welcome this French military presence:

This batallion was on the Tonkin frontier at Laocai, waiting for a favourable opportunity to enter China peacefully and take possession, at the very request of the Chinese authorities, of the province of Yunnan, this metallurgical pearl of the Celestial Empire ... How did so simple and glorious a venture end in a ridiculous and distressing collapse?

Doumer's plans, at the very moment of their success, were wrecked by a denunciation, which would be expected from a German or an Englishman, but whose author was a French official — whom we would like to believe was unconscious of what he was doing [but who] must be considered guilty of a crime of *lèse-patrie*.

The article concluded by naming François as the guilty party and denouncing Delcassé for his support of François.[65] The article was totally fanciful. As François himself repeatedly pointed out, there were less than a hundred men at Laocai and no serious preparations had been made for a march into Yunnan.[66] In his desire to condemn the Quai d'Orsay for its reluctance to carve the Chinese melon, de Pouvourville exposed his ignorance of conditions in the metallurgical pearl which he hoped to see prised from the celestial oyster.

The article made François, who had hitherto sought to conciliate the Governor-General, into a violent enemy of Doumer. He revealed damaging information about Doumer's activities which he had hitherto suppressed. Notably he reported that, following the evacuation of Yunnan, Doumer had offered to pay the fare to France of the pro-vicar of the province, the Reverend Father Maire, who had been in Yunnan since the reign of Louis Philippe. Father Maire held conservative political views and Doumer was anxious for him to tell his superiors and the religious world in general about the Yunnan affair, as he saw it, 'in order to complete, in the religious papers, the campaign being waged in the nationalist newspapers'. Maire's vicar, Escoffier, who was then in Hong Kong, had prohibited him from lending his name to Doumer's propaganda in this way.[67]

In the same personal letter to Delcassé, François revealed that Doumer's anxiety to have the co-operation of the French Consul-General in Kunming had led him to offer François the sum of 1,000 piastres (about 2,500 francs) whenever and as often as François wished. The letter in which Doumer had made this offer had been lost in the attack at Kunming on 10 June 1900. Until François had been himself attacked in Doumer's press campaign he did not mention this virtual attempt at bribery to his superiors. He now reported what, to the best of his memory, Doumer had written in his letter accompanying the 1,000 piastre note:

There is no need to look at a map for very long to see that there is another Fashoda on the Yangzi. Whatever it costs, we must get there before the English. It is obvious that it will be the one of these two nations which advances, keeping the country behind it, that will be master of the situation. France must be that country and it is up to you to prepare the way. Indo-China is rich; its resources in men and money are at your disposal. . . .

I am giving you 1,000 piastres which you need not account for in any way, it is between you and me. It is from a fund outside of the ordinary secret funds and I will renew the sum every time you ask me . . . [68]

The third damaging accusation in François's letter was not so much against Doumer himself, who had not yet returned from France, but against his lieutenants. Clearly they had learned nothing from the events of 1899 and 1900, for, as provided in Article 14 of the 'Cahier des charges' of the convention of 15 June 1901, Indo-China was planning to send yet another mission into Yunnan, this time allegedly to supervise and control the work of the company's engineers. These latter, however, had not yet left France. The mission included no less than sixty men. It head was an Indo-Chinese Public Works Department official named Blin and amongst its members were some of the officers whose conduct had been so compromising prior to the evacuation. The mission was going to construct a telegraph line to Kunming and re-establish the remount service which had earlier caused so much resentment. No attempt had been made to inform the provincial government, let alone obtain permission from Beijing. The acting Governor-General, Broni, and the men around him had clearly learnt nothing from all that had happened in the province so recently.[69]

Delcassé had been told about the Blin mission telegraphically and raised the matter first with Decrais, then in cabinet, where it was decided to order its cancellation.[70] In imitation of the man whose shoes he was filling, Broni disobeyed Decrais's instruction and Blin left on the very day of the arrival in Hanoi of Decrais's cable ordering him to remain. The appearance of Blin and his followers at Laocai led to the ominous reappearance of posters in Kunming telling the population that an invasion had begun and urging it kill and burn the French. François and the Chinese officials agreed not to issue passports to the members of the mission to prevent its entry into Yunnan.[71] Once Doumer had returned to Indo-China in September, he expressed his willingness to reduce the size of the mission to two engineers and ten assistants, and Delcassé asked Paul Beau, now installed as Minister in Beijing, if he would agree to such a mission.[72] Decrais supported Doumer's request in Paris, believing it desirable that the control mission begin its work in Yunnan before the company's first employees arrived in Tonkin on 15 November 1901.[73] Beau and Delcassé, however, refused utterly to consider such a move. Delcassé obtained cabinet support for the proposition that the mission waiting at Laocai be completely dismantled and that the two engineers and their ten assistants, none of whom was to be a military officer, could not enter China until after the company's engineers had begun work.[74] These orders were sent the day after François's damaging personal letter of 20 July reached Delcassé's hand.

Doumer suffered a great defeat on the issue of the Blin mission. A further one was to follow. The accusations contained in François's letter, coupled with Doumer's impetuous behaviour after his return to Indo-China, suggested that a more cautious hand was required to direct France's Indo-Chinese empire. It was decided to recall Doumer as soon as the Hanoi exposition of 1901 was finished. Thus, early in 1902 Paul Beau, a product of the Quai d'Orsay and then Minister in Beijing, was named his successor, and Doumer left Indo-China before the first rail was laid in Yunnan. It was not until 1904 that the company's engineers decided on the exact route the railway was to follow. A separate company, the Socièté de construction de chemin de fer indochinois was formed with a capital of four million francs by the Régie générale de chemins de fer de l'Indo-Chine and the Société des Batignolles to build the line. Guibert, who had done the surveys for the original consortium in 1899 and 1900, was appointed as its engineer-in-chief.[75]

Construction of the Yunnan section did not begin until 1906, when the rails reached Laocai from Haiphong. It was a difficult enterprise. Laocai was at an elevation of 93 metres. From there the line climbed through the malarial valley of the Namti to reach the plateau near Mengzi at an elevation of nearly 2,000 metres. Construction in the humid and unhealthy Namti valley was dangerous and expensive, both in terms of funds and human life. Completion to Kunming, where the first train arrived in 1911, involved the boring of 150 tunnels totalling 18 kilometres in length and building no less than 3,422 bridges on the 464-kilometre section from Laocai. It was France's finest engineering achievement in East Asia. The political aims of the railway, however, were already fading. Even before the line was completed France had agreed that the railway would become the property of the Chinese government in 1935.[76] Although the railway was certainly to be profitable, far beyond the expectations of François for example, and far more profitable than any Indo-Chinese railway,[77] it never secured for France either the economic or political penetration of Yunnan which had been, in the eyes of French imperialists, its *raison d'être*.[78] Ultimately Doumergue was right and the Chinese were the main beneficiaries of this extraordinary French investment, this improbable and indirect child of Fashoda.

The French decision to construct the railway to Kunming in spite of the unhappy events and unencouraging prospects which had attended its surveys, was to put an end to the rival British plan to penetrate Yunnan from Burma. Along with the career of Paul Doumer, the Rangoon to Yunnan railway was the main casualty in the months following the promulgation of the law of 5 July 1901. While the former Governor-General's eclipse was only temporary, the British railway had received a blow from which it would not recover, and by the end of the year had become the butt of ridicule from the highly opinionated Viceroy of India, Lord Curzon. Despite its abandonment in 1901, the proposal for a British railway into Yunnan had been serious and did have substantial official support. The authorization of construction from Mandalay to Kunlong ferry on the Salween in 1895 and the Anglo-Chinese Convention of 4 February 1897 which provided for the connection of the future Burmese and Yunnan railways are proof of these earnest intentions.

For a short while it had appeared as though these intentions might be transformed into reality. By January 1898 there was serious talk in London of extending the Mandalay–Kunlong line into China and Sir Claude Macdonald announced that the right to build such a railway was the precondition of any further British loans to China. Although Curzon, then still in the House of Commons, indicated his lack of enthusiasm for the scheme,[79] *The Times* gave the proposal its support and at the same time accurately analysed the dilemma then confronting the French:

It is merely asking for the Mandalay–Kunlong ferry line a privilege which has been secured for projected extensions from Tonkin; and it will be interesting to see which first reaches the field. The delay on the French side seems due to difference of opinion as to the relative claims of rival routes. There are those who contend now that a line from Langson to Longzhou and Nanning, instead of attracting trade to Tonkin, will prove an additional feeder to the Xi Jiang route, and that the objective of a Tonkin–Yunnan railway should, after all be Sichuan.[80]

Despite the activity of the British in Burma, the officials at the French Foreign Ministry were confident that the Tonkin route would prove to be superior and judged that: 'In so far as ease of penetration of Yunnan is concerned, it is incontestable that, thanks to the geographical position of Tonkin, we are better placed than the English'.[81] This French confidence was amply justified, for, as British politicians became aware of the geographical difficulties confronting a railway builder between the Salween and Kunming, let alone the Yangzi, they cooled markedly to the proposal. In May 1898 the Association of Chambers of Commerce of the United Kingdom told Salisbury that it:

considers it imperative in the interests of British trade that Her Majesty's Government should press for a concession authorizing the construction of a railway to connect British Burma with the upper valley of the Yangzi.[82]

Salisbury replied bluntly that the construction of such a railway would involve enormous difficulties as all the valleys ran at right angles to it. This was a remarkable change of opinion from that he had held two years before when he had described the railway as 'a great benefit to the world'.[83]

Expert opinion in Britain hardened against the proposal. Paul Cambon, the new French Ambassador in London, considered that most English who knew the area believed that the enormous cost of such a railway would never be repaid and that in any case Tonkin was the natural outlet for Yunnan and the Yangzi for Sichuan.[84] In July 1898 the Yunnan Company was established in London to build the railway from Kunlong into Yunnan. The Government of India supported it by seconding two military officers to the company to survey the route. The report of Captain Wingate was so derisive of the whole scheme that it was never taken seriously again.[85] On his return to India he told an audience in Simla that he could see neither strategic nor commercial advantages in building the railway. He claimed that the area the railway would serve was impoverished and that its inhabitants were lazy.[86] He later published his description of the topography of the route:

he estimated the average height of the major ranges the railway would have to cross at 7,000 to 8,000 feet and the average depth of the river beds below the level of the plateau at 2,500 to 3,000 feet. In short, the railway was not a practical proposition.[87] As a French newspaper later put it when describing Wingate's conclusions, 'the construction of the line would involve a succession of tunnels like the Gothard and of bridges like that over the Menai Strait'.[88] The same paper dwelt complacently on Wingate's conclusion that 'our Tonkin is the natural commercial route [to Yunnan]'.[89]

Even the completion of the section of the railway within Burma was in doubt. Its construction involved building the highest bridge in the world across the Gotkeik gorge, a massive fissure over 600 feet deep. The grades on the railway were so steep that it could never handle much traffic and an English journalist was shocked to hear the 'dismal prognostications' of its engineers for the railway's remunerative prospects. Ironically, even the great bridge over the Gotkeik, which at least would bring profits to its manufacturer, was being made by the Pennsylvania Steel Company as the British tenders were all about double the American. The journalist had ample experience of the operating difficulties of the line, but being made of stern stuff reported that 'the return journey, ... barring two small derailments and the breakdown of a locomotive, was accomplished without mishap'.[90]

By 1901 tension between Britain and France had been reduced and the railway building race began to seem absurd. The Yunnan Syndicate, an Anglo-French mining company, had been established and was unsympathetic to the railway's extension into China, as by 1901 construction of a line from Tonkin had begun and the Burmese route offered no commercial advantages and could lead to hostility on the part of Chinese officials and the local population.[91] A British general in Rangoon observed that the railway could produce no commercial benefits and that British political interests, which were in any case protected by the Anglo-French Convention of 15 January 1896, could never justify the enormous cost of its construction, nor indeed the cost of the completion of the Burmese section to Kunlong Ferry.[92] On 11 December 1901 Curzon, now Viceroy, made a celebrated speech in Rangoon in which he ridiculed the possibility of a metre-gauge mountain railway drawing the lucrative trade of Sichuan to Rangoon when navigable rivers such as the Yangzi could move far larger tonnages far more cheaply. He asserted that 'were a bonfire made tomorrow of the prolific literature to which it [the Burma–China railway] has given birth, I do not think anyone in the world would be the loser'.[93] Cambon was later told that Curzon's speech, which effectively killed the proposal, was made as a result of pressure on the British government by the Yunnan Syndicate.[94]

Thus, British rails never reached Yunnan, much less the Yangzi, and in fact construction was suspended before they descended into the Salween valley. The Gotkeik bridge became an expensive and impressive monument on a railway going nowhere. The British had created a Longzhou fiasco of their own, but on a far grander scale and without the excuse that the initiative had been taken by a private company which had not thought through the consequences of its investments. In curtailing the Kunlong railway, the British were doing what the French did in abandoning the Longzhou line. They recognized that the Hong Ha was the natural route of access to Yunnan and that visions of attracting the trade of Sichuan to South-east Asian colonial ports were chimeras.[95]

The involvement of the Yunnan Syndicate in opposition to the British railway is significant as the syndicate represented the beginnings of a tendency to merge French interests in China with those of other powers. It is true that the Beijing–Hankou concession was a joint enterprise with Belgium, but whatever Leopold's pretensions may have been, Belgium was not a great power. Britain was such a power and was widely regarded in France as the nation's principal rival in China and elsewhere. Despite such sentiments, the formation of the Yunnan Syndicate indicated that by 1901 the potential for co-operation anticipated in the Anglo-French Convention of 15 January 1896 was being realized. The area of economic activity in which this was occurring was mining, which by then had been recognized as Yunnan's main attraction for the Western capitalist.

The first serious attempts to exploit Yunnan's mineral resources by Europeans had been undertaken by Emile Rocher early in 1896 when he was leading the Lyon commercial mission in the province. Rocher was a former Consul in Mengzi and had been one of the leading propagators of the myth of Yunnan's wealth. Despite his familiarity with the province and the support of a prominent Chinese banker in his efforts to obtain the concession of a copper mining monopoly in Yunnan, the provincial authorities refused his request. After Rocher's departure, his successor, Brenier, continued to negotiate for such a concession and received the support of Gérard in Beijing, but with no more success than Rocher.[96]

Although the leaders of the Lyon mission were unsuccessful in their attempts to exploit Yunnan's mines with European capital and technology, their reports prompted others to do so. Most notably the Comité des Forges was encouraged by what the Lyon mission had reported. It corroborated what one of its members, Paul Dujardin-Beaumetz, had observed during a visit to China in 1896. In general, Dujardin-Beaumetz had been pessimistic about the possibility of finding a substantial market for the products of French heavy industry in China, but he did see the potential for developing a profitable mining industry in Yunnan. He claimed that the provincial authorities were willing to consider European involvement in mining in order to improve productivity and concluded that 'this Yunnan business merits particular attention'.[97]

Thus, in response to the reports both of Dujardin-Beaumetz and the Lyon mission, early in 1897 the Comité des Forges, in association with the Comité des houillères de France, formed the Société d'études industrielles en Chine. This company was capitalized at 300,000 francs and its chairman was Baron Reille, vice-president of the Comité des Forges. At Reille's request Hanotaux agreed to give the new company official support and on 14 February 1897 Marcel Bélard, a young mining engineer, left for Yunnan as the company's agent.[98] The expectations of rapid success in securing a concession which the Comité des Forges obviously entertained were not realized. After Gérard's success in securing a favourable commercial convention on 12 June 1897, the Comité des Forges approached the Quai d'Orsay about further support for Bélard's efforts.[99] The support was given and Dubail approached the *Zongli Yamen* about the matter in September 1897 and again in January 1898. On both occasions Dubail was informed that the Chinese had not yet decided to commence Western-style mining operations and that the assistance of French engineers and industrialists, as anticipated in the conventions of 20 June 1895 and 12 June 1897, was not yet needed.[100]

After nearly a year without tangible results and in view of Bélard's expenses, which eventually came to 176,000 francs or over a half of the company's capital,[101] the Comité des Forges decided to withdraw its agent from Yunnan.[102] Commercial dispatches from China had convinced Bompard that French industry needed permanent representation if it were to achieve any substantial results there. He therefore prepared a forceful reply for Hanotaux to send to the Comité des Forges urging it to retain Bélard's services in China.[103] Bompard also arranged to have Bélard's work integrated with that of the Guillemoto mission, which was then about to begin its surveys in Yunnan. Reille agreed to do as requested by the Quai d'Orsay.[104] Once again French officials were more enthusiastic supporters of a Chinese investment than French capitalists. As a result of the Quai d'Orsay's intervention Bélard spent most of 1898 in Yunnan, leaving Tonkin for France in November. During that time he worked both independently and with the Guillemoto mission. He impressed Dejean de la Bâtie with his energy in arranging clandestine visits to the wealthy Gejiu tin mines but had no greater success than either Rocher or Brenier in obtaining a concession.[105] Like those of his predecessors, Bélard's report was optimistic: there were large deposits of tin, copper, and coal, but their exploitation by the Chinese was haphazard and Chinese attitudes towards mining would need to change to permit their efficient exploitation on Western lines.[106]

While Bélard was working in Yunnan on behalf of the Société d'études industrielles en Chine, another French mining engineer was in the same area in an official capacity. André Leclère had taught at the St Etienne Ecole des Mines and was added to the Guillemoto mission at the request of the Quai d'Orsay. This decision was not taken until the eve of Guillmoto's departure in October 1897 and it was not until January 1898 that he arrived at Laocai. From then until August he explored the territory along the route of the future railway to Kunming. He spent his remaining eight months in China travelling more widely in Sichuan, Guizhou, and Guangxi, covering a total of 6,000 kilometres in all, and returned to France in June 1899.[107] His reports also were optimistic. He recognized that Yunnan, Guizhou, and Guangxi were amongst the poorest provinces of the Chinese empire and that their penetration 'is not a utopia for France'. Nevertheless, he considered that the exploitation of the exceptionally rich and widespread coal deposits which he had found throughout the three provinces could play a major role in the industrial development of Tonkin.[108] He even suggested that iron ore deposits in Tonkin near Laocai could be exploited in conjunction with Chinese coal and an iron and steel industry created in Indo-China. Such a metallurgical industry, in Leclère's opinion, 'would have to assure Indochina of a real industrial and commercial predominance'.[109] The outlook was also encouraging so far as non-ferrous metals were concerned, provided the technology used could be improved:

So far as metals are concerned, Yunnan, Guizhou, and Guangxi have long been recognized as the most important mining provinces in China. Yunnan especially supplies almost all the copper used in the empire ... Undoubtedly metal deposits are very numerous through a rather vast area, but their exploitation will probably remain a rather uncertain enterprise and will only be successful near coal deposits.[110]

By early 1899 a favourable atmosphere for a Yunnan mining project had been established in Paris. Bélard had returned, without a concession, but with reports of considerable wealth in non-ferrous metals to present to his influential employers. Reports from Leclère encouraged official interest in such a project. Thus, in March 1899 a group of financiers led by the Comte de Bondy approached Delcassé for support in obtaining a 99-year concession of the tin and coal mines within a 200-kilometre radius of Gejiu and the right to build a railway from Gejiu to the future Mengzi station.[111] These were the very mines to which Bélard had made two clandestine visits the previous year and where, within three months, anti-foreign violence was to erupt. The approach was serious: de Bondy was a member of the Comité de direction of the Comité des Forges, a director of le Creusot and chairman of the board of Chantiers et Ateliers de la Gironde. Delcassé was pleased that French capitalists were taking serious interest in a Chinese investment and told de Bondy as much, promising the support of the French legation at Beijing and consulate at Mengzi. At the same time he warned de Bondy that the Chinese would not grant the concession directly to his group but to a Chinese company which would then obtain official permission to use foreign capital, personnel, and technology in its operations. All these arrangements would require the approval of the *Zongli Yamen* and the Central Mines Board in Beijing, so official French support would be necessary.[112]

De Bondy was not discouraged by Delcassé's warning, and on 1 May 1899 the group was formally constituted in Paris as the Syndicat minier du Yunnan. Its president was de Bondy and its vice-president Emile Cellérier, a former *attaché au cabinet* in the Ministry of Finances who had also served as a *chef de service* of the

Indo-Chinese government. Its *secrétaire* was de Francqueville d'Abancourt, who was also *secrétaire* of a company building ironworks in Russia and who represented his uncle, Baron de Nervo, president of the Comité des Forges. The other members of the syndicate were Dujardin-Beaumetz; Bonnaud, also a member of the Comité des Forges and representative of the Société des Aciéries et Forges de Firminy (Loire); Dupuy-Dutemps, a former Minister for Public Works who was then a director of the Compagnie générale de Traction, a Parisian electricity supply company; Chaudey, a former deputy; and the Comte de Gessler, a Russian who was private secretary to the Crown Prince of Bulgaria.[113] Gessler was a friend of the Russian Foreign Minister Muraviev, and Cellérier expected to be able to obtain official Russian support for the concession through this connection.[114] The plan was for Cellérier to go first to Mengzi and sign a contract for mining operations in and around Gejiu with a Chinese man of straw. He would then go to Beijing obtain the ratification of this contract from the *Zongli Yamen* with the support of the French and Russian legations. It was hoped that an Imperial Edict would be issued ordering the co-operation of the Yunnan authorities and that Cellérier could then return to Mengzi and set about establishing mining operations.[115]

Delcassé was enthusiastic about Cellérier's plans and told his agents in Beijing and Mengzi how anxious he was that the quest for the concession be brought to a successful conclusion.[116] This enthusiasm and optimism was not shared by those in China, for in June, while Delcassé's dispatches were on the water, there had occurred the uprising of the Gejiu miners and the riot at Mengzi. Pichon replied that it might not be impossible to obtain the concession, but that it would require 'special precautions, much tact and prudence'.[117] Dejean de la Bâtie was more blunt:

I do not believe that M. Cellérier has at present the least chance of success and I am certain, on the other hand, that if the aim of his trip was known by the Gejiu mining population, very serious troubles would follow which could compromise once again work on the railway and the safety of the European colony.[118]

He described the operations at Gejiu: the mines were not private property and not subject to concession. A mine belonged to the first person to occupy it and its extent was undefined. The provincial government did not regulate mining in any way and intervened only to collect taxes. It would be very difficult, if not impossible, to institute a European concession beside the existing system. Moreover, the 30,000 men who worked the mines were mostly unmarried and a notoriously rough bunch.[119] In the light of this report Pichon wrote to Delcassé urging him to ask the syndicate to postpone Cellérier's departure until security around Mengzi had improved.[120] Under the circumstances the Syndicat minier du Yunnan was informed officially that its proposed exploitation of the Gejiu deposits was impractical for the time being.[121]

This reversal led to a transformation in the aims and constitution of the syndicate. As it appeared to be impossible to commence operations at Gejiu, it was decided instead to solicit a copper mining monopoly for the entire province. As copper deposits were in the northern and western areas of Yunnan, nearer to the Burmese frontier and the Yangzi than to Tonkin, British interests had to be considered and, at the very least, the benevolent neutrality of the British legation in Beijing assured. Moreover, the members of the Syndicat minier du Yunnan seem suddenly to have become aware of the Franco-British Convention of 15 January 1896 and its assertion of the equality of the rights of British and French nationals. It was therefore decided to form an Anglo-French syndicate to pursue the more ambitious concession.[122] It appeared, moreover, that in the wake of the troubles at Mengzi, Cellérier and de Bondy had difficulty maintaining the support of French investors and that the sources of capital needed for the project were evaporating. At any rate Cellérier went to London in search of capital without informing the Quai d'Orsay. Success was quick but the price was high. The Syndicat minier du Yunnan was transformed into the Anglo-French Syndicate, a British company operating under British law.[123]

The new company had six directors, three of each nationality. The English directors were Hubert Bourke, a businessman with connections in the Conservative Party, most notably with Chamberlain, and the founder, with Angelo Luzzati, of the Peking Syndicate; Oakley Maund, chairman of the Matabeleland Mining Company and Bechuanaland Railway Company; and Sir Robert Herbert, permanent Under-Secretary of State at the Colonial Office from 1871 to 1892 and a director of the Peninsular and Oriental Steamship Company. All three were very substantial men in the London commercial and political scene.[124] In contrast the substance which had characterized the Syndicat minier du Yunnan had vanished. Cellérier himself was the only element of continuity on the French side. He had approached various Parisian financiers in his search for two other directors, but had had to make do with MM. Froment-Meurice and d'Andigné, whom Delcassé described as 'young men of very good family, but without the least renown in the Paris market'. The Comte de Bondy, who had retained his links with the group but was unwilling to remain a director, considered that it could not possibly raise any capital in Paris unless Cellérier secured better French directors.[125]

Cellérier retained the services of Dujardin-Beaumetz as a consulting engineer and recruited Bélard from the Société d'études industrielles en Chine. Bélard promptly attempted to transfer a coal mining concession he had obtained in Tonkin from his former to his new employees, thereby provoking prolonged litigation in France and Indo-China. After obtaining financial support in London, Cellérier passed quickly through Paris before leaving for Tonkin where he hoped to join Bélard and use the letters of introduction and promises of support he had obtained from the Quai d'Orsay when vice-president of the Syndicat minier du Yunnan.[126] Cellérier and Bélard, as the company's agents in China, stood to gain greatly from the enterprise. Although there was more English than French capital involved, profits were to be divided equally between the two elements. French shareholders would get 35 per cent of the profits, the French directors five per cent and the company's agents in China, Cellérier and Bélard, ten per cent. English shareholders would receive 45 per cent and the English directors five per cent. All the big names of French finance and industry were absent from the list of French shareholders, but English shareholders included Beit of Werner, Beit and Company; Milton and W. F. Forbes of the Rhodesia Company; A. Walker of Bovril; Standish Grady; the Count of Yarborough and three directors of the British South Africa Company, including its chairman, the Duke of Abercorn, Lord Loch, and Earl Grey.[127]

The reconstitution of the Syndicat minier du Yunnan as the English-dominated Anglo-French Syndicate presented French officials with a dilemma. Should the support that was given to the old syndicate be given to the new? Cellérier's behaviour in not informing the Ministry of Foreign Affairs and his rapid departure for China suggested that he was an adventurer. Feeling at the Pavillon de Flore was decidedly hostile to the possibility of a company with English involvement operating on the borders of Indo-China. Decrais made this quite clear to Delcassé:

> I have no doubt that you will judge like me that it is not only inopportune but also dangerous to allow any company whatever to operate under our auspices, in which the English have significant interests, on the Sino-Vietnamese frontier and especially in Yunnan.[128]

In London Paul Cambon took a very different view. He acknowledged that it was undesirable to create an enterprise on the frontier of Tonkin which the English could easily take over, but concluded that 'this is on the English side a serious enterprise which sooner or later will be implemented, with us, without us, or against us'. As against the one objection to a joint enterprise, he could see much in its favour. Under the terms of the Anglo-French Convention of 15 January 1896, British and French enjoyed equal privileges in Yunnan. Cambon argued that opposition to the Anglo-French Syndicate could put this convention, which was France's best security against a British occupation of the Yangzi valley, into question. He also advanced commercial reasons for supporting the syndicate: if the mine at Gejiu were operated by the British, the ore would be shipped to Hong Kong via the Xi Jiang rather than by the French railway to Haiphong. In any case, he argued, if the mines were profitable French capitalists would invest in them regardless and France would not gain any political benefit from this deployment of her capital. His final argument was the most important: as Anglo-French relations had improved since Fashoda, he believed it was best to cement this improvement by linking the capital of the two nations wherever possible so as to prevent any further estrangement. This was his doctrine of the 'protective mattress'. Cambon therefore felt that the French government should support the syndicate, provided that the French share in its capital and management be at least equal to the English.[129]

Unfortunately for Cambon's hopes, this last condition was not being met and it was becoming increasingly obvious that it could not be unless there was a radical restructuring of the French side of the syndicate. At this crucial stage Delcassé was absent from Paris mourning the death of his father. In the Minister's absence Bompard and Cambon agreed that the English directors should be informed that they could not receive any support from the French government unless French interests were better represented.[130] As no such change occurred, at the end of the year Delcassé ordered his agents to withdraw their support from Cellérier's projects.[131]

This decision, however, was not the end of attempts to form a joint Anglo-French mining company in Yunnan. Once they realized the weakness of Cellérier's backing, Cambon and Delcassé set about creating a French consortium themselves. At least three other French companies, apart from the interests represented by Cellérier, were involved in some way with Yunnan mining proposals. The Société d'études industrielles en Chine, the company established by the Comité des Forges, still existed. Its chairman, Baron Reille, had died and been replaced by the deputy, Georges Berger. Berger took a conservative approach, and, apart from offering Bélard's report to the Anglo-French Syndicate, did not involve the company in the negotiations. One of the companies represented on the board of the Société d'études industrielles en Chine was the Compagnie générale de Traction. Its need for copper wire for its electricity supply and generation activities led its directors to express interest in the copper deposits in Yunnan. France was poor in non-ferrous metals, so its director-general, Obry, approached Delcassé for support in obtaining a concession just at the time that the news broke of the creation in London of the Anglo-French Syndicate.[132] Delcassé offered the support requested and was later to use the interest of the Compagnie générale de Traction in Yunnan's copper to create a new international syndicate.[133] This was facilitated by the fact that the chairman of the Compagnie générale de Traction, Hubert Henrotte, was also a director of the Banque française de l'Afrique du sud, which, despite its name, was an Anglo-French bank established by Werner-Beit, the British financiers whose funds contributed the majority of the capital of Cellérier's Anglo-French Syndicate.

Meanwhile, Cambon also was attempting to involve French capital in a joint mining venture in Yunnan. He approached Noel Bardac, a French director of the Banque Impériale Ottomane, who was visiting London. The Banque Impériale Ottomane was already an Anglo-French organization and both Bardac and Cassel, one of the bank's English directors, responded with interest to Cambon's initiative. Probably at Bompard's request, the Banque Impériale Ottomane and the Compagnie générale de Traction formed a consortium to seek the concession. This consortium was also joined by the Société générale, the Banque Internationale de

Paris and the Banque française de l'Afrique du sud. Additionally, a new British grouping, the Exploration Company, which had been established by the houses of Hambro, Baring, and Rothschild for the purpose, announced its adherence to the consortium. This powerful consortium was represented in its negotiations with the Quai d'Orsay by Henrotte. It was the aim of the French Foreign Ministry to put Henrotte's consortium into contact with Bourke, Maund, and Herbert, the English directors of the Anglo-French Syndicate and, ultimately, to replace Cellérier's weak French element in that syndicate with the powerful French interests represented by Henrotte.[134]

After a number of false starts and despite misunderstandings on both sides, the plan worked. Bourke had been told by Cambon soon after the syndicate had been formed that it could not hope for the support of the French government unless its French interests were more substantially represented. Bourke, therefore, went to Paris and attempted, unsuccessfully, to interest the Banque de Paris et des Pay-Bas in Cellérier's syndicate. He later returned to Paris to discuss what he considered were the French government's objections to the French directors of the syndicate with Bompard at the Quai d'Orsay. Bompard had told him that the objection was not to the directors themselves, but with 'their lack of real power'. Bompard went on:

the Government in fact regards very favourably ... joint Anglo-French projects so that both can exercise in common the equal rights the two countries have in Yunnan, but it is still necessary that in the company formed for this purpose, the French element has the same standing and authority as the English element, that French capital is involved in the same proportion as English capital; in a word that the company works towards profits and costs shared equally by the two nationalities.[135]

The message was clear. Bourke asked if Bompard could suggest a suitable French businessman with whom he could negotiate. Bompard coyly and dishonestly replied that he could not involve the department in business negotiations.[136] In reality the involvement of the Quai d'Osay was intimate and Bompard even drafted what amounted to the by-laws under which he believed the venture should operate.[137]

Reserves about the project had to be overcome in London as much as in Paris. The Foreign Office was initially unsympathetic to the proposal, even though it came from a source as sound as Sir

Robert Herbert. Francis Bertie told Herbert in October 1899 that he could see no objection to the the scheme provided the majority of the capital was British and it was 'placed absolutely under British control'. He feared 'the political importance which France would thereby gain in Yunnan ... with English gold' unless this condition were met. Salisbury agreed and bluntly commented that unless British domination were secured, 'we ought certainly not to assist it'.[138] Salisbury, however, gradually rallied to the project, convinced by Cambon's assertion that 'his object is, by drawing together French and English financial and commercial interests in the Far East, to establish a community of political interests, and thus obviate a conflict of interest which would otherwise become inevitable'.[139] Cambon and Salisbury discussed the matter at least twice, and Salisbury even helped directly in the negotiations by resolving a misunderstanding which had been created by one of the agents of the Exploration Company. Cambon believed that Salisbury's changing attitude to the plan proved that it was an excellent means of ensuring that Anglo-French relations, then improving, continued to do so. He observed:

I attach ... the greatest value to its early realization. The time is favourable; the reverses of English arms in South Africa have silenced for a while the hostile sentiments which reign in this country towards us. This respite may only be temporary and once British troops are again success-ful, the distrust of all French enterprises in China might reappear stronger than ever. We therefore have to hurry if we want to avoid, when the time comes, a conflict of our political interests, to make the 'protective mattress' ... mixing private French and English interests by organizing enterprises jointly founded by the businessmen of both countries.[140]

For Cambon such joint investments were an effective means of encouraging *détente* between the two powers, especially as they involved not just the co-operation of British and French officials and politicians, but of members of the bourgeoisie as well.

Once the groundwork had been laid by Cambon and Bompard, serious negotiations between Henrotte's group and Hubert Bourke of the Anglo-French Syndicate began. Henrotte enlisted as his negotiator Achille Adam, a banker from Boulogne and deputy for Pas-de-Calais who was also a director of the English South-Eastern Railway Company (SER) and had interests in trans-Channel shipping. Adam and Bourke quickly agreed on the terms for amalgamation of the Anglo-French Syndicate and Henrotte's consortium. The terms were essentially those specified by Bompard. However, at the last minute the whole question was thrown open. While insisting on absolute equality of British and French representation in the syndicate, Bompard was prepared to give it official support even though it was going to be a British company with its head office in London. No doubt used to the independence which characterized operations at the Quai d'Orsay during the Third Republic, Bompard, Cambon, and Delcassé had spoken for the French government without consulting other ministers. Eventually the matter had to be brought to Cabinet, which it was on 3 April 1900. Unfortunately for all the efforts of the ministry of Foreign Affairs and despite Delcassé's opinion, Cabinet decided that it was 'absolutely impossible' to use French diplomatic influence in China to assist an English company, even though French interests were involved.[141] In its quest for Anglo-French détente, the Quai d'Orsay had gone beyond what the government was prepared to give, and support for the project had to be withdrawn.

Despite the decision of the French Cabinet, Bourke's and Adam's negotiations came to a successful conclusion just a week later, on 10 April 1900. It was agreed that the three French directors of the Anglo-French Syndicate would be replaced by four new French directors representing the interests involved in Henrotte's group. Bourke and Herbert would remain, but Maund would resign and be replaced by directors nominated by the Exploration Company.[142] These new 'English' directors were Sir Astley Corbett and an Austrian living in Paris named Mandle. The French directors were to be Adam (chairman of the French committee of the board), Henrotte, the Comte de Germiny representing the Banque de Paris et des Pays-Bas, and Ulysse Pila. The group's name was changed to the Yunnan Syndicate, or Syndicat du Yunnan and the old Anglo-French Syndicate was wound up at its meeting in London on 27 April 1900 which ratified the agreement between Bourke and Adam.[143] The capital of the company was increased from 20,000 to 35,000 pounds sterling. Of the 20,000 one pound shares in the former Anglo-French Syndicate, 5,500 had been held by French and 14,500 by English investors. Of the 15,000 new one-pound shares, 3,000 were to be taken by the Exploration Company, increasing the English holding to 17,500 shares or half the total. The remaining 12,000 shares were to be taken by a number of powerful French companies and individuals, including eight banks.[144] It was agreed that a joint Anglo-French mission be sent to China to obtain the concession and that in the meantime one of the deposed French directors already in China, Froment-Meurice, would represent the syndicate.

Despite the Cabinet decision of 3 April 1900, Delcassé felt that French diplomacy could not remain totally indifferent to the fate of the Yunnan Syndicate. He wrote to Pichon in terms which hardly suggest that he was a devotee of the doctrine of Cabinet solidarity, urging him to consider the French interests in the syndicate:

My Department has thought that it could not refuse them its protection in any absolute sense, and I authorize you to lend, to the extent that you judge opportune and in an unofficial way, your good offices to M. Froment-Meurice if he comes to seek them.[145]

If Delcassé discreetly authorized Pichon to give unofficial support to the representatives of the Yunnan Syndicate, his attitude to Cellérier and Bélard was very different. The new Consul in Mengzi, Sainson, was ordered to keep an eye on their activities, which were unauthorized and potentially dangerous. Sainson found the job easy but tedious, as Cellérier was constantly harassing him and requesting introductions and support. None was offered and eventually Cellérier went off to Beijing with a quixotic plan to obtain a concession for a railway from Guangzhou to Fuzhou and Hankou. While Cellérier dreamed of achieving some success, Bélard realized that his hopes for wealth were at an end. After finding no encouragement in Yunnan he went to the coal mine in Tonkin, the concession of which he now claimed as his own. Finding that no one would recognize his claim, he shot himself through the temple.[146] Instead of five per cent of the profits of an international consortium, his defection to the Anglo-French had brought him a lonely suicide's grave far from home. Bélard's death and Cellérier's departure from Yunnan brought to an end the activities of the Anglo-French Syndicate.

Cambon and Delcassé were bitterly disappointed at the Cabinet decision not to support the Yunnan Syndicate. Their desire to reverse it increased during 1900 as the syndicate made further concessions to French interests, notably by deciding that its chairman would be Adam and that the mission to China would be led by a French citizen. Delcassé used these concessions to work on Decrais and succeeded in convincing the Minister for Colonies of the desirability of supporting the syndicate.[147] The strongest French advocate for the Yunnan Syndicate was always Paul Cambon, who believed that it could serve French interests in China in a powerful and effective way. On 11 December 1900 he wrote an important dispatch to Delcassé which was later presented to the French Cabinet in support of a change of attitude to the syndicate. The essence of Cambon's argument was that the syndicate was a real fulfilment of the convention of 15 January 1896, a convention which worked well in French interests:

The convention of 1896 is one of our diplomacy's most profitable arrangements in the Far East; it protects us from temptations to expand our Indo-Chinese empire to the north, which would bring us nothing but a revival of the Tonkin insurrection, a war with China's southern provinces and a conflict with England. It also protects us from the encroachments of this latter power which could pursue the establishment of communications between Burma and the Yangzi Valley, and thus render useless the establishment of the railway we propose to build at great expense between Hanoi and Kunming.

The views of the Government of the Republic in 1896 were no different from those in 1900, they conformed to the policies you have expressed several times, always with the unanimous assent of the Cabinet and the Chambers, to the parliament and to the country.

The formation of an Anglo-French syndicate to survey Yunnan's industrial resources and share in their exploitation is therefore the most complete and fortunate realization of our policies.[148]

Cambon pointed out that the British government was obviously sincere in its desire to see Anglo-French co-operation in Yunnan as it had permitted Sir Robert Herbert to remain on the board of the syndicate. The syndicate itself had indicated its good faith by electing a French chairman in the hope of securing official French support. He speculated, though, that its directors must be wondering what the use of a French chairman was when the French government withheld its support. He expressed enormous regret at this attitude and pointed out that many people in London were beginning to suspect France of trying to overthrow the convention of 1896. This was dangerous as it could lead to the loss of 'an arrangement which gives us our fair share in Yunnan and whose failure will lead the English to do without us or even against us what they were ready to do with us'.[149] In this situation, the French government had three alternative policies:

It may have some second thoughts of conquest to the north of Tonkin. I am ignorant of these and would deplore them. In that case it must know that beyond the perils of a war on the Indo-Chinese frontier, it would be threatened with a conflict with England, and as England armed with the convention of 1896 would be right, it would be exposing itself to a pitiful failure.

It may fold its arms, ignore the new company and let it go its way without lending it its support. In that case the company will be truly English, it will have the support of English agents in China and of the English Minister in Beijing, and the French capital invested in the enterprise will only serve English political interests.

Finally it may, conforming to its own policies, to its reiterated declarations, to the most obvious interests of our influence in China, encourage an enterprise conceived and managed in the spirit of the convention of 1896 and which ... will serve our interests more effectively than English interests in the province of Yunnan.[150]

Cambon's final point in favour of an early French rally to the syndicate's support was that it wished to send Emile Rocher, the former Consul in Mengzi and then Consul in Liverpool, as the head of its mission to China. Rocher wished to accept the position, but would not do so unless his employers enjoyed official French protection.[151]

On 8 January 1901 Delcassé raised the issue of the Yunnan Syndicate in Cabinet once again. He had already won Decrais's support in his attempt to persuade his colleagues to reverse their decision of 3 April 1900. He tabled Cambon's despatch, spoke to it forcefully and ultimately secured Cabinet's assent to Cambon's recommendation.[152] Delcassé immediately placed Rocher at the syndicate's disposition and gave him the rank of Consul-General.[153] Rocher's reputation in England was such that plans to send an English assistant (Sir Edgar Creed, a former Indian Civil Service official) with him to China were abandoned. The Foreign Office instructed Sir Ernest Satow, Macdonald's successor in the British legation in Beijing, to give Rocher every assistance and Delcassé advised Paul Beau to discuss with Satow how best Rocher's success could be assured.[154] Rocher arrived in China in April and by July 1901 was well advanced in his negotiations with the Yunnan authorities for the concession of: firstly, all mineral deposits exploited by the state in the province; secondly, deposits formerly exploited but then abandoned and; thirdly, deposits not yet exploited or not yet discovered. Such a concession left the Gejiu miners undisturbed but assured the syndicate's control of the rich copper mines. Management of the operations would be by a Chinese official representing both the provincial and imperial governments and who would be advised by British and French engineers.[155]

Thus Rocher, one of the earlier proponents of the Yunnan myth, returned to the province where he had spent so much of his life. He was now the agent of an international syndicate representing the capital of both of the powers which had been rivals for so long in the quest for the province's wealth. His return and the decision to support the Yunnan Syndicate were partly the result of the growth of *détente* between Britain and France during the early stages of the South African War. They were also expressions of the ascendancy of commercial over strategic and territorial interests in southern China. This was consistent with the policy which had evolved in the Quai d'Orsay during 1899 and with the repudiation of the expansionist plans of Paul Doumer. These moves were also harbingers of a new era of international financial imperialism where Western capitalists co-operated with each other to exploit what a later age would call the nations of the Third World.

10

Conclusion

DURING the period from 1885 to 1901 French policy in China and on China's frontiers was largely transformed from a territorially based imperialism into an economically based imperialism. The two forms of imperialism co-existed for most of the period, which began with concrete expressions of both, in the forms of the annexation of Tonkin and the railway quasi-monopoly claimed in Article 7 of the 1885 treaty. As the conqueror of a former tributary state on the frontier of China, France was in a position to pursue both a policy of territorial acquisition and a policy which gave primacy to economic gain throughout China. This inherent dualism in French imperialism only began to come under stress in 1898 and 1899 as, on one hand, an unusually zealous annexationist in Indo-China and, on the other, the actions of the other imperialist powers in China, brought to light the potential contradictions of French imperialism. France decided that she could not claim a *de facto* Protectorate and economic monopoly in that part of China near her colonial frontier, while at the same time seeking to extend her influence and place investments throughout the 18 provinces.

For France, at least, imperial endeavour in China was, by 1900, becoming more concerned with investment than trade. For although China continued to be a significant source of raw materials for French industry, especially of raw silk, France was never able seriously to penetrate the Chinese consumer goods market. As an exporter of capital, however, France was extremely competitive and it was in financial imperialism that French efforts were most effective. From these investments there generally flowed industrial advantages of some kind which naturally assisted the French metallurgical industry. The emphasis on investment did not mean that hopes of developing trade had died. Indeed, it was hoped that with the aid of the Indo-Chinese customs regime, the Yunnan railway would turn much of the province into a French commercial monopoly. Certainly most knowledgeable French observers, notably François, were extremely sceptical as to whether this could be achieved. Yet most members of the Chamber of Deputies showed in 1898 and 1901 that they retained the old illusions on which so many of Ferry's ambitions had been based. Indeed, on the former occasion they were so carried away by neo-mercantilist fervour that they resurrected a provision from the old British Navigation Act — the requirement that material for the Yunnan railway be transported in French ships.[1] Despite the survival of such dreams into the early twentieth century, investment became the leading element in France's presence in China.

After the annexation of Tonkin, French diplomatic efforts in northern China were primarily directed towards opening up new areas for investment by French capitalists. These investments, far more than the religious Protectorate for example, were seen as the most effective means of extending French influence. After some early spectacular, but transient, successes, this financial and industrial imperialism faltered badly around 1890, dealing both French prestige and the reputation of French industry a severe blow. During the period when Gérard was in Beijing and Gabriel Hanotaux Minister for Foreign Affairs, there was a recovery. Gérard's energetic policy, supported in Paris by Hanotaux, although at times muddled, did greatly extend French investment in China, both near the Tonkin border and elsewhere in the empire. Although Gérard and Hanotaux were nationalists, they anticipated the twentieth-century trend towards international financial groupings in their willingness to co-operate with Belgium in the Beijing–Hankou project and with Russia in the first indemnity loan and the Russo-Chinese Bank. The Anglo-French agreement on Yunnan and Sichuan of 15 January 1896 indicates that Berthelot and de Courcel at least, like Cambon and Delcassé later, were willing to co-operate with Britain as well. This was despite the fact that throughout the period from 1885 to 1901 there was constant rivalry with Britain — for territorial advantage in south-east Asia, for dominance in Yunnan, and for financial and industrial concessions throughout China. It is true that from 1899 the intensity of this rivalry diminished, but even by 1901 it can hardly be said to have disappeared.

This willingness to fuse French capital with that of other powers suggests that national prestige was far less important in the eyes of French decision makers than economic advantage and real influence. Indeed, they were even willing to underestimate, in public at least, the extent of French involvement in the Beijing–Hankou project in order to achieve real advantages. Whatever may have been the case in Indo-China, in Paris, and in China itself, where the significant decisions and large investments were made, a serious concern for economic gain rather than a sentimental obsession with the putative grandeur brought by territorial conquest was evident. On this point Hanotaux and Delcassé were, for once, in agreement.

The economic rationality and modernity of Delcassé's imperial outlook as manifested in his policy towards China has often been ignored. While Andrew has rightly pointed out that Delcassé was 'not merely indifferent, but actively hostile, to attempts to enlarge France's Asian empire', his concern with Delcassé's alleged Mediterranean vision has distorted his analysis of the reasons for Delcassé's hostility to Doumer's plans.[2] Certainly Delcassé did not want to provoke a conflict with Britain, or indeed with any other power in East Asia, but his concern was not just to leave his hands free for expansion in Africa. Rather he was aware that further expansion of the Indo-Chinese empire would be directly prejudicial to France's own wider economic interests in China itself. Delcassé, like Bompard and Paul Cambon, was extremely solicitous of these interests. He regarded China much as he and his contemporaries regarded Latin America: as a fertile field for investment and exploitation, at times in association with foreign capitalists, but also as an area too vast, too volatile, and too dangerous for annexation. This is, more or less, the attitude of wealthy capitalist powers towards the Third World today. It is thoroughly unsentimental, not at all atavistic, and quite rational.

As something of a modern economic rationalist, Delcassé's conception of imperialism extended beyond territorial acquisition. He had no desire to confine French investment and trade to the colonial empire and actively sought to extend French interests not just in China but in Latin America and the Ottoman empire as well. It is a mistaken and limited view of imperialism which does not embrace activities of this type. While it is understandable that this view could be held, for instance, by Leonard Woolf, obsessed as he was by the memory of controlling the lives behind gaunt Singhalese faces in places like Beddagama; or by William Langer, seated in front of a world map splashed with cartographer's red and purple on the wall of his Washington office; it cannot be sustained in a serious analysis of French policy in East Asia in the late nineteenth century. More satisfactory are the models of Hobson, who, writing at the very time when the Yunnan Syndicate was being created, anticipated such international groupings; and

Lenin, whose emphasis on the need for investment as the primary cause of modern imperialism echoed the repeated concerns of the more economically minded officials at the Quai d'Orsay. Indeed, at times the writings of de Bezaure, Bompard, Gérard and even Delcassé himself, sound like Lenin.

The emphasis which both Hobson and Lenin placed on the role of finance capital in imperialism is supported by the reality of French activity in China. In one important respect, however, French capitalists did not behave as Lenin or Hobson would have imagined they should. For the initiative in developing investment opportunities and arranging the deployment of the necessary capital in China came very often not from the financiers but from the government or its officials. It would appear as though some officials knew more about the role of finance capital in imperialist endeavour than did the capitalists. The views of men such as Gérard, Bompard, and Cambon certainly anticipate those later postulated by Hobson and Lenin.[3] Like Lenin they recognized that territorial control was not necessary to implement an effective imperialist policy.[4] Strangely, it was often their task to rally the capitalists to the imperialist cause. This they did with considerable success. Certainly there were occasions when private industry took the initiative in China, notably when Fives-Lille sought the Long-zhou railway concession and when the Comité des Forges dispatched the Dujardin-Beaumetz and Bélard missions to Yunnan.[5] Nevertheless, on the whole it was the Ministry of Foreign Affairs which set the pace and decided the directions of French business activity in China. This was partly because these activities could only take place within the provisions of international agreements made by diplomats, and partly because elements in the Quai d'Orsay saw more clearly than most the future directions of imperialism.

The central importance of the role of the Ministry of Foreign Affairs was clearly demonstrated in its successful efforts to maintain control of policy in Yunnan. For there was a real attempt to seize the province by a headstrong and well-connected Governor-General and a group of ambitious and talented military officers. They together constituted a relatively powerful peripheral force which, as the experience of the western Sudan demonstrated, could well have determined policy. This challenge, however, was beaten back by the Ministry, and control of policy remained in the Quai d'Orsay. In achieving this victory of the metropolis over the periphery, Delcassé was greatly aided by the existence both of a modest diplomatic network in southern China and of international agreements which obliged France to be circumspect rather than adventurous in pursuing her ambitions on Tonkin's frontiers.

As well as coping with the ambitions of French expansionists, the Quai d'Orsay was constantly engaged in a struggle with Britain for influence, trade, and concessions in China. In the south-west this struggle intensified following the almost simultaneous annexation of Tonkin by France and Upper Burma by Britain. Sino-French customs agreements; the opening of the Xi Jiang to foreign trade; and the multitude of railway-building schemes, some serious, others not, which saw lines drawn on maps from various ports on the Bay of Bengal and the Gulf of Tonkin via diverse and devious routes to improbable and remote destinations in China, were all expressions of this rivalry. Moreover, as events surrounding the construction of the Port Arthur dockyard, the concession of the Beijing–Hankou railway, and the negotiations over the various indemnity loans showed, Anglo-French rivalry was not confined to the southern provinces of China.

It was in the south, though, that the Ministry's most notable efforts to resolve this rivalry were made. On one hand, the favourable customs arrangements negotiated with the Chinese, the mining privileges claimed for French industry, and the decision to construct the Haiphong–Kunming railway were all intended to establish French commercial and industrial dominance. On the other, there were attempts to resolve the conflict by transforming the British from rivals into partners with whom the potential wealth of the region could be exploited jointly. The agreement of 1896 on Sichuan and Yunnan was a first move in this direction. Although it was the product of a short-lived French government, and the principles on which it was based were forgotten in the feverish scramble of 1898, it provided the basis for a more sober reconsideration of French policy in 1899. By the end of that year, in the Quai d'Orsay at least, the trend was towards co-operation with other powers, especially with Britain. Even though the French government was not yet ready to accept the new policy, officials of the Ministry continued to pursue it actively in the business world by helping to create the Yunnan Syndicate. Ultimately, like business, the government rallied to the Ministry's position. Significantly, Anglo-French co-operation in Yunnan was based not on trade, but on investment in resource development.

The new century was clearly bringing a new approach to Chinese affairs.

Thus, both Hanotaux and Delcassé demonstrated that they were willing to co-operate with Russian, Belgian, and British investors in order to further French economic interests. This alignment anticipated that at the outbreak of war in 1914. But Delcassé was also willing to consider, and indeed on one occasion suggested, association with German capital in a Chinese project. During 1899 Sheng Xuanhuai was attempting to obtain finance to build a railway from Kaifeng to Xi'an. This line would bisect the Beijing–Hankou line and, typically, he discussed the possibility of a concession with French as well as other financiers. The Banque de Paris et Pays-Bas was interested in the concession but unwilling to undertake it alone. As the Germans were planning to build a railway from Qingdao to Kaifeng, Delcassé considered that it would be appropriate to consider association with German capital. He considered that such a financial combination would have both economic and political advantages:

From a political point of view, this combination would have the advantage of linking our interests in northern China with those of various powers.[6]

Ultimately Sheng awarded the concession to a Belgian group but the episode illustrates both Delcassé's unsentimental approach to Chinese affairs and his belief that the fusion of French and foreign capital protected rather than threatened French interests.

The flexibility and modernity of Delcassé's attitude to foreign investment and his enthusiasm for fusing French interests with those of other powers suggests that some revision of Thobie's harsh judgement may be needed. Thobie compared the attitudes of Delcassé and Maurice Rouvier to French participation in the Baghdad railway loan in 1902 and 1903. Rouvier, then Minister for Finances, favoured French participation even though the project would be dominated by German capital. Delcassé insisted throughout on absolute equality of French and German participation in both costs and benefits. Thobie had concluded from this difference that Rouvier was, as a businessman, more forward-looking and internationalist in his approach, whereas Delcassé was rather narrow-minded and nationalist:

Rouvier thought that networks of bonds woven amid international companies could create the beginnings of a capitalist solidarity useful for international relations ...

Delcassé, contrary to Rouvier, showed an extreme distruct towards any financial operation, indeed any enterprise, whose capital was not exclusively French.[7]

Delcassé's attitude towards Chinese investments suggests that he was not as rigid, chauvinistic, or exclusive as Thobie would have us believe. Indeed, even in the case of the Baghdad railway, Delcassé was not opposed to French participation but merely insisted on absolute real French equality. This was more than Hanotaux had demanded in the case of the Beijing–Hankou project, but was exactly what Bompard and Delcassé had sought and received in the case of the Yunnan Syndicate.

Delcassé's insistence on equality was partly due to political factors: certainly he did not want French capital employed to extend the influence of foreign powers, although that is exactly what Hanotaux had allowed to happen in the case of the Russo-Chinese Bank. In spite of this well-founded concern on Delcassé's part, Thobie is exaggerating in alleging that:

[Delcassé] was so convinced that political influence abroad was of necessity the product of the creation of economic and financial interests, that he refused all forms for economic and financial collaboration for fear of having to share the political advantages which might flow from it.[8]

In China at least, Delcassé saw a distinct political advantage, security, accruing from sharing the political benefits of a joint investment. The real reason, however, for Delcassé's insistence on equality was his concern for the industrial orders which flowed from such investments. Equality in investment and management of an enterprise would lead to equality in obtaining orders from metropolitan factories. There was the rub. French prices were, in general, the highest in Europe. Many officials in the Quai d'Orsay deplored the high levels of protection which sustained these high prices, but nevertheless consistently ensured that the placement of French capital, often at rather mediocre returns, was rewarded with industrial orders. Bompard and Delcassé were especially insistent on this point. Delcassé's careful monitoring of the proportions of orders from the Beijing–Hankou railway received by French and Belgian industry is evidence of this concern. Without equality in the management of joint enterprises abroad, and without guarantees to that effect, French industry could not be certain that it would receive the rewards commensurate with the French capital involved. The orders were crucial. That unlikely proto-Leninist and social imperialist, the Vicomte de Bezaure, stated the position succinctly when he asserted that:

in these last few years our industry has developed very quickly ... [But] French orders are not enough to assure work for the factories. It is therefore indispensible to look for outlets overseas to succour our national industry and to give bread to the workers.[9]

This concern with orders for French industry, the most tangible of all benefits of imperial endeavour, contributed mightily to the determination of policy. In a very real sense it was the potential industrial orders which decided Delcassé to rally to the policy of the open door in 1899. The integrity of the Chinese empire and adherence to the principles of open competition were the best guarantees of French participation in the benefits of creating China's modern industrial infrastructure.

Delcassé's restrained behaviour during the Boxer Rebellion is consistent with this attitude. Despite the cataclysmic nature of the events and the hysteria prevalent in Europe at the time, Delcassé's greatest concerns were to ensure that the integrity of the Chinese empire be maintained and to prevent any creation of exclusive spheres of influence. The two notable French initiatives undertaken in the wake of the Rebellion, the circularization in October 1900 of the six-point note affirming the continuity of Chinese sovereignty, and the dispatch of French troops to Shanghai when British troops were ordered there, successfully achieved these two aims respectively.

The policy of looking for maximum economic advantage throughout China, sometimes in open competition, sometimes in collaboration with other imperialist powers, survived the strains imposed by the Boxer Rebellion. The ability of the Paris market to absorb loans at relatively low interest rates and the very real commitment of the resources of France's diplomatic network both in East Asia and Europe to the cause of French capital, enabled France to compete effectively with other powers on this basis, despite the high costs of her industrial products. In China, at the very beginning of the twentieth century, French capitalism and the diplomats who fostered and tended its oriental branches already had assumed the roles of the contemporary world. By 1901, thoughts of territorial aggrandisement in East Asia were anachronistic. French imperial capitalism and diplomacy had moved beyond all that.

Notes

Notes to Chapter 1

1. There are three quite good studies of the diplomacy of the period which document French policy with some adequacy. They are Henri Cordier, *Histoire des relations de la Chine avec les Puissances occidentales* (Paris, F. Alcan, 1902), H. B. Morse, *The International Relations of the Chinese Empire* (London, Longmans, Green and Co., 1910–1918), and Pierre Renouvin, *La Question d'Extrême-Orient 1840–1940* (Paris, Hachette, 1946). None of them, however, could be described as either comprehensive or recent and none devotes much attention to economic issues. Michel Bruguière, 'Paul Doumer et la politique d'intervention en Chine (1889–1902)', *Revue d'histoire diplomatique* 77 (1963) is a fine study of one aspect of French policy in China at this time. John F. Cady, *The Roots of French Imperialism in Eastern Asia* (Ithaca, Cornell University Press, 1954), does not discuss the period after 1885 or policy in China in any detail.

2. Paul Boell, *Le Protectorat des missions catholique en Chine et la politique de la France en Extrême-Orient* (Paris, Institut scientifique de la libre pensée, 1899), p. 10.

3. Dép. pol. No. 13 de Nisard à Delcassé, 23 janvier 1899, MAE NS 309, 17–8.

4. Note d'Alphonse Favier, 19 décembre 1899, MAE NS 309, 161.

5. De Favier à Pichon, 19 avril 1900, MAE NS 310, 54–6.

6. Quoted in Wei Tsing-sing, *Le Saint-Siège, la France et la Chine sous le pontificat de Léon XIII, le project de l'établissement d'une nonciature à Pékin et l'affaire du Pei-t'ang, 1880–1886* (Schöneck-Beckenried, Nouvelle Revue de Science missionnaire, 1966), p. vii.

7. De Jules Ferry à Joseph Reinach, 10 août 1886, *Lettres de Jules Ferry 1846–1893* (Paris, Calmann-Lévy, 1914), p. 414; J. L. de Lanessan, *Les Missions et leur protectorat* (Paris, F. Alcan, 1907), p. 33; Boell (1899), see note 2 above, p. 67.

8. Note pour le Ministre, 17 octobre 1899, MAE NS 309, 134–7.

9. Dép. pol. No. 88 de Gérard à Hanotaux, 23 juin 1897, MAE NS 321, 88–98.

10. De Lanessan (1907), see note 7 above, p. 27.

11. Boell (1899), see note 2 above, p. 30.

12. De François à Hanotaux, 21 janvier, 13 septembre 1897, MAE NS 313, 137, 212.

13. De Dautremer à Berthelot, 25 avril 1896; de Dubail à Hanotaux, 25 août 1896; de Bezaure à Hanotaux, 25 novembre 1896; de Gérard à Hanotaux, 18 mars 1897, MAE NS 321, 6, 10–5, 20–7, 48–50.

14. De Pichon à Delcassé, 6 janvier 1900; Note pour le Ministre, 11 avril 1900, MAE NS 310, 2–14, 35–7.

15. Note pour le Ministre, 28 février 1899, MAE NS 309, 31–4.

16. De Pichon à Delcassé, 6 janvier 1900, MAE NS 310, 9.

17. Note pour le Ministre, 15 décembre 1897, MAE NS 313, 229.

18. Note pour le Ministre, 3 mai 1898, MAE NS 314, 75–6. See Chapter 6 below for a discussion of the motives for the railway concession.

19. Note pour le le Ministre, 23 mai 1899; de Pichon à Delcassé, 25 juillet 1899, MAE, NS 309, 71,103.

20. Note sur notre droit d'intervention en faveur des chrétiens en Chine, 15 avril 1900, MAE NS 310, 44.

276   NOTES TO PAGES 9–11

21. This historiography was largely prompted by the work of Robinson and Gallagher on the partition of Africa. See R. E. Robinson and J. Gallagher with A. Denny, *Africa and the Victorians: the Official Mind of Imperialism* (London, Macmillan, 1961), and R. E. Robinson and J. Gallagher, 'The Partition of Africa', *New Cambridge Modern History*, Vol. 2 (Cambridge, Cambridge University Press, 1962). The issue is also discussed in D. K. Fieldhouse, *Economics and Empire* (London, Weidenfeld and Nicolson, 1973), pp. 76–84, and in W. R. Louis (ed.), *Imperialism: the Robinson and Gallagher Controversy* (New York, New Viewpoints, 1976), pp. 1–56. In an important study, Jacques Thobie, *La France impériale 1880–1914* (Paris, Mégrelis, 1982), argues strongly that economic factors dominated French imperial motives and, hence, that it was in Paris that the important decisions were made.

22. On Southern Africa see D. M. Schreuder, *The Scramble for South Africa, 1877–1895* (Cambridge, Cambridge University of Press, 1980). On Australia, New Zealand, and the Pacific, see R. C. Thompson, *Australian Imperialism in the Pacific: the expansionist era, 1820–1920* (Melbourne, Melbourne University Press, 1980), and A. Ross, *New Zealand Aspirations in the Pacific in the nineteenth century* (Oxford, Clarendon Press, 1964). The issues are discussed more generally in Fieldhouse (1973), see note 21 above, pp. 340–62, 437–56.

23. On the role of French military officers in West Africa, see A. S. Kanya-Forstner, *The Conquest of the Western Sudan: a study in French military imperialism* (London, Cambridge University Press, 1969).

24. Ella S. Laffey, 'French Adventurers and Local Bandits in Tonkin: the Garnier Affair in its local context', *Journal of Southeast Asian Studies* 6, 1 (1975), 50–51.

25. J. Kim Munholland, 'Admiral Jauréguiberry and the French scramble for Tonkin', *French Historical Studies*, 11, 1 (1979), 81, 106–7.

26. See Chapter 8 below; also M. Bruguière, 'Le Chemin de fer du Yunnan. Paul Doumer, et la politique d'intervention française en Chine (1889–1902)', *Revue d'histoire diplomatique*, No. 77, 1963, pp. 273–8.

27. Henri Brunschwig, *French Colonialism, 1871–1914, Myths and Realities* (London, Pall Mall Press, 1964).

28. Brunschwig (1964), see note 27 above, p. 89.

29. Brunschwig (1964), see note 27 above, p. 61.

30. Cady (1954), see note 1 above, p. 17.

31. Jean Ganiage, *L'Expansion coloniale de la France* (Paris, Payot, 1968), p. 22.

32. Dieter Brötel, *Französischer Imperialismus in Vietnam* (Freiburg, Atlantis, 1971).

33. Robert Delavignette et Charles André Julien, *Les Constructeurs de la France d'outre-mer* (Paris, Corrêa, 1946).

34. D. V. Mackay, 'Colonialism in the French Geographical Movement', *Geographical Review*, 33, 2 (April 1943), 214–32, and Agnes Murphy, *The Ideology of French Imperialism, 1871–1881* (Washington, Catholic University of America, 1948).

35. John F. Laffey, 'Les Racines de l'impérialisme français en Extrême-Orient. A propos des thèses de J. F. Cady', *Revue d'histoire moderne et contemporaine*, 16 (1969), 282–283. See also John F. Laffey, 'French Imperialism and the Lyon Mission to China' (unpublished Ph.D. thesis, Cornell University, 1966).

36. W. C. Hartel, 'The French Colonial Party 1895–1905' (unpublished Ph.D. thesis, Ohio State University, 1962).

37. Fieldhouse (1973), see note 21 above, pp. 400–1.

38. For the expedition generally, see Milton Osborne, *River Road to China, the Mekong River Expedition 1866–1873* (London, Allen and Unwin, 1975); and

Francis Garnier, *Voyage d'exploration en Indo-Chine, effectué pendant les années 1866, 1867 et 1868*, 2 vols, (Paris, Hachette, 1873).

39. For a brief account with a British emphasis, see Warren B. Walsh, 'The Yunnan Myth', *Far Eastern Quarterly*, 2, 3 (May 1943), 272-85. Also Emile Rocher, *La Province chinoise du Yün-nan*, 2 vols, (Paris, E. Leroux, 1879).

40. An entertaining, highly enthusiastic account of the Garnier affair is given in Paul Gaffarel, *Les Explorations françaises depuis 1870* (Paris, Librairie générale de vulgarisation, 1882), pp. 22-43.

41. See Hsieh Pie-chih, 'Diplomacy of the Sino-French War' (unpublished Ph.D. thesis, University of Pennsylvania, 1968).

42. T. F. Power, *Jules Ferry and the Renaissance of French Imperialism* (New York, King's Crown Press, 1966), p. 190.

43. Power (1966), see note 42 above, p. 197.

44. *J.O.*, 10 décembre 1883, cited in Power (1966), see note 42 above, p. 170.

45. Power (1966), see note 42 above, p. 192. Ferry expressed this concern even more acutely in *Le Tonkin et la mère-patrie* (Paris, V. Havard, 1890), cited in Delavignette et Julien (1946), see note 33 above, p. 293.

46. Ferry (1914), see note 7 above, pp. 202, 189.

47. *J.O.*, 29 juillet 1885, cited Delavignette et Julien (1946), see note 33 above, p. 293.

48. Brunschwig (1964), see note 27 above, p. 73; Power (1966), see note 4 above, p. 193.

49. Delavignette et Julien (1946), see note 33 above, p. 265.

50. *J.O.*, 1 novembre 1883, cited Power (1946), see note 42 above, pp. 168-9.

51. *J.O.*, 19 juillet 1881, cited Brötel (1971), see note 32 above, pp. 78, 397.

52. Delavignette et Julien (1946), see note 33 above, p. 268.

53. Munholland (1979), see note 25 above, 107.

54. La Fortune française en Chine, Note, juin 1900, MAE NS 563, 65-87.

55. Patrick O'Brien and Caglar Keyder, *Economic Growth in Britain and France 1780-1914: Two Paths to the Twentieth Century* (London, George Allen and Unwin, 1978), p. 161. O'Brien and Keyder argue convincingly against the received wisdom that the French economy was retarded compared with that of Britain throughout the nineteenth century. The emphasis they place on French specialization in high value-added products, that is luxury goods, is consistent with this work's explanation of France's poor export performance in China.

56. H. Tyszynski, 'World Trade in Manufactured Commodities, 1899-1950', *The Manchester School of Economic and Social Studies*, No. 3, September 1951, in Paul Blairoch, 'La Place de la France sur les marchés internationaux', in Paul Levy-Leboyer (ed.), *La Position internationale de la France: aspects économiques et financiers, XIX-XX siècles* (Paris, Ecole des Hautes Etudes en Sciences Sociales, 1977).

57. Blairoch (1977), see note 56 above, p. 42.

58. Calculated from O.P. Austin, *Commercial China in 1899: area, population, production, telegraphs, transportation routes, foreign commerce, and commerce of the United States with China* (Washington, Government Printing Office, 1899), pp. 1749-50.

59. Imbault-Huart, Rapport commercial de Canton, 1889, AN FI2 7056.

60. De Spuller à Tirard, 25 septembre 1889, AN FI2 7056.

61. De Haas à Bompard, 15 août 1896, MAE NS 360, 7.

62. De Haas à Bompard, 1 novembre 1896, MAE NS 360, 172-3.

63. De Prince, Fermé et Kingsbourg à Delcassé, 21 octobre 1898, MAE NS 322, 165-6.

64. Dép. com. No. 9 de Claudel à Hanotaux, 19 août 1897; Dep. no. 11 d'Hanotaux à Claudel, 27 novembre 1897, MAE NS 561, 167-8, 200.

65. Quoted in Guillain à Delcassé, 27 mars 1899, MAE NS 562, 155–6.
66. Calculated from table in Laffey (1966), see note 35 above, p. 110.
67. Robert Y. Eng, *Economic Imperialism in China: Silk Production and Exports, 1861–1932* (Berkeley, Institute of East Asian Studies, University of California, 1986), pp. 26–7.
68. Commerce général de la France avec la Chine, Note, juin 1900, MAE NS 563, 63–4.
69. Eng (1986), see note 67 above, p. 31.
70. Calculated from Laffey (1966), see note 35 above, pp. 118–9.
71. So high were Messageries Maritimes' freight rates, that de Bezaure in Shanghai reported that even the most patriotic Frenchmen used British or German shipping companies. (De Bezaure à Delcassé, 15 décembre 1899, MAE NS 562, 91–4.)
72. Imbault–Huart, Rapport commercial de Canton, 1889, AN FI2 7056.
73. De Gaston Kahn (Guangzhou) à Ribot, 20 août 1892, AN FI2 7056.
74. Dép. com. No. 144 de Bezaure à Delcassé, 2 septembre 1898, MAE NS 562, 92.
75. De Dautremer à Ribot, 29 juillet 1892, AN FI2 7056.

Notes to Chapter 2

1. *Le Télégraphe*, 21 mai 1884, Extract in MAE, Chine, Negociations commerciales, IV (mars 1879–juin 1885), 93.
2. Article 7 read in full: 'With a view to developing under the most advantageous conditions the commercial and neighbourly links which it is the aim of this treaty to re-establish between France and China, the Government of the Republic will build roads in Tonkin and encourage railway construction there.
'When, for its part, China decides to build railways, it is understood that she will approach French industry, and the Government of the Republic will give her every possible facility to find in France the personnel she will need. It is also understood that this clause may not be considered to constitute an exclusive privilege in favour of France.' (In H. Cordier, *Histoire des relations de la Chine avec les puissances occidentales 1860–1900* (Paris, F. Alcan, 1902), t. 2, pp. 534–5.)
3. Des Membres délégués du Comité des Forges de France á de Freycinet, 8 février 1886. MAE Negociations commerciales, IV.
4. 'In order to specify the advantages promised to France in the last paragraph of Article 7 of the Treaty of 9 June 1885, it is understood that China will reserve for French industrialists, on equal terms, the construction of 1,000 kilometres of the Empire's railway system.' Annèxe No. 5 à la Dêpêche Commerciale de Pékin, No. 26, 1 avril 1886. MAE Négociations commerciales, V, avril–décembre 1886, 62.
5. Entretien du 28 mars entre M. Cogordan et le Vice-roi Li Hongzhang. Annèxe No. 8 à la Dépêche Commerciale de Pékin (1886), see note 4 above, 75.
6. *Le Matin*, 22 octobre 1886. Extract in Neg. com. V, 233.
7. Dép. com. s.n. de Frandin à Ferry, 15 janvier 1884, MAE, Correspondance commerciale, Tianjin, 2, 46–7.
8. Some of the enterprises forming the Syndicat de la Mission de l'Industrie française en Chine were Fives-Lille, Etablissements Cail, St Charmon, Chantiers de la Loire, Forges et Chantiers de la Méditerranée, Forges de St Nazaire, and Bing et cie.
9. *Hong Kong Daily Press*, 18 August 1892. Cutting in MAE, Chine, Affaires diverses commerciales, Boîte 329 (1892). Jean Marie Thévenet had been born in

1843, and had spent all his career until 1886 in State service, including three years (1878–81) as director of Public Works in Cochin-China. (AN F14 11507)
10. Dép. pol. No. 29 de Bezaure (Tianjin) à Ribot, 10 février 1892, located in MAE (1892), see note 9 above.
11. Dép. com. s.n. de Ristelhueber à Flourens, 27 janvier 1887, MAE, Affaires diverses commerciales, Boîte 328 (1887–9).
12. *Chinese Times*, n.d., Annèxe à la dépêche politique de Lemaire à Goblet, 28 mai 1888, MAE, Correspondance politique, t.72 (1888), 175–6.
13. Zhou Fu (1837–1921) was one of Li's assistants who subsequently became Governor-General at Guangzhou (1906–7). A. W. Hummel, *Eminent Chinese of the Ch'ing Period* (Washington, Government Printing Office, 1943), p. 471.
14. *Chinese Times*, 21 September 1889, Annèxe à la dépêche commerciale s.n. de Lemaire à Spuller, 1 octobre 1889, MAE, Correspondance commerciale, Pékin, t. 6, 447.
15. Dép. pol. No. 29 de Bezaure à Ribot, 10 février 1892, MAE, Affaires diverses commerciales, Boîte 329.
16. De Fliche à Flourens, 5 mars 1887, MAE, Affaires diverses commerciales, Boîte 328.
17. Dép. com. s.n. de Ristelhueber à Flourens, 27 janvier 1887, MAE, Affaires diverses commerciales, Boîte 328 (1887–9).
18. Dép. com. s.n. de Ristelhueber à Flourens, 18 octobre 1887, MAE, Affaires diverses commerciales, Boîte 328 (1887–9).
19. Dép com. s.n. de Ristelhueber a Flourens, see note 18 above.
20. 4,038 taels for 'translations', or 28,000 francs; 20,000 francs paid to Ristelhueber's mother-in-law, Mme Dannet, in Paris; and 5,000 taels or 35,000 francs commission for a turret erected at Weihaiwei. In Paul Boell, *Les Scandales du Quai d'Orsay* (Paris, Savine, 1893), p. 36. Boell at that time was the Republican candidate for the seat of Pont-1'Evèque, and was using his record as an opponent of corruption as part of his campaign. He had informed the Foreign Minister, Ribot, of his charges against Ristelhueber, who by then had been *Chargé d'affaires* at Beijing, in January 1892, and the latter admitted to the payment of 20,000 francs to Mme Dannet. He had been punished by being placed 'available by withdrawal from employment', although he still received an income from the Quai d'Orsay. Boell (1893), see above, pp. 37–8.
21. De Thévenet à Denfert-Rochereau, 8 avril 1887, cited in Boell (1893), see note 20 above, p. 27.
22. De Ristelhueber à Thévenet, 14 janvier 1889, cited in Boell (1893), see note 20 above, p. 35. Denfert's suicide and the collapse of the Comptoir occurred soon after.
23. Boell (1893), see note 19 above, p. 27. The Corps de Ponts et Chaussées was an official body responsible both for registering civil engineers and ensuring the maintenance of roads, bridges, and railways in France.
24. Dép. pol. No. 29 de Bezaure à Ribot, 10 février 1892, MAE, Affaires diverses commerciales, Boîte 329 (1892).
25. Dép. com. confidentielle de Raffray à Ribot, 8 décembre 1892, MAE, Affaires diverses commerciales, Boîte 329 (1892).
26. Dép. com. s.n. de Ristelhueber à Ribot, 18 août 1890, MAE, Correspondance commerciale, Tianjin, 2 (1877–1901), 115–8.
27. Dép. com. s.n. de Bezaure à Ribot, 20 décembre 1890, MAE, Correspondance commerciale, Tianjin, 125.
28. Tél. s.n. de Lemaire à Spuller, 21 avril 1889, MAE, Affaires diverses commerciales, Boîte 329 (1892). Lemaire was in Beijing from July 1887 to March 1894.

29. De Griffon à de Bezaire, 7 mars 1891, MAE, Affaires diverses commerciales, Boîte 328 (1891).

30. Dép. pol. No. 29 de Bezaure à Ribot, 10 février 1892, MAE, Affaires diverses commerciales, Boîte 329 (1892).

31. Dép. pol. No. 29 de Bezaure à Ribot, 10 février 1892, MAE, Affaires diverses commerciales, Boîte 329 (1892).

32. De Thévenet à Li Hongzhang, 14 février 1890, MAE, Affaires diverses commerciales, Boîte 329 (1890).

33. Dép. com. No. 2 de Raffray à Ribot, 21 juin 1892, MAE, Affaires diverses commerciales, Boîte 329 (1892).

34. De Li Hongzhang à Lemaire, 13 avril 1889, MAE, Affaires diverses commerciales, Boîte 328 (1887-9).

35. Dép. pol. s.n. de Lemaire à Spuller, 30 avril 1889, MAE, Affaires diverses commerciales, Boîte 328.

36. Thévenet, 'Note sur la composition du personnel français appelé à diriger l'arsenal de Port Arthur', Tianjin, 20 avril 1889, MAE, Affaires diverses commerciales, Boîte 328. Prosper Griquel had established a naval shipyard near Fuzhou in 1866 on the initiative of the Governor-General of Fujian and Zhejiang, Zuo Zongtang.

37. Du Sénateur Ministre de la Marine à Ribot, 9 mai 1890, MAE, Affaires diverses commerciales, Boîte 328 (1890).

38. De Thévenet à Li Hongzhang, 14 février 1890, MAE, Affaires diverses commerciales, Boîte 328.

39. Dép. com. s.n. de Ristelhueber à Ribot, 24 août 1890, MAE, Affaires diverses commerciales, Boîte 328.

40. Dép. com. s.n. de Ristelhueber à Ribot, 24 août 1890, MAE, Affaires diverses commerciales, Boîte 328.

41. De Cornulier à Besnard, 16 septembre 1890; dép. com. s.n. de Lemaire à Ribot, 4 octobre 1890, MAE, Affaires diverses commerciales, Boîte 328.

42. De Ribot à Rouvier, 8 novembre 1890, MAE, Correspondance politique, 76 (août-décembre 1890), 126.

43. Dép. pol. No. 40 de Ribot à Ristelhueber, 19 décembre 1890, MAE, Correspondance politique, 76 (août-décembre 1890), 198-9.

44. Note du 7 octobre 1891, MAE, Affaires diverses commerciales, Boîte 328 (1891).

45. Dép. com. No. 45 de Lemaire à Develle, 4 juillet 1893, MAE, Affaires diverses commerciales, Boîte 329 (juillet-décembre 1893).

46. Dép. com. No. 12 de Bezaure à Ribot, 18 juin 1891, MAE, Affaires diverses commerciales, Boîte 328 (1891). During this inspection of June 1891, Li also visited Qingdao and Yantai.

47. De Bezaure acidly commented: 'It is very annoying that our industrialists' representatives in China have not been able to exploit better a situation which was very good for them and which could hardly fail, after the completion of the superb works at Port Arthur, to open an important outlet for French industry.' Dép. com. No. 12 de Bezaure à Ribot. 18 juin 1891, MAE, Affaires diverses commerciales, Boîte 328 (1891). De Bezaure was highly critical of the Thévenet group, and late in 1890 expressed the opinion that as a result of its failings 'we now have to walk down the same path again'. Dép. com. s.n. de Bezaure à Ribot, 20 décembre 1890, MAE, Correspondance commerciale, Tianjin 2 (1877-1901), 125.

48. Dép. com. No. 13 de Bezaure à Ribot, 25 août 1891, MAE, Affaires diverses commerciales, Boîte 328 (1891).

49. Dép. pol. No. 29 de Bezaure à Ribot, 10 février 1892; dép. com. confidentielle de Raffray à Ribot, 8 décembre 1892, MAE, Affaires diverses commerciales, Boîte 329 (1892).

50. Dép. com. s.n. de Raffray à Ribot, 11 janvier 1893, MAE, Affaires diverses commerciales, Boîte 329 (1893).

51. Note pour M. Develle, Ministre des Affaires étrangères, au sujet des intérêts français en Extrême-Orient, 15 juin 1893; dép. pol. s.n. de Bezaure à Ribot, 1 avril 1892, MAE, Affaires diverses commerciales, Boîte 329 (janvier-juin 1893 and 1892 respectively). In June 1893 the Groupe français en Chine consisted of ten factories employing a total of 20,000 workers. Its members included the Forges et Chantiers de la Méditerranée, Société de Construction de Levallois-Perret, Forges de Châtillon et Commentry, Sauter, Harlé et cie, and Etablissements Eiffel.

52. De Lemaire à Ribot, 23 mars 1892, MAE, NS 446, 56-8. Lemaire described Taton's and Lavergne's attitude thus: 'With what jealousy and fear does its personnel at Tianjin see the few scraps the Viceroy throws from the table go to the rival organization ... In these conditions the most fatal discord reigns amongst our compatriots in Tianjin; it makes foreigners rejoice and amuses even the Viceroy, who takes a malignant pleasure in fuelling this jealousy which has now reached the point of hatred...'

53. Dép. com. No. 2 de Raffray à Ribot, 21 juin 1892, MAE, Affaires diverses commerciales, Boîte 329 (1892).

54. Dép. com. s.n. de Raffray à Ribot, 27 juin 1892, MAE, Affaires diverses commerciales, Boîte 329 (1892).

55. Boell (1893), see note 20 above, p. 20.

56. Tél. No. 6 de Ribot à Lemaire, 12 avril 1892, MAE, Affaires diverses commerciales, Boîte 329 (1892). Raffray himself analysed the relative prejudices of his two predecessors thus: 'It cannot be denied that the Thévenet syndicate had been favoured by M. Ristelhueber to the exclusion of all others, and the Griffon-Croissade group left, at this time, completely to the side. Under M. de Bezaure's management things seem to have changed, and representatives of the Thévenet syndicate constantly complained, rightly or wrongly, that the consulate was not as impartial as it should be. Accustomed as they were to being favoured, they were certainly bad judges of impartiality, and they may well have taken as partiality against them what was only justice to others.' Dép. com. s.n. de Raffray à Ribot, 17 septembre 1892, MAE, Affaires diverses commerciales Boîte 329 (1892).

57. Dép. com. No. 2 de Raffray à Ribot, 21 juin 1892, MAE, Affaires diverses commerciales, Boîte 329 (1892).

58. Dép. com. No. 2 de Raffray à Ribot, 21 juin 1892, MAE, Affaires diverses commerciales, Boîte 329 (1892).

59. Note pour M. Develle, Ministre des Affaires Etrangères, au sujet des intérêts français en Extrême-Orient, 15 juin 1893, MAE, Affaires diverses commerciales, Boîte 329 (janvier-juin 1893).

60. Boell (1893), see note 20 above, p. 11.

61. Note Pour M. Develle (1893), see note 59 above, MAE, Chine, Affaires diverses commerciales, Boîte 329 (janvier-juin 1893).

62. Boell (1893), see note 20 above, p. 61; France, Ministère des affaires étrangères, *Annuaire diplomatique et consulaire de la République Française* (Paris, Berger-Levrault, 1900).

63. Dép. pol. No. 68 de Lemaire à Ribot, 26 juillet 1892, MAE, Chine, Affaires diverses commerciales, Boîte 329 (1892). Two years later Griffon was still operating without using the services of the consulate and Raffray was only informed of his activities by other Europeans and Chinese. Dép. com. s.n. de Raffray à Casimir-Périer, 2 février 1894, MAE, Affaires diverses commerciales, Boîte 330 (1894).

64. Dép. com. No. 26 de Lemaire à Ribot, 20 juin 1892, MAE, Affaires diverses commerciales, Boîte 329 (1892).

65. *Hong Kong Daily Press*, 2 June 1892. Lemaire thought that the articles had been written by Gustav Detring, a German subject formerly employed by the Imperial Maritime Customs whom Li consulted and used frequently, and Luo Fengluo, an official in the Chinese Admiralty. Dép. pol. No. 68 de Lemaire à Ribot, 26 juillet 1892, MAE, Affaires diverses commerciales, Boîte 329 (1892).

66. Dép. com. s.n. de Raffray à Develle, 10 juillet 1893; dép. com. No. 44 de Lemaire à Develle, 28 juin 1893, MAE, Affaires diverses commerciales, Boîte 329 (janvier-juin 1893).

67. De Raffray à Lemaire, 17 avril 1893, MAE, Affaires diverses commerciales, Boîte 329 (janvier-juin 1893).

68. Note pour M. Develle, 15 juin 1893, MAE, Affaires diverses commerciales, Boîte 329 (janvier-juin 1893).

69. Boell (1893), see note 20 above, p. 60.

70. De Lemaire à Ribot, 6 mai 1890, MAE NS 446, 5-7.

71. Dép. com. No. 10 de Bezaure à Ribot, 10 mai 1891, MAE NS 458, 22-3.

72. De Bezaure à Ristelhueber, 20 juillet 1891, MAE NS 458, 48-9.

73. De Bezaure à Ristelhueber, 20 juillet 1891, MAE, NS 458, 50.

74. De Gaun, Administrateur délégué des Establissements Eiffel, à Clavery, Directeur des affaires commerciales, 20 juillet 1891, MAE NS 458, 32.

75. Dép. com. s.n. de Ristelhueber à Ribot, 12 août 1891, MAE NS 458, 40-1.

76. De Li Hongzhang à Ristelhueber, 20 août 1891, MAE NS 446.

77. De Ribot à Lemaire, 24 octobre 1891, MAE NS 446, 37.

78. De Lemaire à Ribot, 12 janvier 1892, MAE NS 446, 40.

79. Dép. com. No. 2 de Raffray à Ribot, 21 juin 1892, MAE, Affaires diverses commerciales, Boîte 329 (1892).

80. Dép. pol. No. 68 de Lemaire à Ribot, 26 juillet 1892, MAE, Affaires diverses commerciales, Boîte 329 (1892).

81. Dép. com. s.n. de Raffray à Develle, 7 septembre 1893, MAE, Affaires diverses commerciales, Boîte 329 (juillet-décembre 1893).

82. Dép. com. s.n. de Raffray à Develle, 6 octobre 1893, MAE, Affaires diverses commerciales, Boîte 329 (juillet-décembre 1893).

83. Tél. No. 113 de Berthelot à Gérard, 9 décembre 1895, MAE NS 458, 131.

84. Dép. pol. No. 211 de Gérard à Berthelot, 28 décembre 1895, MAE NS 451, 65-8.

85. His scepticism was due to a belief that Hu Yufen's anglophile tendencies would be reinforced by Kinder and by his co-director, Wu Yunpang, who was also a comprador at Tianjin for the Hongkong and Shanghai Banking Corporation. Dép. com. No. 2 de Gérard à Berthelot, 6 janvier 1896, MAE NS 458, 70-1.

86. Dép. com. No. 2 de Gérard à Berthelot, 6 janvier 1896, MAE NS 458, 70-1, Express No. 660 issued by the *Tientsin Press*, 28 December 1895, 'Imperial Railways of North China: Lu-Kou-Chiao Extension — Contracts', MAE NS 458, 104-6.

87. Dép. com. No. 2 de Gérard à Berthelot, 6 janvier 1896, MAE NS 458, 72-3.

88. At this time Griffon represented the following companies: Société des Forges et Chantiers de la Méditerranée; Artillerie Carnet; Compagnie française des métaux; Société de Commentry-Fouchambault; Société centrale du dynamite; Société nouvelle des Etablissements Decauville ainé; Société des aciéries de France; Société des Etablissements Achel; Saulter, Harlé et cie.; and some other smaller, more specialized concerns. In note, s.d. (about Nov. 1896), MAE NS 411, 78-9. The fluctuating composition of the syndicate Griffon represented indicates the lack of willingness of many French companies to support the cost of maintaining a permanent representative in China, and obviously made Griffon's task much more difficult.

89. Dép. com. No. 11 de Gérard à Berthelot, 18 mars 1896, MAE NS 458, 82. Ironically Churchward's cousin, who occupied a similar position on the Great Western Railway in England, soon after began importing French locomotives into England and subsequently acknowledged a considerable debt to French practice in his own designs. However uncompetitive French industry may have been in providing the cheap and robust equipment required in North China, there was nothing absolute about its technical inferiority.

90. Dép com. No. 11 de Gérard à Berthelot, 18 mars 1896, MAE NS 458, 82.

91. Dép. com. No. 20 de Gérard à Hanotaux, 7 mai 1896, MAE NS 458, 98–100.

92. Dép. com. No. 20 de Gérard à Hanotaux, 7 mai 1896, MAE NS 458, 101.

93. D'Hanotaux à Reille, 2 juillet 1896, MAE NS 458, 107–8.

94. Dép. com. No. 74 de Gérard à Hanotaux, 21 novembre 1896, MAE, Correspondance commerciale, Pékin, 8, 84–5.

95. Dép. com. No. 59 de Gérard à Hanotaux, 21 octobre 1896, MAE NS 411, 70–1.

96. Dép. com. No. 8 du Chalyard à Hanotaux, 19 décembre 1896, MAE NS 411, 80–1.

97. Dép. com. No. 8 du Chalyard à Hanotaux, 19 décembre 1896, MAE NS 411, 81.

98. *Chinese Times*, 21 September 1889. Annèxe à la dép. com. s.n. de Lemaire à Spuller, 1 octobre 1889, MAE, Correspondance commerciale, Pékin, 6, 447–8.

99. *Chinese Times*, 21 September 1889, see note 98 above.

100. See A. Murphy, *The Ideology of French Imperialism* (Washington, Catholic University of America, 1948).

101. *Hong Kong Daily Press*, 18 August 1892, MAE, Chine, Affaires diverses commerciales, Boîte 329 (1892).

Notes to Chapter 3

1. Many modern businesses were established in the late Qing period on a basis of 'official supervision and merchant management' (*guandu shangban*). Such enterprises retained the bureaucratic administration of the Confucian state while allowing Chinese capitalists to use Western innovations. See A. Feuerwerker, *China's Early Industrialization: Sheng Hsüan-huai and Mandarin Enterprise* (New York, Atheneum, 1970), pp. 8–12.

2. Feuerwerker (1970), pp. 13–14.

3. A. W. Hummel, *Eminent Chinese of the Ch'ing Period (1644–1912)* (Washington, US Government Printing Office, 1943), p. 29.

4. Tél. s.n. de Lemaire à Spuller, 3 août 1889, MAE Chine C. P., M. Lemaire, t. 73 (juillet–septembre 1889), 82.

5. Tél. confidentiel No. 17 de Spuller à Lemaire, 5 septembre 1889, MAE Chine C. P., M. Lemaire, t. 73 (juillet–septembre 1889), 111–2.

6. Dép. com. s.n. de Lemaire à Spuller, 30 septembre 1889, MAE Chine C. P., t. 73, 130–7.

7. *Chinese Times*, 23 November 1889, annèxe à la dép. com. s.n. de Lemaire à Spuller, 29 novembre 1889, MAE, Correspondance commerciale, Pékin, t. 6 (1886–9), 462–3.

8. Hummel (1943), see note 3 above, p. 29.

9. Hummel (1943), see note 3 above.

10. *North China Herald*, 23 October 1896, 693.

11. De Gérard à Hanotaux, MAE NS 446, 90–4.

12. De Gérard à Hanotaux, MAE NS 446, 90–4.

13. Dép. com. No. 24 de Gérard à Hanotaux, 28 mai 1896, MAE NS 465, 17–8.

14. Dép. com. No. 24 de Gérard à Hanotaux, 28 mai 1896, MAE NS 465.

15. For the details of these attempts see A. A. Fauvel, 'Le Transsinien et les chemins de fer chinois', *Revue politique et parlementaire*, 21 (1899), 463–5.

16. Dép. com. No. 30 de Gérard à Hanotaux, 29 juin 1896, MAE NS 465, 20–9.

17. Dép. com. No. 24 de Gérard à Hanotaux, 28 mai 1896, MAE NS 465, 16.

18. For Sheng's memorial see Dép. com. No. 73 de Gérard à Hanotaux, 19 décembre 1896. MAE NS 465, 84–5; Also P. H. Kent, *Railway Enterprise in China* (London, Edward Arnold, 1907), p. 98.

19. Dép. com. No. 54 de Gérard à Hanotaux, 21 octobre 1896, MAE NS 465, 47–8; Feuerwerker (1970), see note 1 above, p. 67.

20. Dép. com. No. 59 de Gérard à Hanotaux, 5 novembre 1896, MAE NS 465, 57–8.

21. Dép. com. No. 67 de Gérard à Hanotaux, 23 novembre 1896, MAE NS 465, 65–9.

22. Dép. com. No. 67 de Gérard à Hanotaux, 23 novembre 1896, MAE NS 465, 69–70.

23. Tél. No. 82 de Gérard à Hanotaux, 13 décembre 1896; tél. No. 104 d'Hanotaux à Gérard, 26 décembre 1896, MAE NS 465, 77, 106.

24. Note du Ministre, 30 décembre 1896, MAE NS 465, 116.

25. Dép. No. 6 d'Hanotaux à Gérard, 23 janvier 1897, MAE NS 465, 141.

26. De Pinget à Hanotaux, 5 janvier 1897, MAE NS 465, 119. Jean Léonce Frédéric Baron Hély d'Oissel (1833–1920) came from the imperial aristocracy and was a director of many companies.

27. On the connections of the Gouin family see W. C. Hartel, 'The French Colonial Party 1895–1905' (unpublished Ph.D. thesis, University of Michigan, Ann Arbor, 1962), pp. 175–6.

28. Dép. com. No. 3 de Gérard à Hanotaux, 1 février 1897, MAE NS 465, 44–6.

29. Dép. No. 24 de Bezaure à Hanotaux, 18 février 1897, MAE NS 465, 148.

30. Dép. No. 16 de Bezaure à Hanotaux, 19 janvier 1897, MAE NS 465, 135–7.

31. Dép. No. 32 de Bezaure à Hanotaux, 19 mars 1897; du Président de la Société française d'explorations minières en Chine à Bompard, 16 mars 1897, MAE NS 465, 194–6, 176–7.

32. Tél. No. 29 d'Hanotaux à Gérard, 25 mars 1897, MAE NS 465, 205.

33. Note confidentielle, 30 mars 1897, MF F30 372 2/A.

34. The capital of the Société d'études was later increased to 330,000 francs. Note, s.d., MF F30 372 2/A, *Quinzaine coloniale*, 1, 7 (10 avril 1897), 220.

35. Baeyens à de Frondeville, 17 mars 1897, MAE NS 465, 179–80.

36. Tél. No. 9 d'Hanotaux à de Montholon (Bruxelles) 17 mars 1897; Dép. No. 62 de Montholon à Hanotaux, 9 avril 1897, MAE NS 465, 178, 215.

37. Tél. No. 43 d'Hanotaux à Gérard, 29 avril 1897, MAE NS 466, 8.

38. Dép. com. Nos. 34 et 45 de Gérard à Hanotaux, 12, 26 mai 1897, MAE NS 466, 9–10, 23–6.

39. De Claude à Hanotaux, 28 mai 1897, MAE NS 466, 35.

40. Dép. No. 45 de Bezaure à Hanotaux, 3 juin 1897; dép. com. No. 49 de Gérard à Hanotaux, 24 juin 1897; dép. No. 52 de Bezaure à Hanotaux, 30 juin 1897, MAE NS 466, 59–62, 94–5.

41. Macdonald to Salisbury, 25 May 1897, Salisbury to Macdonald, 9 June 1897, Memorandum requesting Railway Concessions in China, September 1898, FO17 1362, 107–8.

42. *The Times*, 3 June 1897.

43. See H. Enselme, *A travers la Mandchourie* (Paris, J. Rueff, 1903).

44. The issue of Leopold's motives is difficult. Certainly profit was the main motive in all his imperialistic activities, but he was also anxious to obtain new territories to exploit. According to Emile Francqui, Leopold considered that the Beijing–Hankou concession would help him acquire the provinces of Hubei and Henan when the Chinese empire was partitioned. See J. Stengers, 'King Leopold's Imperialism', E. R. J. Owen and R. B. Sutcliffe (eds.), *Studies in the Theory of Imperialism* (London, Longman, 1972), p. 257.

45. Tél. No. 16 de Montholon à Hanotaux, 15 juin 1897, MAE NS 466.

46. Dép. com. No. 117 de Montholon à Hanotaux, 17 juin 1897, MAE NS 466, 78–9.

47. Dép. com. No. 119 de Montholon à Hanotaux, 22 juin 1897, MAE NS 466, 83.

48. Dép. com. No. 49 de Gérard à Hanotaux, 24 juin 1897, MAE NS 466, 97.

49. On one occasion de Bezaure congratulated Sheng on exploiting the naïvety of the Belgians when in fact Sheng had agreed to what they were willing to accept. Dép. No. 54 de Bezaure à Hanotaux, 15 juillet 1897, MAE NS 466, 124.

50. Dép. No. 58 de Bezaure à Hanotaux, 28 juillet 1897, MAE NS 466, 132–4.

51. Dép. com. No. 1 de Gérard à Hanotaux, 9 janvier 1897, MAE NS 465, 129.

52. Téls. de Dautremer (Hankou) à Hanotaux, 11, 20 novembre 1897, MAE NS 466, 198, 209.

53. De Montholon à Hanotaux, 10 septembre 1897, MAE NS 466, 160.

54. A. Gérard, *La Vie d'un diplomate sous la Troisième République* (Paris, Plon Nourrit et cie, 1928), p. 300.

55. Gérard (1928), see note 54 above, p. 301. Germain and Duval were old, both born in 1824.

56. Note de la Direction commerciale, 23 septembre 1897, MAE NS 466, 164.

57. Note de la Direction commerciale, 23 septembre 1897, MAE NS 466, 165.

58. W. C. Hartel (1962), see note 27 above, p. 176; Gérard's argument is contained in his 'Note sur l'entreprise du chemin de fer de Hankow à Pékin et sur le contract conclu entre le directeur Sheng et la Société Belge, 27 septembre 1897', MAE NS 466, 170–2.

59. MAE NS 466, 170–2.

60. *North China Herald*, 10 September 1897, 501.

61. *Quinzaine coloniale*, 2, 18 (25 septembre 1897), 184.

62. Gérard (1928), see note 54 above, pp. 311–12.

63. De Gérard à Hanotaux, 25 octobre 1897, MAE NS 466, 185–6.

64. De Gérard à Hanotaux, 25 octobre 1897, MAE NS 466, 185–6. Hanotaux wrote *oui* in the margin beside Gérard's proposal.

65. Tél. No. 29 de Dubail à Hanotaux, 15 novembre 1897, MAE NS 466, 200.

66. Tél. de Dautremer (Hankou) à Hanotaux, 11 novembre 1897; dép. No. 85 de Claudel (Shanghai) à Hanotaux, 17 novembre 1897, MAE NS 466, 199, 206.

67. Tél. No. 89 d'Hanotaux à Dubail, 29 octobre 1897, MAE NS 466, 190.

68. Tél. de Vinck à Favereau, 23 novembre 1897, MAE NS 466, 219. The presence of a Belgian diplomatic telegram in French archives indicates the extent of co-operation at this time.

69. Tél. No. 36 de Dubail à Hanotaux, 24 novembre 1897, MAE NS 466, 222.

70. Personal letter from d'Anethon to Hanotaux, 2 December 1897, MAE NS 466, 232–3.

71. On 17 December 1897 Baeyens (with Stoclet's approval) wrote to de Frondeville as follows: 'To sum up, it is a question of inquiring under what conditions the business is *feasible*... The conditions we have settled on would be wired to Beijing and they would be left with a short time to respond with *yes* or *no*.

By bringing certain influences to play, it is reasonable to hope that *firm* proposals will be accepted there; but they must be firm. You can no longer ask for changes which will be followed by others and which will never end... In short, it is a question of bringing the business to a close; you think about it, you talk about it to your friends and we'll talk about it on Saturday.' MAE NS 466, 236-7.

72. D'Eugène Gouin à Cochery, 15 février 1898, MF F30 372 2/B.

73. Minutes of a meeting of bankers attended by MM. Stern, Baüer, and Villars (all of the Banque de Paris et des Pays-Bas), Rostand and Ullmann (of the Comptoir d'escompte), Dorizon (Société générale), Stoclet and de Frondeville (Belgian Société générale), 22 January, 1898, MF F30 372 2/A.

74. See note 73 above.

75. The conditions are stated in more detail in a letter from Gouin to the French Finance Ministry, 15 February 1898, MAE NS 467, 22-4.

76. V. Stoclet, Note, 13 février 1898. This note contains the text of Stoclet's telegram to Hubert of 8 February 1898, MAE NS 467, 11-12.

77. Tél. No. 30 d'Hanotaux à Dubail, 14 février 1898, MAE NS 467, 16.

78. Dép. com. No. 99 de Bezaure à Hanotaux, 21 février 1898, MAE NS 467, 26.

79. Dép. com. No. 99 de Bezaure à Hanotaux, 21 février 1898, MAE NS 467, 28-31.

80. Dép. com. No. 100 de Bezaure à Hanotaux, 28 février 1898, MAE NS 467, 43.

81. Tél. de Bezaure à Hanotaux, 1 avril 1898; téls. Nos 97, 104 de Dubail à Hanotaux, 31 mars, 6 avril 1898, MAE NS 467, 61, 60, 65.

82. Tél. de Stoclet à Hubert, 4 avril 1898, MAE NS 467, 81.

83. Dép. com. No. 107 de Bezaure à Hanotaux, 11 avril 1898, MAE NS 467, 70-8.

84. Tél. de Stoclet à Hubert, 26 mars 1898, MAE NS 467, 59.

85. It was signed on 14 April. L. K. Young, *British Policy in China 1895-1902* (Oxford, Clarendon Press, 1970), p. 81.

86. Dép. com. No. 108 de Bezaure à Hanotaux, 25 avril 1898, MAE NS 467, 86-91.

87. Tél. No. 117 de Pichon à Hanotaux, 1 mai 1898, MAE NS 467, 104.

88. Tél. No. 90 d'Hanotaux à Pichon, 5 mai 1898, MAE NS 467, 111.

89. Dép. com. No. 3, tél. No. 125 de Pichon à Hanotaux, 9, 20 mai 1898, MAE NS 467, 113-5, 143.

90. Note du 4 mai 1898, MAE NS 467, 110.

91. De Bezaure à Hanotaux, 23 mai 1898, MAE NS 446, 190.

92. Tél. de Vinck à de Favereau, 19 mai 1898, MAE NS 467, 148.

93. Tél. No. 10 de Gérard à Hanotaux, 20 mai 1898, MAE NS 467, 144.

94. From Gérard's account of his presentation of credentials to Leopold, both men spoke of little other than the Beijing–Hankou project. Dép. com. No. 1 de Gérard à Hanotaux, 8 mars 1898, MAE NS 467, 48-9.

95. Tél. No. 128 de Pichon à Hanotaux, 25 mai 1898, MAE NS 467, 156.

96. Tél. No. 13 d'Hanotaux à Gérard, 23 mai 1898, MAE NS 467, 151.

97. Tél. No. 17 d'Hanotaux à Gérard, 2 juin 1898, MAE NS 467, 168.

98. Dép. com. No. 10 de Pichon à Delcassé, 4 juillet 1898, MAE NS 468, 29.

99. D'Hubert à Stoclet, 11 juin 1898, MAE NS 467, 171.

100. Tél. No. 142 de Pichon à Hanotaux, 19 juin 1898, MAE NS 467, 177.

101. Tél. No. 118 d'Hanotaux à Pichon, 17 juin 1898, MAE NS 467, 175.

102. Tél. No. 119 d'Hanotaux à Pichon, 20 juin 1898, MAE NS 467, 178.

103. Tél. No. 152 de Pichon à Delcassé, 1 juillet 1898, MAE NS 468, 1.

104. Tél. de Francqui à Favereau, 23 juin 1898, MAE NS 467, 186.

105. Dép. com. No. 120 de Bezaure à Hanotaux, 27 juin 1898, MAE NS 467, 191–2.
106. The French text of the two contracts is reprinted in its entirety as Appendix B in Kent (1907), see note 18 above, pp. 224–34.
107. *Le Messager de Paris*, 2 juillet 1898. The file of press reaction is located at MF F30 372 2/C.
108. *The Times*, 31 May 1898.
109. *The Times*, 1 June 1898. See Young (1970), see note 85 above, p. 82.
110. Fauvel (1899), see note 15 above, 470–1.
111. Tél. No. 179 de Pichon à Delcassé, 11 août 1898, MAE NS 468, 62.
112. Dép. com. No. 128 de Bezaure à Delcassé, 4 août 1898, MAE NS 468, 56–8.
113. F. Bertie, Note, 17 August 1898, FO17 1362, 197–203.
114. Seymour to Admiralty, 22 August 1898, FO17 1362, 296.
115. Tel. No. 273 from Macdonald to Salisbury 28 August 1898. See also Fauvel (1899), see note 15 above, 471–2, and Kent (1907), see note 18 above, pp. 99–101. The railways conceded to British companies were (1) Shanghai to Nanjing, (2) Shanghai to Fuzhou and Hangzhou, (3) Guangzhou to Kowloon, and (4) from mines conceded to the Peking Syndicate in Hunan to the Yangzi.
116. Dép. com. No. 26 de Pichon à Delcassé, 12 août 1898, MAE NS 468, 70.
117. *Les Débats*, 16 août 1898; *le Rentier*, 27 août 1898, MF F30 372 2/C.

Notes to Chapter 4

1. The Belgian directors were T. Devolder, L. Barbanson, both directors of the Société générale, Chevalier R. de Baüer, manager of the Brussels branch of the Banque de Paris et des Pays-Bas, Carlier of the Société la Métallurgique of Brussels, A. Greiner, a director of John Cockerill of Seraing, and Pochez, treasurer of the Congo Free State. The French directors were Jules Gouin, Duval, L. St Paul de Sinçay, and Frédéric Mallet, all of Paris, Moyeaux, a director of la Société Baume et Marpent de St Pierre and de Schryver of Raismes. MF F30 372 1/A.
2. MF F30 372 1/A. See also Note s. d. du Ministère des Finances, MF F30 372 2/A.
3. *Economiste Français*, 17 septembre 1898.
4. *Le Temps*, 17 avril 1899, MF F30 374.
5. *Le Rentier*, 7 avril 1899, MF F30 372 2/C.
6. *Le Temps*, 10 avril 1899, MF F30 372 2/C.
7. Comptoir national d'escompte à Peytral (Ministre des finances), 20 avril 1899, MF F30 372 2/D.
8. *Le Messager de Paris*, 24 avril 1899, MF F30 372 2/C.
9. De Roblot (Premier adjoint de la Compagnie des Agents de change de Paris) à Caillaux, 31 juillet 1899; de Caillaux à Delcassé, 28 mars 1902, MF F30 372 2/D.
10. Tél. No. 162, dép. com. No. 15 de Pichon à Delcassé, 25, 30 juillet 1898, MAE NS 468, 43, 48–9.
11. Delcassé's marginal notes, MAE NS 468, 43.
12. Dép. com. No. 134 de Gérard à Delcassé, 8 novembre 1898, MAE NS 468, 162.
13. Dép. com. No. 173 de Bezaure à Delcassé, 26 mars 1899, MAE NS 469, 39–41. Also see A. Feuerwerker, *China's Early Industrialization: Sheng Hsüan-huai (1844–1916) and Mandarin Enterprise* (New York, Atheneum, 1970), p. 68.
14. D'Hély d'Oissel à Bompard, 5 mai 1899, MAE NS 469, 68.

15. The contract was first published in the Shanghai *China Gazette*, 16 November 1898. Dép. com. No. 157 de Bezaure à Delcassé, 18 novembre 1898, MAE NS 514, 39–47.

16. Count Cassini, the Russian Ambassador in Washington, even claimed that the concession had been offered for sale to Russia. Dép. com. No. 70 de Thiébaut (*Chargé d'affaires* à Washington) à Delcassé, 16 décembre 1898, MAE NS 514, 67–8.
Despite the lack of capital confronting the syndicate, Senator Brice was publicly extremely confident. In an interview published in the *New York Herald* on 2 October 1898 he described its aspirations in these terms: 'We shall reach all the important points between Hankou and Hong Kong. We are planning branch roads. If our main line does not run through a large city we shall build a branch to it. We shall probably cover the territory completely. If there is enough business to warrant competition, though, perhaps I'll build the competitive road.'
Perhaps, however, Brice's comments should not be taken too seriously. In the same interview he discussed protection: 'Whether the protective system was ever a good one will always be a disputed point. That the law was unjust there is no doubt. Every tax law is.' And on the Philippines he said: 'We never thought of getting here, but now we're here we're going to stay.' Brice died two months later.

17. *New York Herald*, 5 December 1898.

18. Dép. com. Nos. 12, 15 de Pichon à Delcassé, 12, 30 juillet 1898, MAE NS 468, 33–4, 47–8.

19. The draft telegram, dated 9 August 1898, is located at MAE NS 468, 60.

20. Tél. No. 195 de Delcassé à Pichon, 9 septembre 1898, MAE NS 514, 27.

21. Dép. com. No. 77 de Gérard à Delcassé, 1 juin 1899, MAE NS 514, 116–7.

22. Dép. com. No. 87 de Delcassé à Pichon, 17 juillet 1899, MAE NS 514, 133.

23. Dép. com. No. 50 de Jules Cambon (Washington) à Delcassé, 14 juillet 1899, MAE NS 514, 131–2.

24. Dép. com. No. 86 de Delcassé à Gérard, 7 août 1899, MAE NS 514 140–1.

25. Dép. com. No. 50 de Pichon à Delcassé, 11 septembre 1899, MAE NS 514, 155 For the rapid change in Pichon's views, see pp. 211, 223–4, and 226–7.

26. Dép. com. Nos. 101, 104 de Gérard à Delcassé, 11, 16 août 1899, MAE NS 514, 142–3, 147–8.

27. Dép. com. No. 114 de Gérard à Delcassé, 29 août 1899, MAE NS 514, 153–4.

28. Dép. com. No. 137 de Gérard à Delcassé, 10 octobre 1899, MAE NS 514, 166.

29. Dép. com. No. 140 de Gérard à Delcassé, 17 octobre 1899, MAE NS 514, 170–3.

30. Dép. com. No. 57 de Pichon à Delcassé, 22 octobre 1899, MAE NS 514, 186.

31. The first of these agreements bound China not to cede any part of the Yangzi valley to any other power. The second involved mutual British and Russian recognition of their respective spheres of influence on the Yangzi and in Manchuria. See p. 219.

32. Dép. com. No. 140 de Gérard à Delcassé, 17 octobre 1899, MAE NS 514, 174.

33. D'Hély d'Oissel à Bompard, 20 octobre 1899, MAE NS 514, 179.

34. Dép. com. No. 23 de Pichon à Delcassé, 9 avril 1900, MAE NS 514, 209.

35. Dép. com. No. 81 de Gérard à Delcassé, 1 mai 1900, MAE NS 470, 51–2.

36. Dép. com. No. 101 de Delcassé à Gérard, 11 août 1900, MAE NS 514, 224–5.

37. Dép. com. No. 107 de Delcassé à Pichon, 30 novembre 1900, MAE NS 514, 255–60.

38. Dép. com. No. 172 de Gérard à Delcassé, 25 octobre 1901, MAE NS 515, 17–19.

39. Lettre particulière de Bompard à Gérard, 18 novembre 1901, MAE NS 515, 32.

40. De Gouin à Cochery, 15 février, 1898, MF F30 372 2/B.

41. The Chinese Imperial Railways paid the American China Development Company US$6,750,000 by way of indemnification to cancel the concession. See W. R. Braisted, 'The United States and the American China Development Company', *Far Eastern Quarterly*, 11, 2 (February 1952), 147–65.

42. Tél. No. 212 de Pichon à Delcassé, 10 septembre 1898; dép. com. No. 110 de Gérard à Delcassé, 10 septembre 1898; Note, 28 septembre 1898, MAE NS 468, 122, 125, 139.

43. *L'Etoile Belge*, 12 janvier 1901, MAE NS 472, 10.

44. Note, 13 mars 1899, MAE NS 469, 18.

45. Dép. No. 39 de Delcassé à Pichon, 24 mars 1899, MAE NS 469, 31.

46. Dép. com. No. 41 de Gérard à Delcassé, 21 mars 1899, MAE NS 469, 22.

47. Dép. com. No. 52 de Delcassé à Pichon, 22 avril 1899, MAE NS 469, 64.

48. De Bompard à Hély d'Oissel, 19 avril 1899, MAE NS 469, 53.

49. Dép. com. No. 63 de Gérard à Delcassé, 26 avril 1898, MAE NS 469, 65–6.

50. Dép. com. No. 28 de Pichon à Delcassé, 21 avril 1899, MAE NS 469, 62.

51. Dép. com. No. 73 de Delcassé à Gérard, 21 juin 1899, MAE NS 469, 93–4.

52. Dép. com. Nos. 88, 92 de Gérard à Delcassé, 26 juin, 13 juillet 1899, MAE NS 469, 95–7, 99.

53. Dép. com. No. 97 de Gérard à Delcassé, 25 juillet 1899, MAE NS 469, 100–2.

54. Dép. com. No. 56 de Pichon à Delcassé, 22 octobre 1899, MAE NS 469, 132–3. See also *Revue Indo-Chinoise*, 2ème série, 98 (3 septembre 1900), 850–1.

55. De Bompard à Villars, 24 juillet 1899, MAE NS 411, 174.

56. De Villars à Bompard, 15 septembre 1899, MAE NS 411, 176–7.

57. M. Boissonnas, quoted in a letter from Pichon (then Foreign Minister) to the Minister of Finance, 12 July 1909, MF F30 372 1.

58. By comparison, in 1897 the most profitable French railway, the Paris–Lyon–Méditerranée, had a co-efficient of 0.448, the Belgian State Railway 0.599, all British railways 0.560, all Indian broad gauge railways 0.472, and the most profitable railway in North Africa, the Algiers-Oran 0.649. H. Bouillard, 'Note sur l'état de l'avancement du chemin de fer de Pékin à Hankow, novembre 1899', MAE NS 469, 225–9.

59. Dép. No. 65 de Delcassé à Pichon, 1 mai 1900, MAE NS 470, 50.

60. De Sheng à Stoclet, 2 mars 1900, MAE NS 470, 75.

61. De Delcassé à Hély d'Oissel, 9 mai 1900, MAE NS 470, 60.

62. De Stoclet à Sheng, 26 mai 1900, MAE NS 470, 70–3.

63. Dép. No. 20 de Dautremer à Delcassé, 22 octobre 1899, MAE NS 469, 130–1. Dautremer's allegations were subsequently denied by Pichon (tél. No. 117 de Pichon à Delcassé, 24 octobre 1899, MAE NS 469, 136). He was subsequently moved to the less sensitive new post at Nanning (dép. com. No. 56 de Pichon à Delcassé, 22 octobre 1899, MAE NS 134–9). Partly as a result of his involvement in a violent dispute with Francqui, the Belgian Consul, over some of their nationals' land speculation activities in Hankou, and partly as a result of his friendship with the dismissed Petit, Dautremer was extraordinarily anti-Belgian. He made some quite wild accusations and ultimately both he and Francqui left the city in a measure of disgrace. Although trivial in itself, the land speculation row and the polemics that flowed to Paris and Brussels from both consuls greatly concerned Delcassé, de Favereau, Gérard, and Stoclet. Eventually their successors, the Belgian, Siffert, and the Frenchman, Kahn, were given specific orders not to

involve themselves too intimately in the railway's affairs. (Dép. com. No. 83 de Gérard à Delcassé, 9 mai 1900, MAE NS 470, 61-2.)

64. Dép. No. 118 de Delcassé à Pichon, 4 novembre 1898, MAE NS 469, 213.

65. De Delcassé à Jules Gouin, 31 mars 1900, MAE NS 470, 38-39.

66. De Jules Gouin à Delcassé, 21 avril 1900, MAE NS 470, 46.

67. Dép. com. No. 24 de Gérard à Delcassé, 19 février1900, MAE NS 470, 28.

68. The exact figures were 1,769,086.35 francs for March and 1,995,785.1 for April. By comparison Belgian orders for the two months were 840,946.40 francs and 652,232.1 francs respectively. Dép. com. Nos. 42, 92 de Gérard à Delcassé, 13 avril, 18 mai 1900, MAE NS 470, 45, 63.

69. Relatively small orders for specialized items had also gone to some German, American, and British firms. Dép. com. No. 114 de Gérard à Delcassé, 17 juin 1900, MAE NS 470, 110.

70. De Delcassé à Jules Gouin, 3 novembre 1900, MAE NS 471, 92-3.

71. Dép. com. No. 209 de Gérard à Delcassé, 4 décembre 1900, MAE NS 471, 127.

72. De Guillain à Simon (président de la Banque de l'Indo-Chine), 5 août 1909, MF F30 372 1. Guillain listed the 12 French companies which had received orders for the railway. They were the Compagnie des Fives-Lille pour constructions mécaniques et entreprises, the Compagnie Française de Matériel de Chemins de fer, MM. Desouches, David et cie., MM. Schneider et cie., the Société Alsacienne de Constructions Mécaniques, the Société Franco-Belge pour la construction de Machines et de Matériels de Chemins de fer, the Société de Travaux Dyle et Bacalan, the Société de Construction des Batignolles, the Société Française de Constructions Mécaniques (ancien Etablissements Cail), the Société Lorraine des Anciens Etablissements de Dietrich et cie de Lunéville, and the Société Nouvelle des Etablissements de l'Horme et de la Buire.

73. Tél. de Jadot à Stoclet, 24 juillet 1900 quoted in Dép. com. No. 142 de Gérard à Delcassé, 29 juillet 1900, MAE NS 470, 162-8.

74. Lettre particulière de Gérard à Delcassé, 13 août 1900, MAE NS 470, 5-6.

75. Note de M. Saint René Taillandier pour la Direction des Consultats et des Affaires Commerciales, 21 août 1900, MAE NS 471, 19; d'Hély d'Oissel à Delcassé, 17 janvier 1900, MAE NS 470, 148.

76. Tél. de Jadot à Stoclet, 16 septembre 1900, tél. de Marcilly à Delcassé, 9 octobre 1900, MAE NS 471, 40, 56.

77. Tél. de Jadot à Stoclet, 30 mai 1900, MAE NS 470, 89.

78. Tél. de Jadot à Stoclet, 12 juin 1900, MAE NS 470, 107.

79. Dép. com. No. 116 de Maurice Borel (Bruxelles) à Delcassé, 12 septembre 1900, MAE NS 471, 34-5.

80. Tél. de Delcassé à Pichon via Fuzhou, 15 septembre 1900, MAE NS 471, 50.

81. I. C. Y. Hsü, *The Rise of Modern China* (New York, Oxford University Press, 1975, second edition), p. 486.

82. Dép. com. No. 35 de Pichon à Delcassé, 1 octobre 1900, MAE NS 471, 50-4.

83. De Bouillard à Stoclet, 26 octobre 1900, MAE NS 472, 110-13.

84. D'Hély d'Oissel à Delcassé, 13 août 1900; de Delcassé à de Lanessan, 17 août 1900, MAE NS 471, 2, 7.

85. De Lanessan à Delcassé, 28 août 1900; de Delcassé à de Lanessan, 27 octobre 1900, MAE NS 471, 31, 86.

86. Dép. com. No. 1 de Pichon à Delcassé, 12 février 1901, MAE NS 472, 87-9.

87. Dép. No. 34 de Delcassé à Gérard, 19 mars 1901, MAE NS 472, 203.

88. Dép. com. No. 2 de Pichon à Delcassé, 9 avril 1901, MAE NS 472, 275.

89. Dép. com. No. 209 de Gérard à Delcassé, 4 décembre 1900, MAE NS 471, 124.

90. Dép. com. No. 222 de Gérard à Delcassé, 22 décembre 1900, MAE NS 471, 142–143.

91. Dép. No. 17 de Delcassé à Gérard, 1 février 1901, MAE NS 472, 51–2.

92. Dép. No. 17 de Delcassé à Gérard, 1 février 1901, MAE NS 472, 51–2; Tél. No. 7 de Delcassé à Gérard, 11 février 1901, MAE NS 472, 114 bis.

93. De Bompard à Gérard, 11 février 1901, MAE NS 472, 116–17.

94. A. Ruiz, Secrétaire du syndicat financier, à Delcassé, 14 mars 1901, MAE NS 472, 188.

95. Present at the meeting were E. Gouin, president, and J. H. Thors, general manager of the Banque de Paris et des Pays-Bas, François Hottinguer, Hély d'Oissel, vice-president of the Société génerale, A. Rostand, manager, and Ullman, deputy manager of the Comptoir national d'escompte and de Froudeville, a director of the Banque parisienne. (Procès verbal de la séance du Comité du syndicat emprunt chinois 5 per cent 1898, 23 mars 1901, MAE NS 472, 233–6.)

96. D'Hély d'Oissel à Bompard, 28 mars 1901, MAE NS 472, 223.

97. Dép. com. No. 77 de Gérard à Delcassé, 2 avril 1901; tél. de Stoclet à Jadot, 9 avril 1901, MAE NS 472, 250, 285.

98. E. Bousigues, *Chemin de fer de Pékin à Hankéou. Mission technique spéciale en Chine confiée en 1901 à M. Bousigues. Rapport de mission* (Villeneuve-st Georges, Union typographique, 1909), pp. 5–10, 82–5.

99. The loan was floated on 27 December 1905. See MF F30 372 2/D.

100. Bousigues (1909), see note 98 above, postscript, pp. 87–8.

101. Edouard de Laboulaye, *Les Chemins de fer de Chine* (Paris, Larose, 1911), pp. 106–7.

102. Dép. com. No. 110 de Bezaure à Hanotaux, 16 mai 1898, MAE NS 514, 7.

103. Dép. com. No. 110 de Bezaure à Hanotaux, 16 mai 1898, MAE NS 514, 13.

Notes to Chapter 5

1. F. H. H. King, *Money and Monetary Policy in China 1845–1895* (Cambridge, Harvard University Press, 1965), pp. 98–9; A. Feuerwerker, *China's Early Industrialisation* (New York, Atheneum, 1970), p. 226.

2. On the collapse of the Comptoire d'escompte, see M. Lévy-Leboyer, 'La Spécialisation des établissements bancaire', P. Léon and G. Carrier (eds.), *Histoire économique et sociale de la France* (Paris, Colin, 1976), t. 3, p. 460; and G. P. Palmade, *French Capitalism in the Nineteenth Century*, translated by G. M. Holmes (Newton Abbot, David and Charles, 1972), p. 183.

3. Consular Reports, Shanghai, du Consul-général de France à Shanghai au Ministre, 5 juin 1889, AN F12 7058.

4. Palmade estimates total French loans to Russia from 1888 to 1914 at about 10,000 million French francs, or about a quarter of total French foreign investment, Palmade (1972), see note 2 above, p. 196.

5. Consular Reports, Shanghai, du Consul-général de France à Shanghai au Ministre, 5 juin 1889, AN F12 7058.

6. Annèxe, Banque de l'Indo-Chine au Ministre des Finances, 9 décembre 1897, p. 4, MF F30 337 (sous-dossier 3).

7. *DDF*, série 1, t. 12, pp. 10–11.

8. A. Gérard, *Ma Mission en Chine (1893–1897)* (Paris, Plon Nourrit et cie, 1918), pp. 68–69.

9. In June 1897 Prince Ukhtomskii, President of the Russo-Chinese Bank and a director of the Chinese Eastern Railway told Gérard that it had been Witte and the Tsar, who was influenced by his Finance Minister, who had been responsible for organizing the triple intervention. Muraviev and Lobanov, claimed Ukhtomskii, were men of the old school and indifferent to East Asian affairs. Gérard considered Ukhtomskii's enthusiasm may have led him to exaggerate Witte's importance. De Gérard à Hanotaux, *DDF*, série 1, t. 13, p. 429 n.

10. *DDF*, série 1, t. 12, pp. 236-7.

11. Gérard (1918), see note 8 above, p. 247.

12. Dép. pol. No. 87 de Gérard à Hanotaux, 4 juin 1895, MAE NS 350, n. p.

13. Dép. pol. No. 87 de Gérard à Hanotaux, 4 juin 1895, MAE NS 350, n.p; Gérard (1918), see note 8 above, pp.70-2; de Jules Cambon (Washington) à Delcassé, 19 juin 1900, MAE NS 402, 45.

14. For Noetzlin's other connections with companies involved in East Asia see W. C. Hartel, 'The French Colonial Party 1895-1905' (unpublished Ph. D. thesis, Ohio State University, 1962), pp. 236-8.

15. De Noetzlin à Hanotaux, 9 octobre 1895, MAE NS 401, 1-3; Note pour le Ministre, Confidentielle, 18 juillet 1896, MF F30 337.

16. De Gérard à Berthelot, 3 février 1896, MAE NS 401, 10.

17. Dép. No. 149 d'Hanotaux à Gérard, 12 octobre 1895, *DDF*, série 1, t. 12, p. 235.

18. De Gérard à Berthelot, 5 décembre 1895, 3 février 1896, MF F30 337 (sous-dossier 1).

19. De Vauvineux (Chargé d'affaires à St Petersbourg) à Bourgeois, 17 avril 1896, MF F30 337 (sous-dossier 1).

20. Note pour le Ministre, Confidentielle, 18 juillet 1896, MF F30 337.

21. De Gérard à Hanotaux, 27 mai 1896, MF F30 337 (sous-dossier 1).

22. D'Hanotaux à Cochery, 1 juillet 1896, MAE NS 401, 29.

23. Dép. No. 71 de Gérard à Hanotaux, 27 mai 1896, MAE NS 401, 25-6.

24. D'Hanotaux à Doumer, urgent et confidential, 10 février 1896, MAE NS 350, 9.

25. Notes, Emprunt chinois, 15 février, 19 février 1896, MAE NS 350, 14, 19.

26. De Rothstein à Noetzlin (telegram from St Petersbourg to Paris), MAE NS 350, 24.

27. De Gérard à Hanotaux, 19 mai 1896, MAE NS 350, 51/8-9.

28. De Gérard à Hanotaux, 5 mai 1896, MAE NS 350, 51/1-2.

29. Dép. No. 38 de Montebello à Hanotaux, 12 mai 1896, MAE NS 350, 51/4.

30. Note du 11 juillet 1896, MAE NS 401, 31; d'Hanotaux à Cochery, 28 juillet 1896, MF F30 337 (sous-dossier 1).

31. De Nabel (directeur du personnel et de la comptabilité du Ministère des Travaux publics) à Hanotaux, 28 octobre 1896, MAE NS 401, 31.

32. De Pinget (secretary of the Comité des Forges) à Bompard (head of the Direction commerciale of the French Foreign Ministry), 18 janvier 1897, MAE NS 401, 89.

33. De la Direction de la Banque Russo-chinoise à Hivonnait, MAE NS 401, 66.

34. Dép. pol. No. 152 de Gérard à Hanotaux, 13 décembre 1896, MAE NS 401, 71-2.

35. De Gérard à Hanotaux, 6 octobre 1896, MF F30 337 (sous-dossier 1).

36. T. H. von Laue, *Sergei Witte and the Industrialization of Russia* (New York, Columbia University Press, 1963), pp. 151-2.

37. W. L. Langer, *The Diplomacy of Imperialism 1890-1902* (New York, Alfred A. Knopf, 1951, second edition), pp. 401-4.

38. On the Longzhou railway, see Chapter 6 below.

39. See S. Y. Witte, *The Memoires of Count Witte*, translated by A.

Yarmolinsky (London, Heinemann, 1921), pp. 88–93 for Witte's own account of the signing of the secret alliance.

40. De Gérard à Hanotaux, 28 mars 1897, *DDF*, série 1, t. 13, pp. 300–2.

41. V. Chirol, *The Far Eastern Question* (London, Macmillan, 1896), p. 69.

42. Stazzeff had told this to the French Consul in Tianjin, du Chalyard. Dép. pol. No. 57 de Gérard à Hanotaux, 9 mars 1897, NAE NS 401, 99.

43. Langer (1951), see note 37 above, pp. 408–9; de Gérard à Hanotaux, 8 mars 1897, MAE NS 401, 96.

44. On these matters the French archives are silent. This account is taken from Langer (1951), see note 37 above, pp. 409–10, which itself is based on B. A. Romanov, *Russia in Manchuria 1892–1906*, translated by S. W. Jones (Ann Arbor, J. W. Edwards, 1952).

45. De Gérard à Hanotaux, 5 janvier 1897, *DDF*, série 1, t. 13, pp. 103–4.

46. Dép. com. No. 1 de Gérard à Hanotaux, 9 janvier 1897, MAE NS 465, 129.

47. Dép. pol. No. 5 de Gérard à Hanotaux, 5 janvier 1897, MAE NS 401, 75–9.

48. De Gérard à Hanotaux, 17 janvier 1897, F30 337 (sous-dossier 1).

49. D'Hanotaux à Cochery, 24 mars 1897, F30 337 (sous-dossier 1).

50. De Gérard à Hanotaux, 8 mars 1897, MAE NS 401, 96–7.

51. Dép. pol. No. 25 de Bezaure (Shanghai) à Hanotaux, 15 mai 1897, MAE NS 401, 110–13.

52. De Gérard à Hanotaux, 5 juin 1897, *DDF*, série 1, t. 13, pp. 421–2.

53. Undated letter from the Ministry of Finance to the Ministry of Foreign Affairs, about October 1897, F30 337 (sous-dossier 1).

54. Note de M. Gérard, *DDF*, série 1, t. 14, pp. 89–90.

55. See p. 128.

56. Note pour le ministre sur la Banque russo-chinoise, la Banque franco-chinoise, l'Etablissement de la Banque de l'Indo-Chine en Chine, 15 juin 1898, p. 3, F30 337, I/1; Note pour le Ministre, 22 juillet 1898, MAE NS 401–50.

57. Annèxe à une lettre de la Banque de l'Indo-Chine à Cochery, 9 décembre 1897, F30 337 (sous-dossier 3).

58. Note pour le Ministre, 15 juin 1898, pp. 3–4, F30 337, I/1.

59. Note pour le Ministre, see note 58 above, p. 4; de Rothstein à Cochery, 16 avril 1898, F30 337 (sous-dossier 3).

60. Note pour le Ministre, 15 juin 1898, p. 5, F30 337, I/1.

61. Dép. No. 116 de Cochery à Hanotaux, 18 juin 1898; Note, s.d., MAE NS 401, 138–9.

62. Dép. No. 125 de Cochery á Hanotaux, 15 juillet 1898, MAE NS 401, 142.

63. Note de la Direction commerciale, 21 novembre 1899, MAE NS 401, 194–7.

64. Delcassé à Raffalovich, 21 novembre 1899, MAE NS 401, 192–3.

65. De Witte à Raffalovich, 20 novembre 1899, MAE NS 401, 201.

66. Dép. com. Nos 158, 191 de Verstraete à Delcassé, 25 juin, 1 décembre 1900, MAE NS 402, 50, 97; Note, 20 octobre 1899, F30 337, I/3.

67. De Noetzlin à Verstraete, 10 février 1901, MAE NS 472, 124–130.

68. De Verstraete à Bénac, Directeur du Mouvement général des fonds, Ministère des Finances, 4 juillet 1903, F30 337 (dossier 4).

69. Dép. com. No. 37 de Montebello à Delcassé, 26 février 1902, MAE NS 402, 181–2.

70. De Delcassé à Caillaux, 29 juin 1903, F30 337 (dossier 5).

71. De Delcassé à Caillaux, 9 décembre 1904, F30 337 (dossier 4).

72. De Boutiron (Chargé d'affaires à St Petersbourg) à Bourgeois, 5 mai 1906, F30 337 (dossier 7).

73. The bank applied to float 80,000 shares in Paris. The Foreign Minister, Bourgeois, willingly assented to the request in view of the bank's transformation

since Rothstein's death and the Russian defeat. De Bourgeois au Ministre des Finances, 29 mai 1906, F30 337 (dossier 7).

74. Dép. com. No. 147 de Gérard à Hanotaux, 25 novembre 1896, MAE NS 350, 70/16-17.

75. Dép. pol. No. 162 de Gérard à Hanotaux, 28 décembre 1896, MAE NS 350, 76-7.

76. Dép. com. No. 221 de Geoffray (Londres) à Hanotaux, 9 juillet 1896; dép. com. No. 160 de Courcel (Berlin) à Hanotaux, 21 septembre 1896, MAE NS 350, 60/1, 70/12.

77. Note de M. Bompard, 13 juillet 1897, MAE NS 350, 87.

78. Note pour M. Bompard, 29 juillet 1897, MAE NS 350, 87.

79. Dép. pol. No. 422 de Geoffray à Hanotaux, 20 septembre 1897; dép. pol. No. 12 de Dubail à Hanotaux, 29 septembre 1897; tél. No. 23 de Dubail à Hanotaux, 10 octobre 1897, MAE NS 351, 22-4, 47-8, 56.

80. Note de M. Bompard, 17 novembre 1897, MAE NS 351, 98.

81. Dép. confidentielle de Cochery à Hanotaux, 17 septembre 1897, MAE NS 351, 18.

82. Bompard à Delatour (directeur-général des fonds, Ministère des Finances), 1 décembre 1897, Delnormandie à Cochery, 3 décembre 1897, MAE NS 351, 109-10.

83. Note du 30 décembre 1897; Dép. pol. No. 3 de Geoffray à Hanotaux, 5 janvier 1898; tél. No. 2 de Geoffray à Hanotaux, 8 janvier 1898, MAE NS 351, 116, 123, 130; The Times, 5 January 1898.

84. The Times, 17 January 1898.

85. Tél. No. 10 d'Hanotaux à Dubail, 20 janvier 1898, MAE NS 351, 202.

86. MAE NS 352, 40.

87. Dép. pol. No. 11 de Montebello (St Petersbourg) à Hanotaux, 31 janvier 1898, MAE NS 351, 236-7.

88. Tél. No. 3 de Dubail à Hanotaux, 5 février 1898, MAE NS 351, 19.

89. Dép. pol. No. 53 de Bezaure à Hanotaux, 7 février 1898; dép. pol. No. 211 de Dubail à Hanotaux, 11 février 1898, MAE NS 352, 27, 38-9.

90. The bankers concerned were Delnormandie of the Banque de l'Indo-Chine and the Comptoir national d'escompte, Gouin of the Banque de Paris et des Pays-Bas, Hély d'Oissel, and a representative of the Crédit industriel, Note de M. Bompard, 18 février 1898; Note, emprunt chinois, 23 février 1898, MAE NS 352, 46-7, 53.

91. Dép. No. 43 de Cochery à Hanotaux, 25 mars 1898, MAE NS 352, 150.

92. Dép. No. 118 de Delcassé à Courcel, 10 octobre 1898, MAE NS 352, 204.

93. Dép. com. No. 23 de Gérard à Hanotaux, 13 avril 1897, MAE NS 401, 107-8.

94. Fives-Lille's price represented a cost per kilogram of 1.315 francs, compared with the German price of 1.13 francs. per kilo and the price of 1.75 francs per kilo which the French railway companies had paid for their most recent orders. P. Hivonnait, 'Compte rendu confidentiel des négociations entamées par la Banque Russo-chinoise auprès de l'administration du Chemin de fer de l'Est chinois pour obtenir des commandes en faveur de l'industrie française, adressé à M. Bompard de New York le 4 juin 1897', MAS NS 401, 117-22.

95. Dép. pol. No. 61 de Montebello á Hanotaux, 3 août 1897, MAE NS 401, 125. Witte further told de Montebello that the cost of the locomotives would be 35,250 roubles each, compared with 30,853 for German and 27,325 for American machines.

96. D'Hivonnait à Bompard, 31 juillet 1900, MAE NS 402, 72.

97. Dép. com. No. 21 de Dubail à Hanotaux, 27 septembre 1897, MAE NS 490, 1-2.

98. De Verstraete à Bompard, 30 novembre 1897, MAE NS 490, 6–7.
99. D'Hivonnait à Pokotilov, Compte-rendu sommaire d'une reconnaisance entre Chengting et T'aiyuen, 25 novembre 1897, MAE NS 490, 15–22.
100. Dép. com. No. 33 de Dùbail à Hanotaux, 15 décembre 1897, MAE NS 490, 12–13.
101. L. K. Young, *British Policy in China, 1895–1902* (Oxford, Clarendon Press, 1970), pp. 89–9.
102. De Pokotilov à Rothstein, 29 avril 1898, MAE NS 490, 76–80.
103. 'Memorandum of agreement made between D. D. Pokotilov and Angleo Luzzati on 26 April 1898 at Peking', MAE NS 490, 82–3.
104. Dép. com. No. 8 de Pichon à Hanotaux, 21 juin 1898, MAE NS 490, 61–3.
105. De Pokotilov à Pichon, 10 mai 1900, MAE NS 490, 125–8.
106. Réponse officielle du *Zongli Yamen* à Pichon, 17 mai 1900, MAE NS 490, 131–2.
107. Note sur la ligne de chemin de fer de Chengting a T'aiyuen, s.d. [about October 1902], MAE NS 490, 166–72; E. de Laboulaye, *Les Chemins de fer de Chine* (Paris, Larose, 1911), pp. 122–3.
108. Langer (1951), see note 37 above, pp. 411–2.

Notes to Chapter 6

1. *Le Temps*, 11 octobre 1892. Vietnamese resistance to French occupation is discussed in D. G. Marr, *Vietnamese Anticolonialism, 1885–1925* (Berkeley, University of California Press, 1971), pp. 72–5, and in J. K. Munholland, '"Collaboration Strategy" and the French Pacification of Tonkin, 1885–1887', *Historical Journal* 24, 3 (1981), 631–6, 647–8. On the Black Flags, see H. McAleavy, *Black Flags in Vietnam: the Story of a Chinese Intervention, the Tonkin War of 1884–1885* (London, Allen and Unwin, 1968).
2. *North China Herald*, 31 December 1897, 1177.
3. *Revue Indo-Chinoise*, 1, 6 (janvier 1894), 378; *le Temps*, 10 novembre 1893.
4. *Revue Indo-Chinoise*, 1, 10 (mai 1894), 18–20.
5. Léon Escande, *Etude sur la navigabilité du fleuve Rouge, voie de pénétration commerciale vers l'intérieur de la Chine* (Paris, Imprimérie nationale, 1894), pp. 11–13.
6. Escande (1894), see note 5 above, p. 19.
7. Escande (1894), see note 5 above, p. 5.
8. R. Lefèvre, *Les Chemins de fer de pénétration dans la Chine méridionale* (Paris, Thèse pour le doctorat, 1902), p. 62.
9. De Rocher à Spuller, 12 octobre 1889, AN, F12 7057, III.
10. De Lallemant-Dumoutier à Ribot, 9 février 1892, AN, F12 7057, III.
11. De Rocher à Ribot, 26 janvier 1893, AN, F12 7057, III.
12. De Rocher à Ribot, 23 janvier 1892, AN, F12 7057, III.
13. De Spuller à Tirard, 4 octobre 1889, AN, F12 7057, I.
14. Annèxe à dép. No. 18 de Bons d'Anty à Ribot, 5 février 1892, AN, F12 7057, I.
15. Annèxe à dép. No. 34 de Bons d'Anty à Develle, 15 février 1891, AN, F12 7057, I.
16. Annèxes aux dép. Nos. 33 et 44 de Bons d'Anty à Develle, 9 février, 16 octobre 1893, AN, F12 7057, I.
17. De Ribot à Biche, 3 février 1891; de Develle à Riche, 23 avril 1892; d'Hanotaux à Lourties, 23 octobre 1894, AN, F12 7057, I.
18. *North China Herald*, 22 December 1893, 1000–1.
19. *Quinzaine Coloniale*, 1, 6 (25 mars 1897), 176.

20. Marcel Monnier in *le Temps*, 22 février 1896.

21. Henri Cordier, *Histoire des Relations de la Chine avec les puissances occidentales 1860-1902* (Paris, F. Alcan, 1902), v. 3, pp. 106-35; W. L. Langer, *The Diplomacy of Imperialism* (New York, Alfred A. Knopf, 1951, second edition), pp. 43-5.

22. Cordier (1902), see note 21 above, v. 3, pp. 138-9.

23. E. Cavaglion quoted in Louis Pichon, *Un Voyage au Yunnan* (Paris, Plon Nourrit et cie, 1893), p. 266. Cavaglion argued that the French should follow the British example in Burma, and build a railway parallel to the Hong Ha.

24. L. K. Young, *British Policy in China, 1895-1902* (Oxford, Clarendon Press, 1970), p. 34.

25. Henry Norman, *The Peoples and Politics of the Far East* (New York, Scribner, 1895), p. 503.

26. Jules Ferry, *Le Tonkin et la Mère-patrie* (Paris, V. Havard, 1890).

27. Henri d'Orléans, *Around Tonkin and Siam*, translated by C. B. Pitman (London, Chapman and Hall, 1894), p. 425.

28. D'Orléans (1894), see note 27 above, pp. 20-2.

29. Joseph Chailley-Bert, *La Colonisation de l'Indo-Chine* (Paris, A. Colin, 1892).

30. Emile Rocher, *La Province chinoise du Yün-nan* (Paris, E. Leroux, 1897).

31. Pichon (1893), see note 23 above, pp. 191-2.

32. See W. B. Walsh, 'The Yunnan Myth', *Far Eastern Quarterly*, 2 (1943), 272-85.

33. Although a familiar practice in the Fifth Republic, this was the last time a Minister was appointed who was not either a Deputy or Senator during the history of the Third Republic. See A. A. Heggoy, *The African Policies of Gabriel Hanotaux 1894-1898* (Athens, University of Georgia Press, 1972), pp. 2-5.

34. MAE, Correspondance commerciale, Pékin, t. 7 (1890-6), 145-6; Cordier (1902), see note 21 above, v. 3. pp. 161-2.

35. Auguste Gérard, *Ma Mission en Chine (1894-1897)* (Paris, Plon Nourrit et cie, 1918), pp. 55-61; S. M. Meng, *The Tsungli Yamen: its organization and functions* (Cambridge, Harvard University Press, 1962), pp. 55-6.

36. The contract of 19 February 1895 was never implemented, as at the end of that year the new Resident-General, Rousseau, obtained a loan of some 80 million francs in Paris and resolved that the government would perform the work itself. Compagnie de Fives-Lille, Entreprise du Chemin de fer de Dong-Dang à Longzhou (Guangxi) Note, Paris, 15 juillet 1898, MAE NS 483, 28-29.

37. Gérard (1918), see note 35 above, pp. 61-3; Cordier (1902), see note 21 above, v. 3, pp. 162-9. The industrial provisions of the Commercial Convention of 20 June 1895 were contained in its fifth Article which read: 'It is understood that China, for the operation of its mines in the provinces of Yunnan, Guangxi and Guangdong, will first approach French industrialists and engineers, the operations remaining under the jurisdiction of regulations enacted by the Imperial government concerning national industry.

'It is agreed that railways either already existing or plannel in Annam [Vietnam] may, after mutual agreement and under conditions yet to be defined, be extended into Chinese territory.' Cordier (1902), see note 21 above, 168.

38. O'Conor to Salisbury, 20 August 1895; London Chamber of Commerce to Salisbury, 28 August 1895, FO17 1270, 42 349; Beauclerk to Salisbury, 14 November 1895, FO17, 1272, 33.

39. Tel. No. 99 from O'Conor to Salisbury, 22 August 1895, FO17 1270, 88; India Office to Foreign Office, 4 October 1895, FO17 1271, 185.

40. Tel. No. 79 from Salisbury to Macdonald, 11 November 1896; Tel. No. 102 from Macdonald to Salisbury, 7 December 1896, FO17 1296, 140-1, 280-2.

41. Gérard (1918), see note 35 above, pp. 66-7.

42. Gérard (1918), see note 35 above, pp. 81, 130.

43. French metallurgical companies received less than half the quantity of orders for railway equipment in 1895 compared with 1885. Le Comité des Forges, *La Sidurgie française, 1864–1914* (Paris, Comité des Forges, 1919), p. 201.

44. Gérard (1918) see note 35 above, pp. 81, 131.

45. Gérard (1918), see note 35 above, pp. 83, 130; de Berthelot à Gérard, 22 novembre 1895, cited MAE NS 481, 4.

46. Gérard (1918), see note 35 above, pp. 131-2; Dép. com. No. 4 de Gérard à Berthelot, 12 janvier 1896, MAE NS 481, 16.

47. De Gérard à Berthelot, 4 decembre 1895, MAE NS 481, 4-5.

48. Dép. pol. de Gérard à Bourgeois, 31 mars 1896, MAE NS 481, 60.

49. Tél. de Gérard à Berthelot, 20 février 1896, MAE NS 481, 28.

50. De Guieysse à Berthelot, 22 février 1896; de Guieysse à Bourgeois, 17 avril 1896, MAE NS 481, 38, 79-80.

51. See MAE NS 481, 42-4.

52. De Bourgeois à Guieysse, 23 avril 1896, MAE NS 481, 87.

53. De Lebon à Hanotaux, 9 mai 1896, MAE NS 481, 98.

54. D'Hanotaux à Lebon, 19 et 22 mai 1896, MAE NS 481, 101, 107-8.

55. Note, MAE NS 481, 109.

56. Note sur le type de voie à adopter pour le Réseau de Chemin de fer qui doit être construit dans les provinces méridionales de la Chine, 26 mai 1896, MAE NS 481, 110-16. The note is also interesting in that it reveals that Fives-Lille's management had no illusions as to the quantity of traffic that would use the new line.

57. De Gérard à Hanotaux, 28 mai 1896, MAE NS 481, 119.

58. De Gérard à Hanotaux, 9 juin 1896, MAE NS 481, 123.

59. Gérard (1918), see note 35 above, p. 134.

60. De Gérard à Hanotaux, 18 août 1896, MAE NS 481, 150-1.

61. François described Renault, who lived in Nanning, as the only priest in Guangxi who was preoccupied with French national interests as well as religious matters. Renault later commented about possible extension of the railway to Nanning and beyond: 'All that would be rightly to the profit of the English'. Dép. No. 10 de François à Hanotaux, 3 avril 1897, MAE NS 482, 2.

62. François wrote his criticisms in red ink on the margins of Fives-Lille's 'Instructions relatives a l'Entreprise du Chemin de fer de la Frontière de Chine à Longzhou (Guangxi)', 8 septembre 1896, MAE NS 481, 152-64.

63. Dép. No. 1 de François à Hanotaux, 26 décembre 1896, MAE NS 481, 191.

64. Dép. No. 2 d'Hanotaux à François, 24 avril 1897, MAE NS 482, 17.

65. Dép. pol. No. 13 de Gérard à Hanotaux, 25 janvier 1897, MAE NS 481, 197-200.

66. By inserting question marks on the margins of Dép. pol. No. 13 de Gérard à Hanotaux, 25 janvier 1897, MAE NS 481, 197-200.

67. Dép. pol. No. 5 de François à Hanotaux, 3 février 1897, MAE NS 481, 205-11. Three days later François put the military arguments against the railway in even stronger terms: 'If these works must be undertaken, it will be only to meet General Su's desire to have a railway completing his defences on the frontier, and the compensation for France's agreement to the opening of the Xi Jiang to the benefit of English merchants against Tonkin's trade will be reduced to the concession of an absolutely Chinese military job directed against this same Tonkin.' Dép. pol. No. 6 de François à Hanotaux, 6 février 1897, MAE NS 481, 216.

68. Dép. pol. No. 10 de François à Hanotaux, 3 avril 1897, MAE NS 482, 4-6.

69. D'Hanotaux à Lebon, 20 juillet 1897; de Lebon à Hanotaux, 11 septembre 1897, MAE NS 482, 70, 120.

70. Dép. pol. No. 16 de François à Hanotaux, 14 juin 1897, MAE NS 482, 49-52.

71. Dép. pol. No. 16 de François à Hanotaux, 14 juin 1897, MAE NS 482, 46-48.

72. Dép. com. No. 31 de Gérard à Hanotaux, 27 avril 1897, MAE NS 482, 21.

73. De Gérard à Hanotaux, 19 mai 1897, MAE NS 482, 33-4.

74. De Duval à Hanotaux, 27 septembre 1897, MAE NS 482, 127-128.

75. Tél. No. 80 d'Hanotaux à Dubail, 29 décembre 1897; dép. pol. No. 36 de François à Hanotaux, 6 octobre 1897, MAE NS 482, 130, 133.

76. G. P. Bertrand, *Dix Ans sur les frontières du sud de la Chine et du Tonkin de 1896 à 1906* (Mayenne, C. Colin, 1905), pp. 1-8; Dép. Nos. 6, 7, 8, 11, et 30 de François à Hanotaux, 6 février, 10 et 23 mars, 20 avril et 5 septembre 1897, MAE NS 481, 215-16, 225-8; NS 482, 14, 109.

77. Dép. pol. Nos. 20, 27 de François à Hanotaux, 5 juillet, 27 août 1897, MAE NS 482, 62-4, 87.

78. Dép. pol. No. 30 de François à Hanotaux, 5 septembre 1897, MAE NS 482, 109.

79. De Duval à Hanotaux, 16 janvier 1897, MAE NS 482, 194-5.

80. D'Hanotaux à Lebon, 30 janvier 1897, MAE NS 481, 202-3.

81. Dép. No. 14 d'Hanotaux à Gérard, 8 février 1897, MAE NS 481, 217.

82. De Lebon à Hanotaux, 6 mars 1897, MAE NS 481, 224. For the Lyon mission generally see J. F. Laffey, 'French Imperialism and the Lyon Mission to China' (Ph. D. thesis, Cornell University, 1966).

83. D'Hanotaux à Gérard, 21 mai 1897; d'Hanotaux á Lebon 25 mai 1897, MAE NS 482, 40, 41.

84. Note de M. Bompard, 24 juin 1897, MAE NS 482, 57.

85. De Lebon à Hanotaux, 5 juin 1897, MAE NS 494, 88-95.

86. Bompard, Note pour la Direction politique, 15 juin 1897, MAE NS 494, 110-12. Bompard's arguments are similar to those used by Curzon when he rejected the building of a British railway from Burma into Yunnan. See Chapter 9 above.

87. *North China Herald*, 18 June 1897, 1106.

88. Dép. com. No. 7 de Souhart à Flourens, 20 octobre 1887, MAE négociations commerciales, VI (1887-97), 27-8.

89. Dép. pol. Nos. 12, 18, 19 de Gérard à Hanotaux, 24 janvier, 4 février 1897, MAE NS 191, 29, 64, 67-71.

90. *Quinzaine coloniale*, 1, 10, 25 mai 1897, 320.

91. Dép. pol. No. 17 de Gérard à Hanotaux, 2 février 1897, MAE NS 191, 55.

92. Tél. No. 16 d'Hanotaux à Gérard, 5 février 1897; dép. pol. No. 51 de Gérard à Hanotaux, 31 mars 1897, MAE NS 191, 73, 164.

93. Question mark in the margin of tél. No. 13 de Gérard à Hanotaux, 7 février 1897; tél. No. 17 d'Hanotaux à Gérard, 9 février 1897, MAE NS 191, 76, 88.

94. Dép. pol. Nos. 43 et 51 de Gérard à Hanotaux, 15 mars, 31 mars 1897, MAE NS 191, 137-8, 163-5.

95. Annèxe No. 3 à la dép. pol. de Gérard à Hanotaux, 18 juin 1897, MAE NS 482, 56.

96. D'Hanotaux à Lebon, 30 juin 1897, MAE NS 494, 114.

97. De Lebon à Hanotaux, 24 mars 1898, MAE NS 482, 216. Later in the year, when the company was even more insistent, Lebon argued that national interests had to be determined before allowing Fives-Lille to increase its investment: 'Essentially it is a question of determining the basis of the future rail system in southern China before the extension of the Longzhou line. In any case there can be no question ... of favouring the immediate realization of the Company's projects, at least until the Guillemoto mission has finished its job and

we know the results.' Quoted in Note pour le Ministre au sujet du chemin de fer de Longzhou et son prolongement eventuel, 4 août 1898, MAE NS 483, 66–7.

98. De Lebon à Hanotaux, 6 octobre 1897, MAE NS 494, 136.

99. 'Programme des travaux de la mission d'étude des chemins de fer à établir dans la Chine méridionale' (Paris, 7 octobre 1897), MAE NS 494, 143–9. For the Lyon mission, see pp. 175–7.

100. Guillemoto, 'Mission d'études techniques en Chine, Division de la Mission' (Hanoï, 6 janvier 1897), MAE NS 495, 44.

101. Despite his hostility Su provided a guard for Wiart's group as instructed by the *Zongli Yamen*. (Dép. com. de Guillien à Hanotaux, 25 janvier 1898, MAE NS 482, 179–80.) The *Chargé d'affaires* in Beijing, Dubail, had obtained Chinese consent and protection for the mission by omitting to mention that it would be surveying possible railway routes and merely indicating that it would consist of engineers and hydrographers. (Dép. com. No. 5 de Dubail à Hanotaux, 1 août 1897, MAE NS 494, 121.)

102. Guillemoto. 'Mission d'études techniques en Chine. Division de la Mission', MAE NS 494, 45–6.

103. Dép. pol. No. 50 de Dubail à Hanotaux, 2 février 1898; tél. inexpédié d'Hanotaux à Pichon, 16 février 1898, MAE NS 193, 33, 50; L. K. Young (1970), see note 24 above, pp. 88–9.

104. De Lebon à Hanotaux, 23 avril 1898, MAE NS 449, 38.

105. Note de la Direction politique, 'Chemin de fer de Pakhoi', 4 mai 1898, MAE NS 449, 43.

106. Annèxes Nos 1 et 2 à la dép. pol. No. 18 de Pichon à Hanotaux, 7 juin 1898; dép. pol. No. 21 de Pichon à Hanotaux, 14 juin 1898, MAE NS 449, 58–60, 63.

107. *The Times*, quoted in *Journal des Débats*, 18 septembre 1898 cutting in AN, F30, 374; analyse de l'article du *Hong Kong Daily Press* du 30 juin 1898, pièce-jointe à la lettre No. 46 adressée à la Direction politique par le Consul de France à Hong Kong le 30 juin 1898, MAE NS 449, 81–2.

108. Note, s.d. (about 1 July 1898) MAE NS 449, 83.

109. *Hong Kong Daily Press*, 3 December 1889; Dép. com. No. 9 de Gauthier à Spuller, 15 décembre 1889; dép. pol. No. 4 de Pichon à Hanotaux, 7 mai 1898, MAE NS 449, 6, 48.

110. De Guillain à Delcassè, 16 novembre 1898; dép. pol. No. 7 de Pichon à Delcassé, 26 janvier 1899, MAE NS 449, 109, 131.

111. De Guillien à Dubail, 24 décembre 1897; dép. pol. Nos. 4 et 5 de Guillien à Hanotaux, 3 et 5 février 1898, MAE NS 482, 174, 184–5, 187–8.

112. De Krantz à Hanotaux, 10 février 1898, MAE NS 482, 190–3.

113. Dép. pol. No. 15 de Guillien à Hanotaux, 13 mai 1898, MAE NS 482, 252–4; Compagnie de Fives-Lille, 'Entreprise du chemin de fer de Dong-Dang à Longzhou (province du Guangxi), Note' (Paris, 15 juillet 1898, MAE NS 483, 39–44.

114. De Krantz à Delcassé, 26 août, 22 septembre 1898; Notes pour le Ministre, 30 août, 18 novembre 1898, MAE NS 483, 76, 86, 78, 98.

115. Tél. No. 260 de Pichon à Delcassé, 10 novembre 1898; dép. pol. No. 93 de Pichon à Delcassé, 19 novembre 1898, MAE NS 483, 94, 101.

116. *Texte de l'Echo des Mines et de la Metallurgie*, about 2 septembre 1898, 4141, cutting in MAE NS 483.

117. *Texte de l'Echo des Mines et de la Metallurgie* (1898), see note 116 above, 4143.

118. De Krantz à Delcassé, 23 novembre 1898, MAE NS 483, 107–8.

119. De Guillain à Delcassé, 5 janvier 1899, MAE NS 483, 114.

120. The arbitrators' estimate was 20,800,000 francs, whereas Vallière's had been 18,138,690 francs. Tribunal arbitral institué par le contrat d'arbitrage du 29

mars 1899 pour la fixation du prix forfaitaire d'établissement de la ligne, 'Sentence du Tribunal arbitral' (Hanoï, 8 mai 1898), MAE NS 483, 167.

121. Dép. pol. No. 120 de Pichon à Delcassé, 20 septembre 1899, MAE NS 482, 209-13.
122. De Thoulon à Pichon, 5, 19 mars 1900, MAE NS 483, 255, 261.
123. De Duval à Delcassé, 2 avril, 10 mai 1900, MAE NS 483, 264-7, NS 484, 7-8.
124. Dép. Nos. 24, 33 du Vaur à Pichon, 19 janvier, 23 mars 1901, MAE NS 484, 81-3, 103-4.
125. De Siegfried à Delcassé, 17 mai 1901, MAE NS 484, 119-20.
126. De Delcassé à Siegfried, 7 juilllet 1902, MAE NS 484, 201.
127. De Doumergue à Delcassé, 4 octobre 1904, MAE NS 484, 247-50.
128. De Guillain à Delcassé, 29 mai 1899, MAE NS 483, 175-6.
129. Marginal note to Guillain à Delcassé, (1899), see note 128 above.
130. Dép. pol. No. 104 de Pichon à Delcassé, 8 août 1899, MAE NS 483, 201-4.
131. Note pour le Directeur de la Direction politique, 16 octobre 1899, MAE NS 483, 232-3.
132. H. Develle, Inspecteur des Chemins de fer de l'Indo-Chine, in *La Politique Indo-Chinoise*, 13 mars 1909, cutting in MF, F30 374.
133. Wang Wen-yuan, *Les Relations entre l'Indo-Chine française et la China* (Paris, P. Bousset, 1937), p. 81.
134. Bertrand (1905), see note 76 above, pp. 15-16.
135. Dép. pol. No. 30 de François à Hanotaux, 5 septembre 1897, MAE NS 482, 110-11.
136. De Krantz à Hanotaux, 10 février 1898, MAE NS 482, 190.
137. Dép. pol. No. 30 de François à Hanotaux, 5 septembre 1897, MAE NS 482, 111.
138. Quoted in dép. pol. No. 36 de François à Hanotaux, 6 octobre 1897, MAE NS 482, 134.

Notes to Chapter 7

1. See p. 143.
2. Dép. com. No. 40 d'Imbault-Huart (Guangzhou) à Hanotaux, 17 novembre 1894, MAE, Affaires diverses commerciales, Boîte 330 (1894); Annual report for 1898 from Mengzi, AN, F12 7057. These figures in fact understate the growth in trade as the value of silver, and hence of the tael in which the trade was measured, was depreciating quite rapidly during this period.
3. The British Consul at Guangzhou, Byron Brennan, stated this in his annual report for 1893 cited in the *Hong Kong Daily Press*, 6 November 1894, annèxe à la dép. com. No. 40 d'Imbault-Huart à Hanotaux, 17 novembre 1894, MAE, Affaires diverses commerciales, Boîte 330 (1894).
4. De Dejean de la Bâtie (Mengzi) à Hanotaux, 15 avril 1898, MAE NS 495, 87-8. See also J. K. Munholland, '"Collaboration Strategy" and the French Pacification of Tonkin, 1885-1897', *Historical Journal* 24, 3 (1981), 643-7.
5. Of course French governments were concerned with the promotion of trade in China, as Hanotaux's commercial instructions to Gérard (see pp. 147-8) indicate, but generally they confined their activities to facilitating trade through diplomatic measures and ensuring that the infrastructure necessary for trade, notably port improvements and railways, was provided.
6. J. F. Laffey, 'French Imperialism and the Lyon Mission to China', (unpublished Ph. D. thesis, Cornell University, 1966), p. 260.

7. Laffey (1966), see note 6 above, pp. 254–6.

8. Ulysse Pila, 'Rapport, Mission d'exploration commerciale en Chine', Chambre de Commerce de Lyon, *Compte rendu, 1895*, pp. 297–304, cited Laffey (1966), see note 6 above, pp. 258–9.

9. Pila in Laffey (1966), see note 6 above, p. 362.

10. The modest and business-like nature of the Lyon mission's conclusions is constantly emphasized by Laffey in his account of the mission. His fine account of the mission, its antecedents and results, is very complete and it is not proposed to repeat his analysis in these pages.

11. Laffey (1966), see note 6 above, pp. 370–1.

12. Laffey (1966), see note 6 above, pp. 351, 407.

13. Laffey (1966), see note 6 above, pp. 408–11.

14. *Rapport sur la création d'une société commerciale au Tonkin et dans la Chine méridionale* (Lyon, A. H. Storck, 1897), p. 9.

15. See note 14 above, pp. 9–11.

16. See note 14 above, p. 12.

17. Dép. com. No. 16 de Dejean de la Bâtie à Delcassé, 1 août 1899, MAE NS 562, 187–8.

18. J. L. de Lanessan, *L'Expansion coloniale de la France* (Paris, F. Alcan, 1886), p. 576.

19. Situation commerciale de Pakhoi au 31 décembre 1890, de Gauthier à Ribot, 25 mars 1891, AN F12 7057, IV (Beihai, 1888–1903).

20. Dép. com. No. 3 de Rocher à Spuller, 27 mai 1889, MAE, Affaires diverses commerciales, Boîte 328 (1887–9).

21. Charles Saglio, 'Les Intérêts français dans la Chine méridionale', *La Chine Nouvelle*, 4 (15 november 1899), 552–3.

22. Dép. com. No. 7 de Dejean de la Bâtie à Hanotaux, 26 mars 1898, MAE NS 495, 57.

23. Dép. com. No. 151 de Paul Cambon à Delcassé, 22 novembre 1899, MAE NS 441, 151.

24. Raymond Lefèvre, *Les Chemins de fer de pénétration dans la Chine méridionale* (Paris, thèse pour le doctorat, 1902), pp. 80–3.

25. Lefèvre (1902), see note 24 above, p. 1.

26. Lefèvre (1902), see note 24 above, p. 57.

27. De Rocher (Mengzi) à de Freycinet, 24 juillet 1891, AN F12, 7057, III (Mengzi, 1889–1904).

28. De Clavery (Consul en mission) à Delcassé, 10 juin 1903, AN F12, 7057, V (Beijing, 1884–1906).

29. *London and China Telegraph*, 25 May 1903, AN F12, 7057, V (Beijing, 1884–1906).

30. De Sainson (Mengzi) à Delcassé, 29 octobre 1903, AN F12, 7057, III (Mengzi, 1889–1904).

31. See pp. 148–9.

32. On the background to the convention see J. D. Hargraves, 'Entente manquée; Anglo-French relations 1895–1896', *The Cambridge Historical Journal*, 11 (1953), 65–92; L. K. Young, *British Policy in China 1895–1902* (Oxford, Clarendon Press, 1970), pp. 31–3; and M. Bruguière, 'Le chemin de fer du Yunnan. Paul Doumer et la politique d'intervention française en Chine (1889–1902)', *Revue d'histoire diplomatique* 77, (1963), 45–6. The text of Article 4 of the convention is printed in *DDF*, série 1, t. 12, pp. 408–9: 'IV. The two Governments agree that all commercial and other privileges and advantages conceded in the two Chinese provinces of Yunnan and Sichuan either to France or Great Britain, in virtue of their respective Conventions with China of the 1st March, 1894, and the 20th June, 1895, and all privileges and advantages of any nature which may in the

future be conceded in these two Chinese provinces, either to France or Great Britain, shall, as far as rests with them, be extended and rendered common to both powers and to their nationals and dependents, and they engage to use their influence and good offices with the Chinese Government for this purpose.'

33. In France Berthelot's coup was dismissed as that of a chemist not a diplomat, while in England those two enthusiastic promoters of the Yunnan myth and of the Burma–Yunnan railway, Holt Hallet and Archibald Colquhoun, both derided the convention and insisted that Britain should retrieve the situation by railway construction. Holt Hallet, 'Recent Frontier Alterations on the Upper Mekong', annèxe à Dép. com. No. 48 de Surrel (Manchester) à Berthelot, 15 mars 1896; A. Colquhoun in *The Standard*, 28 May 1898, MAE NS 485, 7–8, 117.

34. Holt Hallet, 'France and Russia in China', *Nineteenth Century*, March 1897, 500–1.

35. 'Article 12 of the Agreement between Great Britain and China modifying the Convention of March 1, 1894 relative to Burmah and Thibet signed at Peking on February 4, 1897', MAE NS 485, 50. This agreement also bound China not to cede any futher territory along the Mekong to a third power and opened the lower Xi Jiang to foreign trade. See p. 150.

36. See pp. 150–1.

37. De Rocher à Hanotaux, 24 mars 1897, MAE NS 494, 41–2.

38. Note de M. Liébert, 12 mars 1897, MAE NS 494, 34.

39. Note de M. Liébert, 12 mars 1897, MAE NS 494, 29.

40. Voies d'accès à frayer du Tonkin en Chine en vue de l'établissement d'un réseau ferré dans les deux pays. Rapport de M. le commandant d'Amade, ancien attaché militaire de France en Chine, 29 avril 1897, MAE NS 446, 24–5. (The report is also located at MAE NS 494, 46–81.)

41. MAE NS 446, see note 40 above, 54.

42. MAE NS 446, see note 40 above, 55 *et seq*.

43. Henri d'Orléans, 'L'insurrection des Boxeurs', *Questions diplomatiques et coloniales*, 10 (15 juillet 1900), 83.

44. De Dautremer (Hankou) à Ribot, 29 juillet 1892, AN F12 7056, I (Hankou 1870–1906).

45. M. Dejean de la Bâtie, 'Note sur le commerce du Sichuan et sur les routes commerciales qui conduisent en cette province' (Fuzhou, 1 juillet 1893), MAE Correspondance commerciale V (Fuzhou) 4 (1893–1901), 269.

46. De Dautremer (Hankou) à Ribot, 29 juillet 1892, AN F12 7056, I (Hankou 1870–1906).

47. *Le Temps*, 22 février 1896, MAE NS 481.

48. Emile Rocher, *La Province chinoise du Yün-nan* (Paris, E. Leroux, 1879), Vol. 1, p. xi.

49. De Duval à Hanotaux, 16 janvier 1897, MAE NS 481, 194.

50. De Lebon à Hanotaux, 6 mars 1897, MAE NS 481, 224.

51. Bruguière (1963), see note 32 above, 133.

52. Tél. no. 67 d'Hanotaux à Gérard, 9 février 1897, MAE NS 191, 88.

53. Annèxe à dep. pol. de Gérard à Hanotaux, 18 juin 1897, MAE NS 482, 56.

54. De Gérard à Hanotaux, 26 février 1897, *DDF* série 1, t. 13, 234.

55. On the diplomacy of this period see H. Cordier, *Histoire des Relations de la Chine avec des puissances occidentales 1860–1902* (Paris, F. Alcan, 1902), vol. 3, pp. 349–9; H. B. Morse, *The International Relations of the Chinese Empire* (London, Longmans, Green and Co., 1918), vol. 3, pp. 106–21; and Young (1970), see note 32 above, pp. 69–73.

56. The Ministry of the Marine had considered the question of the best site for a coaling station on the south China coast as early as 1895, when Admiral de Beaumont had expressed a favourable opinion of Guangzhouwan. Beaumont

ordered Captain Boutet to survey the bay in the *Alger* in October and November 1896. Boutet was less than impressed and reported that most of the bay was shallow and that the channel was not accessible to ships drawing more than six metres. This would exclude many naval vessels from using the port, although most merchant vessels on the China coast could, but probably would not want to bother as there was practically no commercial activity there. (Note au Ministre au sujet de la baie de Guangzhouwan . . . , 28 mars 1898, MAE NS 193, 151–3.) Well aware that, because of its proximity to Haiphong, Guangzhouwan offered the navy only a marginal strategic benefit, the Minister of the Marine, Admiral Besnard, told Hanotaux that a coaling station was needed further north, and suggested Sansha Bay, some fifty kilometres north of Fuzhou. (De Besnard à Hanotaux, 2 avril 1898, MAE NS 194, 7–8.) By then, however, the Chinese had already agreed to the lease of Guangzhouwan; Sandu'ao, the main town on Sansha Bay had been declared a treaty port; and the Japanese were successfully pushing for a declaration of the inalienability of Fujian. (D'Hanotaux à Besnard, 7 avril 1898, MAE NS 194, 22.) Thus, first the French navy and then the government of Indo-China were stuck with Guangzhouwan, an acquisition of negligible strategic or commercial significance which nevertheless represented the only leased French territory in China.

57. Cordier (1902), see note 55 above, v. 3, p. 368; *The Times*, 26 February 1898, annèxe à la dép. pol. No. 113 de Courcel à Hanotaux, 26 février 1898, MAE NS 193, 55. The use of declarations of inalienability to preserve a power's position had been pioneered by France the previous year. In response to fears of British and German encroachment on the island of Hainan, which although part of the province of Guangdong, enclosed the Gulf of Tonkin, Gérard obtained a verbal declaration of the inalienability of the island to third powers in February 1897. Tél. de Gérard à Hanotaux, 14 février 1897, *DDF*, série 1, t. 13, p. 169 note.

58. Telegram from the *Zongli Yamen* to Jing Zhang, 4 April 1898, MAE NS 194, 16.

59. Tél. No. 43 d'Hanotaux à Dubail, 7 mars 1898, MAE NS 193, 62–3.

60. *Hong Kong Daily Press*, 30 March 1898, MAE NS 193, 167–8; *North China Herald*, 14 March 1898.

61. Tél. No. 49 de Geoffray à Hanotaux, 21 mars 1898; dép. pol. No. 325 d'Hambourg, 28 mars 1898, MAE NS 193, 124, 154.

62. Dép. pol. No. 78 de Dubail à Hanotaux, 11 avril 1898, MAE NS 194, 31.

63. MAE 194, see note 62 above, 30. In fact the next Postal Secretary appointed was a French citizen, Théophile Piry, who took up this position in 1901. When the post office was separated from the customs service in 1911, Piry became China's first Postmaster General. See Morse (1918), see note 55 above, pp. 65–71.

64. Dép. pol. No. 78 de Dubail à Hanotaux, 11 avril 1898, MAE NS 194, 28–30.

65. Tél. no. 69 d'Hanotaux à Dubail, 5 avril 1898, *DDF*, série 1, t. 14, p. 204.

66. He annotated 'Pas d'inconvenient. La distance est en notre faveur' on tél. No. 96 de Dubail á Hanotaux, 31 mars 1898, *DDF*, série 1, t. 14, 182.

67. Auriac, 'Aperçu politique et commerciale', Hanoï, 15 octobre 1896, MAE NS 228, 144–5.

68. Dép. pol. No. 1591 de Doumer à Lebon, 31 juillet 1897, MC IC Carton 32 B11 (27).

69. D'Hanotaux à Lebon, 4 octobre 1897, MC IC Carton 32 B11 (27).

70. D'Hanotaux à Lebon, 13 novembre 1897, MC IC Carton 32 B11 (27).

71. Indo-Chine, *Journal officiel*, 25 novembre 1897, MC IC Carton 32 B11 (27).

72. D'Hanotaux à Lebon, 25 november 1897, MC IC Carton 32 B11 (27).

73. Tél. de Doumer à Lebon, 27 novembre 1897, MC IC Carton 32 B11 (27).

74. De Dejean de la Bâtie à Dubail, 14 janvier 1898, MAE NS 495, 8–13.

75. De Pennequin à Dejean de la Bâtie, 9 janvier 1898, MAE NS 495, 14–15.

76. Dép. No. 262 d'Angoulvant à Dubail, 5 février 1898, MAE NS 229, 10–11.

77. Dép. No. 254 d'Angoulvant à Dubail, 31 janvier 1898, MAE NS 229, 6–7.
78. D'Hanotaux à Lebon, 10 décembre 1897, MC IC Carton 32 B11 (25).
79. Dép. com. No. 7 de Dejean de la Bâtie à Hanotaux, 26 mars 1897 (see also Guillemoto's report which was attached as an annex to this dispatch), MAE NS 495, 54–65.
80. De Dejean de la Bâtie à Pichon, 27 juin 1898, MAE NS 495, 101–3.
81. De Dejean de la Bâtie à Pichon, 4 août 1898, MAE NS 495, 105.
82. De Dejean de la Bâtie à Pichon, 18 août 1898, MAE NS 495, 106–7.
83. De Dejean de la Bâtie à Pichon, 28 mai 1898, MAE NS 495, 100.
84. MAE 495, see note 83 above, 101.
85. De Guillemoto à Dejean de la Bâtie, 15 septembre 1898, MAE NS 495, 147.
86. Dép. pol. No. 5, très confidentiel, de Dejean de la Bâtie à Delcassé, 14 octobre 1898, MAE NS 495, 137–9.
87. MAE 495, see note 86 above, 140.
88. MAE 495, see note 86 above, 141–3. Dejean de la Bâtie defended himself from such allegations by pointing out that he was himself of colonial origin, had four brothers working in Tonkin and had shown such hospitality to the various missions which had come from Tonkin that his consulate had become 'a vast caravanserai'. MAE 495, see note 86 above, 144.
89. Quinzaine coloniale, 25 août 1898, 488–90.
90. Marginal note to dép. pol. No. 5, très confidentiel, de Dejean de la Bâtie à Delcassé, 14 octobre 1898, MAE NS 495, 137.
91. Dép. pol. No. 5, confidentiel, de Delcassé à Dejean de la Bâtie, 30 décembre 1898, MAE NS 495, 151.
92. Instructions générales remises à M. Pichon, Ministre Plénipotentaire se rendant en Chine, 10 mars 1898, MAE NS 193, 73–6.
93. Tél. de Doumer à Guillain, 2 mars 1899, MC IC Carton 32 B11 (28).
94. Tél. de Doumer à Guillain, see note 93 above.
95. Tél. No. 52 de Guillain à Doumer, 4 mars 1899, MC IC Carton 32 B11 (28).
96. Tél. de Doumer à Guillain, 8 mars 1899, MC IC Carton 32 B11 (28).
97. Tél. de Doumer à Guillain, 8 mars 1899, MC IC Carton 32 B11 (28).
98. Tél. de Doumer à Guillain, 8 mars 1899, MC IC Carton 32 B11 (28).

Notes to Chapter 8

1. See M. Bruguière, 'Le Chemin de fer du Yunnan. Paul Doumer et la politique d'intervention française en Chine (1889–1902)', Revue d'histoire diplomatique 77 (1963), 144.
2. J.O., Ch. de D., 25 novembre 1898, 2445–6.
3. J.O., Ch. de D., 25 novembre 1898, 2449.
4. J.O., Ch. de D., 25 novembre 1898, 2451.
5. J.O., Ch. de D., 25 novembre 1898, 2442.
6. J.O., Ch. de D., 25 novembre 1898, 2450.
7. J.O., Ch. de D., 25 novembre 1898, 2452.
8. J.O., Ch. de D., 25 novembre 1898, 2454–5.
9. Bruguière (1963), see note 1 above, 147, 151; Note, chemin de fer de Laocai à Yunnansen, décembre 1900, MAE NS 496, 82–3.
10. De Pichon à Delcassé, 1 septembre 1899, annèxe à Delcassé à Decrais, 7 novembre 1899, MC IC Carton 32 B11 (17).
11. MC IC Carton 32 B11 (17), see note 10 above.
12. De Dejean de la Bâtie à Pichon, 9 avril 1899, MAE NS 229, 76. Pichon and Delcassé were so concerned with the possible effects of such behaviour that

the matter was raised at ministerial level in Paris. De Delcassé à Decrais, 21 juillet 1899, MC IC Carton 33 B11 (30).

13. Tél. de Doumer à Pichon, 25 janvier 1899; De Delcassé à Guillain, 6 avril 1899, MC IC Carton 32 B11 (28).

14. De Delcassé à Guillain, 14 avril 1899, MC IC Carton 32 B11 (28).

15. See pp. 198–9 for these telegrams.

16. De Delcassé à Guillain, 16 mai 1899, MC IC Carton 32 B11 (28).

17. Tél. de Guillain à Doumer, 20 mai 1899, MC IC Carton 32 B11 (28).

18. De Dejean de la Bâtie à Delcassé, 20 septembre 1899 (annèxe à Delcassé à Decrais, 16 novembre 1899), MC IC Carton 32 B11 (17); de François à Delcassé, 29 septembre 1899, MAE NS 230, 78–9. François, who blamed the hostile reception he received on Doumer's actions, described the visit thus: 'M. Doumer came in ahead of everyone else. He arrived in Kunming on horseback, unannounced, quite impromptu. These circumstances allowed the mandarins (who had, nevertheless, been very well warned) not to receive him on his arrival; he went into an empty house which he had to find himself and where he had to look after the details of moving in himself. He received the welcome given to a man of no importance and the Viceroy would have made it clear to him that he recognized only the authority of Beijing.' MAE NS 230, 79.

19. H. Michael Metzgar, 'The Crisis of 1900 in Yunnan: Late Ch'ing Militancy in Transition', *Journal of Asian Studies* 35, 2 (1976), 186. A translation of Songfan's written reply to Doumer was made by the apostolic pro-vicar in Yunnan, Father Maire, and given to Dejean de la Bâtie. Dép. pol. No. 14 de Dejean de la Bâtie à Delcassé, 16 juillet 1899, MAE NS 229, 194.

20. Déclaration du père Maire, provicaire de la Mission catholique du Yunnan, au sujet des entretiens qui ont eu lieu entre M. le Gouverneur général de l'Indo-Chine et le Viceroi de Yunnan-sen, annèxe à la dép. pol. No. 4 de François à Delcassé, 4 juin 1901, MAE NS 233, 120.

21. Dép. personnelle et confidentielle de François à Delcassé, 20 juillet 1901, MAE NS 233, 181.

22. De Dejean de la Bâtie à Delcassé, 20 septembre 1899, MC IC Carton 32 B11 (17).

23. G. Estassy, 'Incidents à Tsi-kay', Mengzi, 3 juillet 1899, MAE NS 496, 23.

24. See for example, Moreau de Champguillaume, 'Rapport sur les incidents …', Mengzi, 15 juillet 1899 and Martial, 'Rapport sur le travail fait et les divers incidents …', Mengzi, 16 juillet 1899, MAE NS 496, 25–30.

25. Guillemoto, 'Rapport du Directeur gènéral des Travaux publics de l'Indo-Chine sur les études du Chemin de fer de Laocai à Yunnan-sen', Saigon, 7 novembre 1899, MAE NS 496, 64–9; de Dejean de la Bâtie à Delcassé, 20 septembre 1899, MC IC Carton 32 B11 (17).

26. Tél. No. 100 de Doumer à Decrais, 27 juin 1899, MC IC Carton 32 B11 (17).

27. Tél. No. 103 de Doumer à Decrais, 29 juin 1899, MC IC Carton 32 B11 (17).

28. Tél. No. 131 de Decrais à Doumer, 29 juin 1899, MC IC Carton 32 B11 (17).

29. De Dejean de la Bâtie à Delcassé, 26 juin 1899, MC IC Carton 32 B11 (17).

30. De Pichon à Delcassé, 1 septembre 1899, MC IC Carton 32 B11 (17).

31. Dép. pol. No. 87 (confidentielle) de Pichon à Delcassé, 1 juillet 1899, MAE NS 229, 122–4.

32. Tél. de Dejean de la Bâtie à Delcassé, s.d. about 28 June 1899, annèxe à Delcassé à Decrais, 2 juillet 1899, MC IC Carton 32 B11 (17).

33. Tél. No. 105 de Doumer à Decrais, 1 juillet 1899, MC IC Carton 32 B11 (17).

34. Tél. No. 107 de Doumer à Decrais, 4 juillet 1899, MC IC Carton 32 B11 (17).

35. Marginal note to tél. de Doumer à Dejean de la Bâtie, 2 juillet 1899, MAE NS 230, 66.

36. De Delcassé à Decrais, 12 juillet 1899, MC IC Carton 32 B11 (17).

37. Tél. No. 116 de Doumer à Decrais, 15 juillet 1899, MC IC Carton 32 B11 (17).

38. Tél. No. 134 de Decrais à Doumer, 2 juillet 1899, MC IC Carton 32 B11 (17). Delcassé's draft is located at MAE NS 226, 132.

39. Dép. No. 508 de Delcassé à P. Cambon, 6 juillet 1899, MAE NS 229, 157–9.

40. Tél. de Doumer à Guillain, 18 juin 1899, MC IC Carton 32 B11 (28).

41. The decree establishing the commission was rather pretentiously dated at Kunming on 7 June 1899 and published in the *Journal officiel* of Indo-China on 27 July 1899. MC IC Carton 33 B11 (30).

42. Tél. de Dejean de la Bâtie à Delcassé, 26 juin 1899, MC IC Carton 32 B11 (17).

43. De Dejean la Bâtie à Pichon, 22 juillet 1899, MC IC Carton 32 B11 (17); Dép. pol. No. 13 de François à Delcassé, 7 octobre 1899, MAE NS 496, 43–5.

44. MAE NS 496, see note 43 above.

45. De Pichon à Delcassé, 1 septembre 1899, annèxe à Delcassé à Decrais, 7 novembre 1899, MC IC Carton 32 B11 (17). Nearly all correspondence on the Yunnan question was forwarded from the Quai d'Orsay to the Pavillon de Flore, largely, it would appear, to urge Decrais to restrict Doumer's freedom of action as much as possible.

46. Dép. pol. No. 106 de Pichon à Delcasse, 13 août 1899, MAE NS 496, 31–3. These views were also stated the previous day in a telegram from Pichon to Delcassé. He wired: 'The more we retain the railway's industrial nature, the greater its value to us. The more we give an official and political nature to our involvement in Yunnan, the more we will provoke distrust and difficulties. We will displease the people, arouse the fears of the authorities who will create embarrassment for us, and risk leading China to support strongly English opposition to check the plans for conquest which they attribute to us. As a result I believe that the railway commission must remain purely technical and abstain from discussing political and diplomatic matters . . .' MC IC Carton 33 B11 (30).

47. De Delcassé à Decrais, 18 août 1899, MC IC Carton 33 B11 (30).

48. Note du 16 août 1899, MAE NS 496, 36–7.

49. MAE NS 496, see note 48 above.

50. De Dejean de la Bâtie à Pichon, 22 juillet 1899, MC IC Carton 33 B11 (30).

51. De Delcassé à Decrais, 10 septembre 1899, MC IC Carton 33 B11 (30).

52. De Decrais à Delcassé, 13 septembre 1899, MC IC Carton 33 B11 (30).

53. Dép. No. 1685 de Doumer à Decrais, 2 octobre 1899, MC IC Carton 33 B11 (25).

54. De Dejean de la Bâtie à Delcassé, 17 mars 1899, MC IC Carton 32 B11 (25).

55. De Delcassé à Decrais, 23 juin 1899, MC IC Carton 32 B11 (25).

56. Dép. No. 1848 de Doumer à Decrais, 15 novembre 1899, MC IC Carton 33 B11 (30).

57. De Delcassé à Decrais, 13 décembre 1899, MC IC Carton 33 B11 (30).

58. De François à Delcassé, 29 septembre 1899, MAE NS 230, 77.

59. De Pichon au *Zongli Yamen*, 6 décembre 1899, MAE NS 230, 192.

60. Tél. No. 221 de Decrais à Doumer, 2 décembre 1899, MAE NS 230, 180.

61. De Delcassé à Decrais, 30 novembre 1899, MAE NS 230, 165–6.

62. De Dejean de la Bâtie à Pichon 28 juillet 1899, MAE NS 229, 243–5.

63. Dép. pol. No. 16 de François à Delcassé, 5 novembre 1899, MAE NS 230, 132.

64. De Dejean de la Bâtie à Pichon, 28 octobre 1899, MAE NS 230, 141.

65. Rabaud, directeur de la Compagnie Lyonnaise Indo-Chinoise (Laocai) à Dejean de la Bâtie, 10 août 1899, MAE NS 230, 64.

66. Dép. pol. No. 13 (confidentielle) de François à Delcassé, 7 octobre 1899, MAE NS 496, 42.

67. Dép. pol. No. 15 de François à Delcassé, 30 octobre 1899, MAE NS 230, 124–5.

68. Tél. No. 107 de Pichon à Delcassé, 2 octobre 1899, MAE NS 230, 83.

69. Guillemoto, 'Rapport du Directeur des travaux publics de l'Indo-Chine sur les études du Chemin de fer de Laocai à Kunming', Saigon, 9 novembre 1899, MAE NS 496, 70–1.

70. Note de la Direction politique, 3 octobre 1899, MAE NS 230, 85.

71. De Delcassé à Decrais, 8 octobre 1899, MC IC Carton 33 B11 (30).

72. Dép. No. 378 de Decrais à Doumer, 29 septembre 1899, MC IC Carton 33 B11 (30).

73. Dép. No. 1732 de Doumer à Decrais, 22 octobre 1899, MC IC Carton 32 B11 (17).

74. Tél. No. 282 de Decrais à Doumer, 15 décembre 1899, MC IC Carton 33 B11 (30).

75. Tél. No. 222 de Doumer à Decrais, 19 décembre 1899, MC IC Carton 33 B11 (30).

76. Procès-verbal de la première séance de réunion des Commissions supérieures du chemin de fer tenue au Yamen de Directeur des Télégraphs, 9 octobre 1899, MAE NS 496, 47–8.

77. Songfan, who was about to retire, wrote: 'O you, who with a sincere heart observe the rules of morality — Who practise virtue, in fulfilling the rites — Who act with circumspection and reason as your mottos!' Annèxe à la dép. No. 2184 de Doumer à Decrais, 28 décembre 1899, MC IC Carton 33 B11 (30).

78. Dép. pol. No. 19 de François à Delcassé, 27 décembre 1899, MAE NS 230, 201.

79. Tél. No. 12 de Pichon à Delcassé, 24 janvier 1900, MAE NS 231, 4.

80. De Decrais à Delcassé, 3 février 1900; de Delcassé à Decrais, 9 février 1900, MC IC Carton 33 B11 (30).

81. Téls. de Pichon (Hanoi) à Delcassé, 2, 8 mars 1900, MAE NS 231, 18, 19.

82. Dép. pol. No. 24 de Pichon à Delcassé, 8 avril 1900, MAE NS 231, 37–40.

83. MAE NS 231, see note 82 above, 34–7.

84. MAE NS 231, see note 82 above, 41.

85. Dép. pol. No. 22 de François (Hekou) à Delcassé, 1 avril 1900, MAE NS 231, 22–4.

86. Tél. No. 89 de Doumer à Decrais, 5 mai 1900, MC IC Carton 33 B11 (30).

87. See L. K. Young, *British Policy in China, 1895–1902* (Oxford, Clarendon Press 1970), pp. 94–5.

88. Note pour le Ministre de la Direction commerciale, 2 mai 1899, MAE NS 195, 64.

89. MAE NS 195, see note 88 above, 61–68.

90. Dép. pol. No. 124 de P. Cambon à Delcassé, 11 mai 1899, MAE NS 195, 76.

91. MAE NS 195, see note 90 above, 80–81.

92. Dép. pol. No. 233 de Delcassé à Noailles (Berlin), 31 mai 1899, MAE NS 195, 99–100.

93. Dép. pol. No. 192 de Noailles à Delcassé, 11 juin 1899, MAE NS 195, 128–9.

94. Dép. pol. (lettre chiffrée) de Delcassé à Noailles, 20 juin 1899, MAE NS 195, 133.

95. Dép. pol. No. 260 de Boutiron (Chargé d'affaires à Berlin) à Waldeck-Rousseau (Ministre des affaires étrangères per interim), 7 août 1899, MAE NS 195, 169.

96. Dép. pol. No. 64 de Pichon à Delcassé, 22 mai 1899, MAE NS 195, 91-5.

97. Delcassé à Lockroy (Ministre de la Marine), 4 octobre 1899, MAE NS 194, 120-121.

98. Note pour le Ministre, difficultés avec le Gouvernement chinois, 3 octobre 1898; tél. de Beaumont à Lockroy, 3 octobre 1898; dép. pol. No. 108 de Pichon à Delcassé, 23 décembre 1898, MAE NS 194, 115-7, 124, 192-3.

99. Delcassé was not enthusiastic about the religious Protectorate as anything more than an excuse to pursue other advantages, and rejected Pichon's idea of continuing it in an area where French economic and political interests were not involved. Pichon's suggestion for maintaining the religious Protectorate in the British sphere of influence attracted an emphatic ministerial 'Non' in the margin. Dép. pol. No. 71 de Pichon à Delcassé, 3 juin 1899, MAE NS 195, 112-4.

100. MAE NS 195, see note 99 above, 114.

101. Dép. pol. No. 96 de Pichon à Delcassé, 27 juillet 1899, MAE NS 195, 141-52.

102. Dép. pol. No. 184 de Cambon à Delcassé, 30 juillet 1899, MAE NS 195, 155-6.

103. MAE NS 195, see note 102 above, 156.

104. The article is disccussed by Cambon in Dép. pol. No. 190 de Cambon à Delcassé, 11 août 1899, MAE NS 195, 173.

105. Note pour La Direction politique de M. Bompard, 31 juillet 1899, MAE NS 195, 159.

106. MAE NS 195, see note 105 above, 160-3.

107. Note pour le Ministre de la Direction politique, 18 septembre 1899, MAE NS 195, 198.

108. This analysis of Hay's motives is that of Thiébaut, the French *Chargé d'affaires* in Washington. See Dép. pol. No. 128 de Thiébaut à Delcassé, 26 octobre 1899, MAE NS 195, 214.

109. Dép. pol. No. 133 de Pichon à Delcassé, 20 octobre 1899, MAE NS 195, 204. See pp. 89-90, 211, 217, and 222-4 for the vacillations in Pichon's views.

110. MAE NS 195, see note 109 above, 206.

111. MAE NS 195, see note 109 above.

112. Dép. com. de Cambon à Delcassé, 30 octobre 1899, MAE NS 195, 216-7.

113. MAE NS 195, see note 112 above. See 255-61.

114. *J.O.*, Ch. de D., 24 novembre 1899, 499-500.

115. De Delcassé au général Porter, 16 décembre 1899, MAE NS 195, 257.

116. Dép. pol. No. 25 de Jules Cambon (Washington) à Delcassé, 28 mars 1900, MAE NS 196, 31-2.

117. Dép. pol. No. 322 de Cambon à Delcassé, 11 décembre 1899, MAE NS 195, 245-6.

Notes to Chapter 9

1. Dép. pol. No. 18 de François à Delcassé, 27 décembre 1899, MAE NS 230, 197-8.

2. Etat-major des Troupes de l'Indo-Chine, 2ème Bureau, Renseignements,

'Notice sur le Yunnan', Hanoï, 15 décembre 1899, MC IC Carton 32 B11 (29), pp. 126–7.
  3. MC IC Carton 32 B11 (29), see note 2 above, pp. 132–4.
  4. H. Michael Metzgar, 'The Crisis of 1900 in Yunnan: Late Ch'ing Militancy in Transition' *Journal of Asian Studies* 35, 2 (1976), 187.
  5. Dép. No. 17 de François à Delcassé, 3 avril 1900, MAE NS 231, 28.
  6. Dép. pol. No. 23 de François à Delcassé, 27 avril 1900, MAE NS 231, 50–3.
  7. De Delcassé à Decrais, 16 mai 1900, MAE NS 231, 69–70.
  8. Dép. pol. Nos. 20, 21 de François à Delcassé, 18, 19 mai 1900, MAE NS 231, 77–8, 82–9.
  9. De *Zongli Yamen* à Pichon, 15 mai 1900; de Pichon au *Zongli Yamen*, 19 mai 1900; tél. No. 38 de Delcassé à Pichon, 17 mai 1900, MAE NS 231, 105, 108, 72.
  10. Tél. No. 72 de Pichon à Delcassé, 24 mai 1900, MAE NS 231, 110.
  11. Tél. No. 41 de Delcassé à Pichon, 26 mai 1900, MAE NS 231, 119.
  12. De Decrais à Delcassé, 2 juin 1900, MAE NS 231, 129.
  13. Tél. de François à Delcassé, 7 juin 1900, MAE NS 231, 133.
  14. Tél. de Sainson à Delcassé, 14 juin 1900, MAE NS 231, 151–2.
  15. Tél. de François à Delcassé, 14 juin 1900, MAE NS 231, 151–2.
  16. Tél. de Sainson à Delcassé, 19 juin 1900, MAE NS 231, 150.
  17. Note: Conversation du Ministre avec le Ministre de Chine, 18 juin 1900, MAE NS 231, 165.
  18. Annèxe à François (Hanoi) à Delcassé, 12 juillet 1900, MAE NS 232, 54. This was a bluff as French forces in Tonkin were severely depleted. See pp. 235–8.
  19. Tél. de Sainson à Delcassé, 20 juin 1900, MAE NS 231,190.
  20. Annèxe à Yu Geng à Delcassé, 26 juin 1900, MAE NS 231, 191; Metzgar (1976), see note 4 above, 195.
  21. See Doumer's account of Su's requests for further subsidies in Tél. personnel et confidentiel No. 264 de Doumer à Decrais, 18 août 1900, MAE NS 232, 101.
  22. Journal de M. François, annèxe à François à Delcassé, 12 juillet 1900, MAE NS 232, 29–47.
  23. Tél. No. 138 de Doumer à Decrais, 17 juin 1900, MAE NS 231, 163.
  24. Tél. de Decrais à Doumer, 17 juin 1900, MAE NS 231, 161.
  25. On 30 June Doumer wired François: I ask you earnestly to stay at Mengzi with all our agents and officers and ask me to send you troops for protection. You are in a good position at Mengzi where you would resist until our arrival which would be prompt ...' Notes et télégrammes, annèxes à François (Hanoï) à Delcassé, 12 juillet 1900, MAE NS 232, 54.
  26. Tél. de Doumer à Sainson, 30 juin 1900, MAE NS 232, see note 25 above.
  27. Tél. de François (Yenbay) à Delcassé, 7 juillet 1900, MAE NS 232, 10.
  28. Notes et télégrammes, annèxes à François (Hanoï) à Delcassé, 12 juillet 1900, MAE NS 232, 55.
  29. MAE NS 232, see note 28 above.
  30. De François (Hanoï) à Delcassé, 17 juillet 1900, MAE NS 232, 77.
  31. MAE NS 232, see note 30 above. 75.
  32. Tél. de Decrais à Doumer, 6 juillet 1900, MAE NS 232, 5.
  33. Notes et télégrammes, annèxes à François (Hanoï) à Delcassé, 12 juillet 1900, MAE NS 232, 55.
  34. Tél. No. 5 de Delcassé à Sainson, 6 juillet 1900, MAE NS 232, 6.
  35. Téls. de Sainson à François, 11, 12 juillet 1900, MAE NS 232, 58.
  36. Tél. de Sainson (Laocai) à Delcassé, 17 juillet 1900, MAE NS 232, 68.
  37. François described Guillemoto's activities in these terms: 'They wanted an

expedition at any price. They strived to provoke it, to make it inevitable. I was frightened by the lack of conscience among the men who were sustaining this illusion in M. Doumer's mind, when he could not bring help in time, when no troops were assembled and when it would have taken some time to defeat the serious obstacles the soldiers would encounter.

'I saw clearly that they had decided to launch the Governor-General on an adventure whose consequences he did not perceive and which is absolutely contrary to the Government's views. They flatter a project dear to M. Doumer, hiding from him the responsibilities he would incur, not only in adopting the project but even more by lacking all the material means essential for its execution. They were only looking for the opportunity to oblige the Government to accept the inevitability of an expedition which would only serve the ambitions of a few persons.' De François à Delcassé, 17 juillet 1900, MAE NS 232, 73–4. François's analysis of the ambitions current in Hanoi was accurate, but he was probably overly solicitous of Doumer's reputation in ascribing these ambitions to his lieutenants rather than to the Governor-General himself.

38. MAE NS 232, see note 37 above.

39. Tél. No. 9 de Delcassé à François, 14 juillet 1900, MAE NS 232, 61–2. The telegram was drafted in Delcassé's own hand and is the clearest indication of his thoughts on the Yunnan question in the summer of 1900.

40. De Delcassé à Yu Geng, 19 juillet 1900, MAE NS 232, 82.

41. Tél. No. 91 de Delcassé à Pichon, 24 octobre 1900, MAE NS 232, 125.

42. Guillemoto, Rapport No. 263 du Directeur Général des travaux publics de l'Indo-Chine, 25 février 1901, MC IC Carton 32 B11 (17).

43. 'Conditions qui pourraient être imposées à la Chine pour assurer la reinstallation de nos agents et nationaux au Yunnan; et réparations qu'il serait avantageux de reclamer à la suite des troubles de 1899 et 1900'. Note de M. François, 20 décembre 1900, MAE NS 232, 163–4.

44. Tél. No. 17 de Pichon à Delcassé, 24 janvier 1901, MAE NS 233, 20.

45. De Sainson (Laocai) à Pichon, 4 février 1901, MAE NS 237, 28; Metzgar (1976), see note 4 above, 198.

46. Instructions de la Direction politique à M. François, 15 février 1900, MAE NS 233, 40–3.

47. Tél. de François (Mengzi) à Delcassé, 29 avril 1901, MAE NS 233, 98.

48. De François à Delcassé, 1 décembre 1900, MAE NS 232, 143–4.

49. 'Positions que la France pourrait prendre au Yunnan, suivant les différentes éventualités qui peuvent se présenter'. Note de M. François, 20 décembre 1900, MAE NS 232, 157–9.

50. Note, chemin de fer de Laokay à Yunnansen, décembre 1900, MAE NS 496, 83.

51. De François à Delcassé, 17 juillet 1900, MAE NS 232, 73.

52. M. Bruguière, 'Le Chemin de fer du Yunnan. Paul Doumer et la politique d'intervention française en Chine (1889–1902)', Reveue d'histoire diplomatique, No. 77, 1963, 261–2.

53. Note, chemin de fer de Laokay à Yunnansen, décembre 1900, MAE NS 496, 82–84.

54. Bruguière (1963), see note 52 above, 267–9.

55. Concession du chemin de fer de Haiphong à Yunnansen, Convention, Paris, 15 juin 1901, MAE NS 496, 91.

56. De Delcassé à Beau, 29 juin 1901, MAE NS 496, 150.

57. Cahier des charges, MAE NS 496, 95–106.

58. Rapport fait au nom de la Commission des Colonies chargée d'examiner le projet de loi ayant pour object d'approuver la convention conclue par le Gouvernement général de l'Indo-Chine pour la construction partielle et

l'exploitation du chemin de fer de Haiphong à Yunnansen, par M. Maurice Ordinaire, député. *J.O.*, Ch. de D., Annèxe au procès-verbal de la séance du 24 juin 1901, 115-16. It is curious that a project costing 101 million francs should provide *more* than 100 millions in orders to French industry.

59. *J.O.*, Ch. de D., 2ème séance du 27 juin 1901, 1617-8.

60. *J.O.*, see note 59 above, 1607.

61. Bruguière (1963), see note 52 above, 272.

62. Bruguière (1963), see note 52 above, 266.

63. Etienne, the deputy from Oran in Algeria, was an ardent, indeed promiscuous supporter of colonial causes. He was also founder of the Comité de l'Afrique française in 1890 and of the Comité du Maroc in 1904. He had also been Sous-secretaire d'Etat des Colonies in three ministries in the period 1887 to 1892.

64. *Bulletin du Comité de l'Asie française*, 1 (avril 1901), 1-10.

65. *Avenir du Tonkin*, 27 avril 1901.

66. Note à l'annèxe No. 1 de la dép. pol. No. 4 de François à Delcassé, 4 juin 1901; dép. pol. No. 9 de François à Delcassé, 21 juin 1901, MAE NS 233, 118, 161-2.

67. Dép. confidentielle et personnelle de François à Delcassé, 20 juillet 1901, MAE NS 233, 180.

68. MAE NS 233, see note 67 above, 185.

69. MAE NS 233, see note 67 above, 183-4; de Delcassé à Decrais, 10 juillet 1901, MAE NS 496, 164-5.

70. Tél. de Decrais à Broni, 18 juillet 1901, MAE NS 233, 176.

71. Dép. pol. No. 20 de François à Delcassé, 29 août 1901, MAE NS 233, 244.

72. Tél. de Delcassé à Beau, 17 septembre 1901, MAE NS 496, 182.

73. De Decrais à Delcassé, 19 septembre 1901, MAE NS 496, 185-8.

74. Tél. de Decrais à Doumer, 12 octobre 1901, MAE NS 496, 191.

75. Note: Compagnie française des chemins de fer de l'Indo-Chine et du Yunnan et Société de construction de chemins de fer indo-chinois, 31 décembre 1904, MAE NS 496, 204-8.

76. Wang Wen-yuan, *Les Relations entre l'Indo-Chine française et la Chine* (Paris, P. Bousset, 1937), p. 68.

77. Between 1910 and 1921 the average working co-efficient (ratio of costs to expenses) of the Yunnan section was 0.723, compared with an average working co-efficient of Indo-Chinese railways of 0.829 over the same period. (Calculated from the table 'Chemins de fer du Nord de l'Indo-Chine' in A. Sarraut, *La Mise en valeur des colonies françaises* (Paris, Payot, 1923), p. 489.) Even more dramatic are the figures for working profit per kilometre in 1932-3. In that year this sum for the Yunnan line was 1,012 piastres whereas from all other Indo-Chinese railways it was only 278 piastres. Wang Wen-yuan (1937), see note 76 above, p. 74.

78. The railway's prosperity was not based on a large international trade. Throughout its history most traffic was local, confined either to Tonkin or Yunnan, as this table of freight traffic (in tonnes) shows:

	1917	1926	1930	1933
Haiphong Docks–Yunnan	10,996	24,227	29,339	24,703
Yunnan–Haiphong Docks	19,719	11,856	9,434	10,644
Tonkin–Yunnan	1,802	39,622	1,860	16,538
Yunnan–Tonkin	1,878	958	953	648
Local Tonkin	29,403	111,107	108,833	82,676
Local Yunnan	61,322	136,591	130,524	116,723

Wang Wen-yuan (1937), see note 76 above, p. 74.

79. Dép. com. No. 23 de Geoffray (Chargé d'affaires à Londres) à Hanotaux, MAE NS 485, 74.

80. *The Times*, 25 January 1898, MAE NS 485, 55.

81. Note sur les chemins de fer birmans: leur prolongation vers le Yunnan, et leur jonction avec le réseau d'Hindoustan, 2 février 1898, MAE NS 485, 61–2.

82. *Daily Graphic*, 28 May 1898, MAE NS 485, 55.

83. *Rangoon Gazette*, quoted in dép. com. No. 17 de Pina (Rangoon) à Delcassé, 13 juillet 1898, MAE NS 485, 122–3.

84. Dép. pol. No. 11 de Cambon à Delcassé, 20 janvier 1899, MAE NS 485, 153–4.

85. L. K. Young, *British Policy in China, 1895–1902* (Oxford, Clarendon Press, 1970), p. 89.

86. Dép. pol. No. 19 du Consul-général à Calcutta à Delcassé, 4 octobre 1899, MAE NS 485, 170.

87. A. M. S. Wingate, 'Recent journey from Shanghai to Bhamo through Hunan', *Geographical Journal*, 1899, 642–4, MAE NS 485, 190–1.

88. *Journal des Débats*, 10 mai 1899, MF F30 374.

89. *Journal des Débats*, 20 juin 1899, MF F30 374.

90. William Sherriff, 'The Burmah-China Railway', *Manchester Guardian*, n.d. [March 1900], MAE NS 485, 193.

91. See pp. 261–3 for the Yunnan Syndicate.

92. Dép. pol. No. 13 de Claine (Rangoon) à Delcassé, 10 novembre 1901, MAE NS 485, 217.

93. Dép. pol. No. 350 de Cambon à Delcassé, 11 décembre 1901, MAE NS 485, 223. See also Young (1970), see note 85 above, p. 35.

94. Dép. pol. No. 19 de Cambon à Delcassé, 22 janvier 1902, MAE NS 485, 232.

95. On British plans to connect Burma and Yunnan by rail see Jeshurun Chandran, *The Burma–Yunnan Railway: Anglo-French rivalry in mainland Southeast Asia and South China, 1895–1902* (Athens, Ohio University Center for International Studies, 1971).

96. A. Leclère, Rapport de l'Ingénieur des mines, 10 décembre 1899, MAE NS 418, 159–62.

97. Such was Dujardin-Beaumetz's pessimism on the industrial prospects for China that he asserted that talk of a Beijing–Hankou railway would come to nothing. See M. Bruguière (1963), see note 52 above, 55–6.

98. De Reille à Hanotaux, 3 février 1897; d'Hanotaux à Reille, 8 février 1897, MAE NS 441, 1, 3.

99. Note de M. Bompard pour la Direction politique, 8 août 1897, MAE NS 441, 4–5.

100. Dép. com. No. 23 de Dubail à Hanotaux, 26 septembre 1897; de *Zongli Yamen* à Dubail, 14 janvier 1898, MAE NS 441, 12, 20.

101. A. Leclère, Rapport de l'Ingénieur des mines, 10 décembre 1899, MAE NS 418, 160.

102. De Reille à Hanotaux, 9 février 1898, MAE NS 441, 24.

103. D'Hanotaux à Reille, 17 février 1898, MAE NS 441, 27–8.

104. De Reille à Hanotaux, 18 février 1898, MAE NS 441, 29.

105. Dép. com. No. 29 de Dejean de la Bâtie à Hanotaux, 20 avril 1898, MAE NS 441, 35–8.

106. M. Bélard, *Rapport sur son voyage de mission au Yunnan* (Paris, Comité des Forges, 1899).

107. A. Leclère, Rapport de l'Ingénieur des mines, 31 juillet 1899, MAE NS 418, 131–6; Leclère file, AN F14 11630.

108. MAE NS 418, see note 107 above, 138.

109. A. Leclère, Rapport de l'Ingénieur des mines, 24 août 1899, MAE NS 418, 143.

110. A. Leclère, Rapport de l'Ingénieur des mines, 31 juillet 1899, MAE NS 418, 138.

111. Note de M. le Comte de Bondy, 15 mars 1899, MAE NS 441, 77-83.

112. De Delcassé à de Bondy, 8 avril 1899, MAE NS 441, 84-5. Delcassé sent de Bondy copies of three contracts for Chinese projects to indicate to him the kind of formula which was acceptable to the Chinese.

113. De Cellérier à Bompard, 9 mai 1899, MAE NS 441, 92.

114. De Cellérier à Bompard, 30 juin 1899, MAE NS 441, 102. This Russian support was expected in spite of the provisions of the Anglo-Russian agreement of 28 April 1899, by which Russia promised not to make such interventions in central or southern China. See p. 219 on this agreement.

115. De Bondy à Bompard, 8 mai 1899, MAE NS 441, 88-9.

116. Dép. de Delcassé à Pichon (No. 62) et Dejean de la Bâtie (No. 2), 20 mai 1899, MAE NS 441, 96-7.

117. Dép. com. No. 44 de Pichon à Delcassé, 7 juillet 1899, MAE NS 441, 105.

118. De Dejean de la Bâtie à Pichon, 22 août 1899, MAE NS 441, 117-18.

119. MAE NS 441, see note 118 above, 120.

120. De Pichon à Delcassé, 3 octobre 1899, MAE NS 441, 137-8.

121. Note de MM. Emile Cellérier et le Comte de Bondy, 20 novembre 1899, MAE NS 441, 160-3.

122. MAE NS 441, see note 121 above.

123. Bruguière (1963), see note 52 above, 253.

124. Note, s.d., MAE NS 442, 52-9.

125. Dép. No. 27 de Delcassé à Cambon, 5 février 1900, MAE NS 442, 60-2.

126. MAE NS 442, see note 125 above; note de MM. Emile Cellérier et le Comte de Bondy, 20 novembre 1899; tél. de Cellérier à Bélard (Poste restante, Singapore), 31 octobre 1899, MAE NS 441, 163, 140.

127. Note, s.d., MAE NS 442, 52-9.

128. De Decrais à Delcassé, 4 décembre 1899, MAE NS 441, 169.

129. Dép. com. No. 151 de P. Cambon à Delcassé, 23 novembre 1899, MAE NS 441, 152-3.

130. Tél. No. 103 de Delcassé à P. Cambon, 11 décembre 1899, MAE NS 441, 170.

131. Tél. No. 118 de Delcassé à Pichon, 30 décembre 1899, MAE NS 441, 176.

132. D'Obry (directeur général de la Compagnie générale de Traction) à Delcassé, 2 novembre 1899, MAE NS 441, 142-3.

133. De Delcassé à Obry, 16 novembre 1899, MAE NS 441, 148-9.

134. Dép. No. 27 de Delcassé à P. Cambon, 5 février 1900, MAE NS 442, 60-7.

135. MAE NS 442, 64, see note 134 above.

136. MAE NS 442, 64, see note 134 above; note, 6 avril 1900, MF F30 373 2 (sous-dossier 5).

137. Bompard expressed his attitude in these terms: 'The Government will give its support to the Anglo-French Syndicate if this company's by-laws (*statuts*) are modified and the necessary measures taken to achieve the following conditions whose object is to ensure equality between the two nationalities:

Ordinary and deferred shares will be equally divided between French and English;

The head office being in London, the central administration will be in Paris;

There will be a Paris Committee and a London Committee, each composed of the same number of directors;

Each Committee will elect its own President;
Board meetings will take place alternatively in Paris and London, chaired by the President of the London Committee;
The Board may only deliberate if both committees are represented at the meeting;
The General Meetings will take place in London, chaired alternatively by the Presidents of the London and Paris committee;
The companies created by the Anglo-French Syndicate will have their head office in France or England following an equal allotment between the two countries, on the base of taking half each and taking account of the site of their operations for those which will function near the Burma or Tonkin frontiers.' Note du 31 mars 1900, MAE NS 442, 128.

138. F. Bertie, Minute, 14 October 1899 and Salisbury's marginal note thereon, FO17 1405, 406.

139. Sir Vincent Caillard (quoting Cambon) to Bertie, 15 December 1899 and Salisbury's marginal note thereon, FO17 1407, 270–3.

140. Dép. com. Nos. 29, 35 de Cambon à Delcassé, 22 février, 2 mars 1900, MAE NS 442, 76–9, 86–7.

141. De Delcassé à Bompard, 3 avril 1900, MAE NS 442, 129.

142. Dép. No. 72 de Delcassé à Cambon, 11 avril 1900, MAE NS 442, 133–4.

143. Copie de la convention conclue entre M. Bourke et M. Adam le 10 avril 1900, MAE NS 442, 136–42.

144. The 12,000 new French shares were issued as follows: Banque de Paris et des Pays-Bas 1,500; Comptoire national d'escompte 1,500; Société générale 1,500; Crédit industriel et commercial 1,000; Banque de l'Indo-Chine 1,000; M. Henrotte 1,000; M. Adam 1,000; Société d'Orient (Banque Ottomane) 900; Banque française de l'Afrique du sud 700; Banque internationale de Paris 600; Ulysse Pila and the Lyon group 600; M. Gouin (Société des Batignolles) 250; M. Vitali (Société des Constructions) 250; and M. Verneuil, syndic des agents de change 200. Note, s.d., MAE NS 442, 132.

145. Dép. No. 97 de Delcassé à Pichon, 20 octobre 1900, MAE NS 442, 153–4.

146. Dép. com. No. 1 de Sainson (Mengzi) à Delcassé, 5 mai 1900, MAE NS 442, 143–5.

147. De Delcassé à Decrais, 22 décembre 1900, MAE NS 442, 167–8.

148. Dép. com. No. 179 de Cambon à Delcassé, 11 décembre 1900, MAE NS 442, 157–8.

149. MAE NS 442, see note 147 above, 160.

150. MAE NS 442, see note 147 above, 161–2.

151. MAE NS 442, see note 147 above, 162–3.

152. Note de M. Delcassé, 8 janvier 1901, MAE NS 442, 156.

153. Tél. No. 1 de Delcassé à Cambon, 11 janvier 1901, MAE NS 442, 175.

154. Tél. No. 12 de Delcassé à Beau, 24 janvier 1901, MAE NS 442, 176.

155. Dép. com. No. 3 de François à Delcassé, 5 juillet 1901, MAE NS 442, 189–91.

Notes to Chapter 10

1. See pp. 201–3, 242–3.

2. Christopher Andrew, *Théophile Delcassé and the making of the Entente Cordiale, a reappraisal of French foreign policy 1898–1905* (London, Macmillan, 1968), pp. 256–7.

3. See J. A. Hobson, *Imperialism: A Study* (London, Allen and Unwin, 1902);

and V. I. Lenin, *Imperialism: the Highest Stage of Capitalism* (Moscow, Progress Press, 1966).

4. This aspect of Lenin's analysis is discussed in Eric Stokes, 'Late nineteenth century colonial expansion and the attack on the theory of economic imperialism: a case of mistaken identity?', *Historical Journal*, 12, 2 (1969), 291–5.

5. See Chapter 6 and pp. 251–2 respectively.

6. Dép. com. No. 92 de Delcassé à Pichon, 29 juillet 1899, MAE NS 523, 3–4.

7. Jacques Thobie, *Intérêts et impérialisme français dans l'Empire Ottoman (1895–1914)* (Paris, Publications de la Sorbonne, 1977), pp. 606–60.

8. Thobie (1977), see note 7 above, p. 607.

9. Dép. com. de Bezaure à Berthelot, 27 décembre 1895, MAE, Chine, Affaires diverses commerciales, Boîte 331 (juillet–décembre 1895).

Glossary of Chinese Names

ALL Chinese names appear in the text in *pinyin* form. This is a list of all those *pinyin* forms and their equivalent Wade-Giles or Post Office forms. In addition the equivalent French forms in common use about 1900 have been listed if they were widely divergent from the Post Office or Wade-Giles forms then normally in use in English.

pinyin	*Wade-Giles or Post Office*	*French about 1900*
Baoding	Paoting	Paotingfou
Beihai	Pakhoi	
Beihe	Pei-ho	fleuve de Pékin
Beijiang	Pei-chiang	Peh-kiang
Beijing	Peking	Pékin
Beiyang	Peiyang	
Bohai Sea	Gulf of Peichihli	
Bose	Poseh	Posé
Chengdu	Chengtu	Tcheng-tu
Chongqing	Chungking	Tchong-king
Dagu	Taku	
Dali	Tali	
Dalian	Talienwan	
Daotai	Taot'ai	
Daye	Ta-yeh	
Ding Baozhen	Ting Pao-chen	
Ding Chenduo	Ting Ch'en-to	
Foshan	Fatsan	
Fujian	Fukien	
Fuzhou	Foochow	Fou-tchéou
Gaobeiding (Xincheng)	Kaopeiting	
Gejiu	Kochiu	Ko-kieou
Prince Gong (Yi Xin) (1833–98)	Prince Kung (I Hsin)	prince Kong or Koung
Gong Yuheng	Kung Yu-hêng	
Guangdong	Kwangtung	Kouangtoung

pinyin	Wade-Giles or Post Office	French about 1900
Guangxi	Kwangsi	Kouangsi
Guangxu (1875–1908)	Kuang-hsü	Kouang-Siu
Guangzhou	Canton	
Guangzhouwan	Kwangchowan	Kouang-tchéou-wan
Guiyang	Kweiyang	Kouei-yang
Guizhou	Kweichow	Kouei-tchéou
Hangzhou	Hangchow	Hang-tchéou
Hankou (Wuhan)	Hankow	Hankéou
Hanyang (Wuhan)	Hanyang	
Hekou	Hok'ou	Hék'ou
Henan	Honan	
Hu Yufen (died 1906)	Hu Yü-fen	
Huang He	Yellow River	fleuve Jaune
Hubei	Hupei	Hupeh
Hunan	Hunan	
Jiangfu	Chiangfu	Kiang-fu
Jiangxi	Kiangsi	
Jiaozhou	Kiaochow	Kiao-tchéou
Jing Zhang	Ching Chang	K'ing Tch'ang (also Tching Tchang)
Kaifeng	K'aifeng	
Kaiping	K'ai-p'ing	
Kunming	K'unming (Yunnanfu)	Yunnanfou (Yunnansen)
Li Hongzhang (1824–1901)	Li Hung-chang	Li Houng-tchang
Li Jingxi	Li Ching-hsi	
Liaodong	Liaotung	Liaotoung
lijin	lichin	likin
Linxi	Lin-hsi	
Liu Chunlin	Liu Ch'un-lin	
Liu Wenquan	Liu Wen-chüan	
Liu Yongfu (1837–1917)	Liu Yung-fu	Lieou Yong-fou
Longzhou	Lungchow	Long-tchéou
Lugouqiao	Lukouch'iao	
Luo Fengluo	Lo Fêng-lo	
Luzhou	Luchow	Lu-tchéou
Mengzi	Mengtze	Mong-tsé

pinyin	Wade-Giles or Post Office	French about 1900
Nanjing	Nanking	
Niuzhuang	Newchwang	Nieou-tchouang
Pingding	P'ing-ting	
Pingxiang	P'ing-hsiang	
Qing	Ch'ing	K'ing
Prince Qing	Prince Ch'ing	prince K'ing
(Yi Kuang)	(I K'uang)	
(1836–1916)		
Qingdao	Tsingtao	
Sandu'ao	Santuao	
Sansha Bay	Samsah Bay	
Sanshui	Samshui	
Shandong	Shantung	Chan-toung
Shanghai	Shanghai	Chang-haï
Shanhaiguan	Shanhaikwan	Chan-haï-kouan
Shanxi	Shansi	Chansi
Sheng Xuanhuai	Sheng Hsüan-huai	
(1844–1916)		
Sichuan	Szechüan	Sseu-tchouan
Simao	Ssumao	
Songfan	Sung-fan	
Su Yuanchuan	Su Yüan-ch'uan	Sou Youan-tch'ouan
Taiyuan	T'ai-yüan	
Tang Zhoumin	T'ang Chou-min	
Tangshan	T'angshan	Tang-chan
Tianjin	Tientsin	
Tongzhi	T'ung-chih	T'oung-Tche
(1856–75)		
Tongzhou	T'ungchou	T'ong-tchéou
Wang Wenshao	Wang Wên-shao	
Weihaiwei	Wei-hai-wei	
Wenshan	Wenshan	Kaihwa
Wu Tingfang	Wu T'ing-fang	
(1842–1922)		
Wu Yunpang	Wu Yün-p'ang	
Wuchang	Wuchang	
(Wuhan)		
Wusong	Woosung	Wou-song
Wuzhou	Wuchow	Wou-tchéou
Xi Jiang	West River	Si-kiang
	(Hsi-chiang)	
Xi'an	Hsian (Sian)	

pinyin	Wade-Giles or Post Office	French about 1900
Xinjian	Hsin-chien	
Xinyang	Hsinyang	Sinyang
Yangzi Jiang	Yangtse River	fleuve Bleu
Yantai	Chefoo	
Yibin	Suifu	
Yichang	I-ch'ang	
Yu Geng	Yü Kêng	
Yuan Jiang	Red River	fleuve Rouge
Yuanjiang	Yüanchiang	Youan-kiang
Zhang Zhidong (1837–1921)	Chang Chih-tung	Tchang Tché-tong, Tchang Tchi-toung
Zhejiang	Chekiang	Tché-kiang
Zhengding	Chengting	
Zhili (Hebei)	Chihli	Tchihli
Zhou Fu (1837–1921)	Chou Fu	
Zhou Wen	Chou Wen	
Zongli Yamen	Tsungli Yamen	Tsongli Yamen
Zuo Zongtang (1812–85)	Tso Tsung-t'ang	

Bibliography

French Government Archives

Archives of the Ministry of Foreign Affairs

Conserved in the Ministry's building on the Quai d'Orsay, these archives have been the most valuable single source for this work. They not only contain material relevant to the determination of official policy, but also some correspondence and records of discussions with and between businessmen. A number of series were consulted.

1. Correspondance politique, 1871–1896; série Chine
Volumes 67 to 96, covering correspondence to and from the Beijing legation for the period 1885–1896.

2. Correspondance politique des consuls
Série Chine

v. 5	1882–1892	Fuzhou
v. 6	1884–1892	Tianjin
v. 7	1889–1892	Longzhou
v. 8	1887–1889	Guangzhou
v. 9	1890–1892	Guangzhou
v. 10	1889–1892	Mengzi
v. 11	1883–1893	Hankou, Beihai
v. 12	1893	Guangzhou, Longzhou, Mengzi, Tianjin
v. 13	1894	all Consulates except Shanghai
v. 14	1895	Guangzhou, Fuzhou, Hankou
v. 15	1895	Longzhou, Mengzi, Beihai, Tianjin

Série Shanghai

v. 13	1886–1890
v. 14	1891–1892
v. 15	1893–1895

3. Correspondance consulaire et commerciale de 1793 à 1901

Guangzhou	v. 4	1878–1889
	v. 5	1890–1901
Fuzhou	v. 3	1884–1892
	v. 4	1893–1901
Hankou	v. 2	1883–1901
Hong Kong	v. 3	1878–1888
	v. 4	1889–1892
	v. 5	1893–1896

	v. 6	1897–1901
Longzhou	v. 1	1889–June 1894
	v. 2	July 1894–1901
Mengzi	v.s.n.	1889–1901
Beihai	v. 1	1888–1895
	v. 2	1896–1901
Beijing	v. 6	1886–1889
	v. 7	1890–1896
	v. 8	1896–1898
	v. 9	1898–1901
Rangoon	v. 1	1883–June 1893
	v. 2	July 1893–1901
Simao	v.s.n.	1896–1901
Yantai	v.s.n.	1897–1901
Chongqing	v.s.n.	1896–1901
Tianjin	v. 2	1877–1901

4. Négociations commerciales: Chine
v. 3   March 1879–June 1885
v. 4   July 1885–March 1886
v. 5   April–December 1886
v. 6   1887–1897

5. Série affaires diverses commerciales antérieures à 1902. Chine, affaires diverses commerciales, industrielles
Boîte 328
1   1862–1869, 1887–1889
2   1890
3   1891
Boîte 329
4   1892
5   January–June 1893
6   July–December 1893
Boîte 330
7   January 1894–June 1895
Boîte 331
8   July–December 1895
9   January–July 1896
10   August 1896–December 1898

6. Papiers d'Agents
These holdings were only of marginal value. The most extensive in this series, those of Auguste Gérard are, unfortunately, practically illegible.
P.R. Bons d'Anty
vv. 1–5   Longzhou   1889–1896
v. 6   Simao   1896–1898

J.P.P. Casimir-Périer
v. 4   folio 80   Défense du Tonkin, 30 décembre 1894
A. Gérard
vv. 1–7   Chine 1894–1897
v. 13   Lettres et documents
v. 16   Note sur la Chine

7. Correspondance politique et commerciale 1897–1918. Nouvelle série, Chine
Volumes in this series are grouped according to topic. The following volumes were of considerable value.
Défense nationale: armée et marine
NS 82   1896–1897
NS 83   1898–1904
Relations avec les puissances
NS 95   February 1897–June 1899
NS 96   July 1899–May 1900
NS 97   June 1900
Relations avec la France
NS 191   January–March 1897
NS 192   April–December 1897
NS 193   January–March 1898
NS 194   April–December 1898
NS 195   1899
NS 196   1900–1901
Yunnan
NS 228   1896–1897
NS 229   January 1898–August 1899
NS 230   September–December 1899
NS 231   January–June 1900
NS 232   July–December 1900
NS 233   January–August 1901
NS 234   September–December 1901
NS 235   1902
NS 243   1900–1910 (supplementary)
Protectorat réligieux de la France
NS 309   1899
NS 310   1900
NS 311   1901
NS 321   1896–1897
NS 322   1898
Finances publiques
NS 343   1878–1902   Dossier général
NS 350   June 1895–August 1897   Emprunts

NS 351	September 1897–January 1898	Emprunts
NS 352	February–December 1898	Emprunts
NS 353	January 1899–1903	Emprunts

Finances privées

NS 401	1895–1899	Banques, bourses
NS 402	1900–1902	Banques, bourses
NS 409	1898–1900	Sociétés françaises et étrangères
NS 410	1901–1906	Sociétés françaises et étrangères

Industrie

NS 411	1896–1900	Dossier général
NS 412	1901–1902	Dossier général
NS 418	1897–1906	Mines — dossier général
NS 420	1895–1898	Mines — Chine centrale
NS 421	1899	Mines — Chine centrale
NS 422	1900–1904	Mines — Chine centrale
NS 425	1897–1898	Mines — Guangxi et Guangdong
NS 426	1899–1906	Mines — Guangxi et Guangdong
NS 428	1896–1902	Mines — Chine du nord
NS 433	1895–1898	Mines — Sichuan
NS 434	1899	Mines — Sichuan
NS 435	1900	Mines — Sichuan
NS 436	1901	Mines — Sichuan
NS 441	1897–1899	Mines — Yunnan
NS 442	1900–1901	Mines — Yunnan

Chemins de fer

NS 446	1890–1898	Dossier général
NS 447	1899–1906	Dossier général
NS 449	1888–1902	Lignes du Sud
NS 458	1890–1906	Lignes de la Mandchourie
NS 465	October 1895–April 1897	Beijing–Hankou
NS 466	May–December 1897	Beijing–Hankou
NS 467	1898	Beijing–Hankou
NS 468	1898	Beijing–Hankou
NS 469	1899	Beijing–Hankou
NS 470	1900	Beijing–Hankou
NS 471	1900	Beijing–Hankou
NS 472	1901	Beijing–Hankou
NS 473	1901	Beijing–Hankou
NS 474	1902	Beijing–Hankou
NS 481	November 1895–March 1897	Ligne de Longzhou
NS 482	April 1897–June 1898	Ligne de Longzhou
NS 483	July 1898–April 1900	Ligne de Longzhou
NS 484	May 1900–1905	Ligne de Longzhou
NS 485	1896–1906	Ligne Sino-birmane

NS 486	1897–1901	Lignes du centre et de l'Est
NS 490	1897–1902	Ligne du Shanxi
NS 494	1897	Ligne du Yunnan
NS 495	1898	Ligne du Yunnan
NS 496	1899–1901	Ligne du Yunnan
NS 497	1902	Ligne du Yunnan
NS 514	1898–1900	Ligne Hankou–Guangzhou
NS 515	1901–1903	Ligne Hankou–Guangzhou
NS 523	1899–1902	Ligne Kaifeng–Xi'an

Relations commerciales

NS 553	1896–1901	Dossier général
NS 559	1896	Relations avec la France
NS 560	1896	Relations avec la France
NS 561	1897	Relations avec la France
NS 562	1898–1899	Relations avec la France
NS 563	1900–1901	Relations avec la France

Enseignement
NS 581   1897–1902
Postes et télégraphes
NS 607   1896–1906
Affaires industrielles
NS 614   1898

Holdings of the National Archives

1. Série F12, Commerce

F12 6499 Négociations commerciales avec la Chine, 1881–1898

I	a) Négociations avec la Chine. Note générale [s.d., about 1885]
	b) Voeux du commerce étranger en Chine, 1882
II	Traité de commerce entre la France et la Chine, 1885
III	Dossier de M. Delourne, 1884
IV	Dossier Radiguet, 1885
V	Négociations avec la Chine, documents parlementaires, 1885, 1886
VI	1889
VII	1890
VIII	1891
IX	1884
X	Rapport de M. von Brandt, Ministre allemand à Pékin, 1878
XI	1891–1896
XII	1898
XIII	1892 [sic — in fact material dates from 1896 and 1897]
XIV	1884–1888

XV Négociations 1884–1885 [sic — in fact contains press clippings from 1881 and 1885]

XVI 1884 [sic — material dates from 1877 and 1885: includes an interview with de Freycinet on Article 7 of the 1885 treaty and its implications for Chinese railway construction]

F12 7056 Rapports des Consuls
I   Hankou   1870–1906
II   Guangzhou   1866–1906
III Fuzhou   1869–1904
IV   Hekou   1897–1904

F12 7057 Rapports des Consuls
I   Longzhou   1889–1893
II   Ningbo   1866–1867
III Mengzi   1889–1904
IV   Beihai   1888–1903
V   Beijing   1884–1906
VI   Shanghai   1864–1872

2. Série F14, Travaux Publics
The personnel files of this series were consulted for biographical details of French engineers.
Dossiers individuels de fonctionnaires des travaux publics
F14 2154–3284   Ingénieurs des Ponts et Chaussées, Mines, Chemins de fer, Ports, etc., XIXe siècle
F14 11459–11510 Ingénieurs des Ponts et Chaussées, XIXe siècle à 1910
F14 11623–11635 Ingénieurs des Mines, XIXe siècle à 1910

Archives of the Ministry for Finances

The following boxes, in the series F30, Finances, then located in the rue de Rivoli, were examined.

F30 337	Banque Russo-chinoise	1895–1909
F30 370	Indemnité des Boxeurs	1901–1914
F30 371	Situation économique et financière	1897–1914
F30 372	Emprunts: chemins de fer	
1	Hankou–Guangzhou	1898–1903
2	Hankou–Beijing	1897–1909
3	Guangdong–Hankou	1909
4	Kowloon–Guangzhou	1907
5	Beihai–Nanning	1898–1899
F30 373/2		
1	Documents statistiques re dette publique et produit des douanes impériales	1900–1913

2	Banques	1887–1908
3	Sociétés, mines	1893–1914
4	Entreprises industrielles et financières au Yunnan	1900–1902
5	Banque industrielle de Chine	1906–1914
F30 1913	Emprunts chinois d'avant la guerre 1914–1918	
6	Emprunt chinois 4% 1895	1895–1938
7	Emprunt chinois 7% 1902 Shanxi	1926–1934
F30 1918	Chemins de fer	
F30 2422	Mouvement général des fonds	1881–1907
F30 2423	Statistique générale	1881–1907

National: Archives: Section Outre-Mer

These holdings in the Pavillon de Flore are those of the former Ministry of the Marine and Colonies and the former Ministry of Colonies. They are indicated in the notes by the abbreviation MC (Ministère des Colonies). Two Indo-China series; A, concerned with general political matters and B, concerned with relations with other states, have been consulted. Within each series, there are two groups of holdings: the older deposits contained in numbered cartons, and the *nouveaux fonds* (abbreviated as NF). The most valuable material was contained in the older deposits in series B.

Série A: Généralités

A00 35	Carton 2	Dismemberment [sic] of China	1894
A0 NF 2378		Guangzhouwan	1899–1914
A11 3–9	Carton 3	Instructions to Governors	1879–1897
A20 38–39	Carton 7	Reports of de Lanessan	1897
A50 21	Carton 23	Piracy on the Langson railway	1892–1893
A50 25	Carton 23	Piracy on the Langson railway	1894
A50 NF 2377		Troubles at Guangzhouwan	1903–1905
A60 1–2	Carton 24	Dupuis file	
A60 3–9	Carton 25	Dupuis file	
A75 NF 969		Sino-French Treaties	1858–1916
A82 4–12	Carton 27 bis	Press clippings	1883–1894

Série B: Relations étrangères

B04 NF 681	Border Consulates	1895
B04 NF 318	Indo-China aid to French Institutions in China	1895–1908

B04 NF 320		Indo-China personnel at Longzhou	1896–1901
B04 9	Carton 30	Hekou and Kunming Consulates	1897–1899
B10 1	Carton 31	Hainan	1897–1907
B10 NF 627		Guangzhouwan	1897–1907
B11 11	Carton 32	Treaty of Tianjin, 9 June 1885	1887–1897
B11 NF 694		Frontier delimitation	1887–1888
B11 14–5	Carton 32	Frontier troubles and piracy	1887–1889
B11 17	Carton 32	Troubles in Mengzi and Kunming	1889–1904
B11 NF 682–3		Frontier situation: General Su	1895–1897
B11 24	Carton 32	Chinese railways — military view	1896
B11 25	Carton 32	Relations between consular and colonial authorities in Yunnan	1896–1899
B11 27	Carton 32	Pennequin mission	1897–1899
B11 28	Carton 32	Doumer's visits to Yunnan and Guangxi	1897, 1899
B11 29	Carton 32	Yunnan	1897–1908
B11 NF 788		Capture of Beijing	1900
B11 30	Carton 33	Correspondence between MC and MAE *re* penetration of Yunnan; Doumer/Pichon meeting	1899–1900
B11 31	Carton 33	Return of François	1900–1901
B11 32	Carton 33	Commercial routes in Guangdong and Guangxi	1900–1902
B11 34	Carton 33	Dispatch of Vietnamese troops to Shanghai	1901
B12 1–6	Carton 34	Frontier delimitation	1885–1889
B12 7–13	Carton 35	Frontier delimitation	1889–1894
B12 NF 691		Convention of 20 June 1895	1895–1897
B12 NF 692–5		Frontier delimitation	1893–1896
B12 14	Carton 35	Frontier delimitation	1892–1894
B12 15	Carton 36	Frontier delimitation	1894
B12 NF 682		Chinese troops on frontier	1895–1896
B12 16–19	Carton 36	Monthly reports of Sino-French mixed frontier police	1899–1914
B12 NF 678		Mixed frontier police discussions	1896–1912

B13 1	Carton 37	Commercial information — Yunnan	1897–1908
B13 2	Carton 37	Public works in Longzhou	1898
B13 NF 317		Indo-China aid to French schools in China	1898–1907
B13 3	Carton 37	Creation of a French rival to the Burma–Yunnan Trading Co.	1898

British Government Archives

The Foreign Office Archives, conserved in the Public Record Office at Kew, were a valuable source on British reaction to French initiatives in China. The most important file is the General Correspondence, China file, F017.

Within that file the following volumes were the most interesting.

F017

1236	O'Conor dispatches	1895
1253	Various diplomatic	1895
1254	Various diplomatic	1895
1265–72	Burma, Siam, and French proceedings, etc.	1895
1292–96	Burma, Siam, and French proceedings, etc.	1896
1362	Various	1898
1405	Yunnan Syndicate	1899
1407	Various diplomatic	1899
1449	Various diplomatic	1900
1450	Various diplomatic	1900
1499	Various diplomatic	1901
1500	Various diplomatic	1901

Printed Sources

Official and Semi-official Publications

China, Imperial Maritime Customs, *Decennial reports on the trade, navigation, industries, etc., of the ports open to foreign commerce in China and on the condition and development of the treaty port provinces* (Shanghai, 1892–1901, 1902–1911).

—— *Reports of foreign trade of China and abstract of statistics of returns of trade at the treaty ports and trade reports* (Shanghai, 1882–1904).

—— *Treaties, conventions, etc., between China and foreign states*, (Shanghai, 1917, second edition), vols 1–2.

France, Assemblée nationale, Chambre des Députés, *Journal officiel. Débats parlementaires* (Paris, IN, 1885–1902).

France, Ministère des affaires étrangères, *Annuaire diplomatique et consulaire de la République Française* (Paris, Berger-Levrault, 1885–1901).
—— *Documents diplomatiques, affaires du Tonkin* (Paris, IN, 1884).
—— *Documents diplomatiques, affaires de Chine et du Tonkin* (Paris, IN, 1885).
—— *Documents diplomatiques, affaires de Chine 1894–1898* (Paris, IN, 1898).
—— *Documents diplomatiques, affaires de Chine, 1898–1899* (Paris, IN, 1900).
—— *Documents diplomatiques, affaires de Chine, 1899–1900* (Paris, IN, 1900).
—— *Documents diplomatiques, affaires de Chine, 1900–1901* (Paris, IN, 1901).
—— Commission de publication des documents relatifs aux origines de la guerre de 1914, *Documents diplomatiques français (1871–1914)*, 1ière série (1871–1900), vols 1–16. (Especially vols 4, 5, 10–16.)
—— Commission de publication des documents relatifs aux origines de la guerre de 1914, *Documents diplomatiques français (1817–1914)*, 2ième série (1900–1911), vols 1, 2.
Great Britain, Foreign Office, *Treaty series No. 7 (1897). Agreement between Great Britain and China, modifying the convention of 1 March 1894 relative to Burma and Thibet, 4 February 1897* (London, Her Majesty's Stationery Office, 1898).
—— *China no. 3 (1898). Despatch from Her Majesty's Minister at Peking, forwarding a report by the Acting British Consul at Ssumao on the trade of Yunnan* (London, Her Majesty's Stationery Office, 1899).
—— *Correspondence between Her Majesty's Government and the Russian Government with regard to their respective railway interests in China* (London, Her Majesty's Stationery Office, 1899).
—— *Exchange of notes between the United Kingdom and Russia with regard to their respective railway interests in China* (London, Her Majesty's Stationery Office, 1899).

Contemporary Newspapers and Periodicals

Some of these newspapers and periodicals have been consulted systematically and thoroughly, others more selectively, during the years indicated.
Avenir du Tonkin (1886–1902).
Bulletin de l'Union Coloniale Française (1894–1896). Renamed *Quinzaine Coloniale* in 1897.
Bulletin du Comité de l'Asie Française (1901–1903).
La Chine Nouvelle, Revue illustrée de l'Extrême-Orient (1899–1905).

Chinese Times (November 1886–March 1891). Renamed *Peking and Tientsin Times.*
l'Echo de Chine (1897–1902).
Economiste Français (1873–1902).
Imperial and Asiatic Quarterly Review (1892–1901).
North China Herald and Supreme Court and Consular Register (1885–1901).
Peking and Tientsin Times (March 1894–October 1902).
Questions Diplomatiques et Coloniales (1897–1902).
La Quinzaine Coloniale (1897–1904).
La Revue d'Asie (1901).
Revue des Deux Mondes (1873, 1885–1902).
Revue Indo-Chinoise (1893–1894, 1899–1905).
Revue Politique et Parlementaire (1897–1902).
Le Temps (1892–1901).
The Times (London) (1885–1901).

Contemporary Books and Pamphlets

Allier, R., *Les Troubles de Chine et les missions chrétiennes* (Paris, Fischbacher, 1901).
Anthouard, Baron d', *La Chine contre l'étranger: les Boxeurs* (Paris, Plon Nourrit et cie, 1902).
Austin, O. P., *Commercial China in 1899: area, population, production, railways, telegraphs, transportation routes, foreign commerce, and commerce of the United States with China* (Washington, Government Printing Office, 1899).
Ballero, Eugène, *Ouverture de la Chine à l'influence française au cours des XIXe et XXe siècles* (Thèse pour le doctorat de la Faculté de droit de Paris, Paris, 1902).
Bélard, Marcel, *Rapport sur son voyage de mission au Yunnan* (Paris, Comité des Forges, 1899).
Beresford, Lord Charles, *The Break-up of China with an account of its present commerce, currency, waterways, armies, railways, politics and future prospects* (New York and London, Harper, 1899).
Bertrand, Georges-Pierre, *Dix Ans sur les frontières du sud de la Chine et du Tonkin de 1896 à 1906* (Mayenne, C. Colin, 1905).
Bertrand, Pierre, *Les Atrocités de la guerre en Chine* (Paris, Bellais, 1901).
Boell, Paul, *Le Protectorat des missions catholiques en Chine et la France en Extrême-Orient* (Paris, Institut scientifique de la libre-pensée, 1899).
Boell, Paul, *Les Scandales du quai d'Orsay. Les pots de vin du consul, la trahison Bourée, le case de M. Lemaire* (Paris, A. Savine, 1893).
Bonnetain, Paul, *L'Extrême-Orient* (Paris, Quantin, 1887).

Bousigues, E., *Chemin de fer de Pékin à Hankéou. Mission technique spéciale en Chine confiée en 1901 à M. Bousigues, ... rapport de mission* (Villeneuve St Georges, Union typographique, 1909, 2ième édition).

Brenier, Henri Valéry, *Chambre de commerce de Lyon. Rapport général sur l'origine, les travaux et les conclusions de la mission lyonnaise d'exploration commerciale en Chine* (Lyon, A. Rey, 1897).

—— *La mission lyonnaise d'exploration commerciale en Chine, 1895–1897* (Lyon, A. Rey, 1898).

—— *Note sur le développement commerciale de l'Indo-Chine de 1897 à 1901 comparé avec la période quinquennale 1892–1896* (Hanoï, Imprimérie de l'Extrême-Orient, 1902).

Chailley-Bert, Joseph, *La Colonisation de l'Indochine* (Paris, A. Colin, 1892).

Chailley, Joseph, *Paul Bert au Tonkin* (Paris, G. Charpentier, 1887).

Chambre de commerce de Lyon, *La Mission lyonnaise, d'exploration commerciale en Chine, 1895–1897. Récits de voyages. Rapports commerciaux et notes diverses* (Lyon, A. Rey, 1898).

Chardin, le P. Pacifique-Mairie, *Les Missions franciscaines en Chine, notes géographiques et historiques* (Paris, A. Picard, 1915).

Chéradame, André, *La Colonisation et les colonies allemandes* (Paris, Plon Nourrit et cie, 1905).

—— *Le Monde et la guerre russo-japonaise* (Paris, Plon Nourrit et cie, 1906).

Chirol, Valentine, *The Far Eastern Question* (London, Macmillan, 1896).

Chollot, J.-J., *Intérêts industriels français en Chine (27 octobre 1905)* (Mirecourt, A. Chassel Jeune, 1905).

Clements, P. H., *An Outline of the politics and diplomacy of China and the powers, 1894–1902* (New York, Columbia University Press, 1914).

Collas, Henry, *La Banque de Paris et des Pays-Bas* (Dijon, L. Marchal, 1908).

Colquhoun, Archibold Ross, *China in Transformation* (London, Harper, 1898).

—— *The Overland to China* (London, Harper, 1900).

Cordier, Henri, *Conférence sur les relations de la Chine avec l'Europe* (Rouen, L. Gy, 1901).

—— *Le Conflit entre la France et la Chine. Etude d'histoire coloniale et de droit international* (Paris, L. Cerf, 1883).

—— *Histoire des relations, de la Chine avec les puissances occidentales, 1860–1902* (Paris, F. Alcan, 1902).

—— *Les Origines de deux établissements français dans l'extrême-orient; Changhaï–Ning-po. Documents inédits, publiés avec une introduction et des notes par Henri Cordier* (Paris, F. Alcan, 1896).

Cunningham, Alfred, *The French in Tonkin and South China* (Hong Kong, Hong Kong Daily Press, 1902).

Curzon, George Nathaniel, Marquis of Kedleston, *Problems of the Far East ... Japan, Korea, China* (London, Longman, 1894).

Davies, H. R., *Yunnan. The Link between India and the Yangtze* (Cambridge, Cambridge University Press, 1909).

Deschamps, Léon, *Histoire de la question coloniale en France* (Paris, Plon Nourrit et cie, 1891).

Devéria, Gabriel, *La Frontière sino-annamite. Description géographique et ethnographique, d'après des documents officiels chinois traduits pour la première fois par G. Devéria* (Paris, E. Leroux, 1886).

—— *Histoire des relations de la Chine avec l'Annam-Viêtnam du XVIe au XIXe siècle, d'après des documents Chinois traduits pour la première fois par G. Devéria* (Paris, E. Leroux, 1880).

Douglas, Sir Robert Kennaway, *Europe and the Far East* (Cambridge, Cambridge University Press, 1904).

Doumer, Paul, *l'Indo-Chine française, souvenirs* (Paris, Vuibert et Nony, 1905).

Dujardin-Beaumetz, P., *La Chine dans ses rapports actuels avec l'Europe* (Paris, Société d'économie sociale, 1897).

Dupuis, Jean, *Les Origines de la question du Tonkin* (Paris, Challemel, 1896).

—— *Le Tonkin et l'intervention française, souvenirs* (Paris, Vuibert et Nony, 1905).

Enselme, Capitaine Hyppolyte-Marie-Joseph-Antoine, *A Travers la Mandchourie: le chemin de fer de l'Est chinois, d'après la mission du capitaine H. de Bouillane de Lacoste et du capitaine Enselme* (Paris, J. Rueff, 1903).

Escande, Léon, *Etude sur la navigabilité du fleuve Rouge, voie de pénétration commerciale vers l'intérieur de la Chine* (Paris, Imprimérie nationale, 1894).

L'Entente franco-anglo-russe, par un patriote français (Paris, P. Ollendorff, 1897).

Famin, P., *Au Tonkin et sur la frontière du Kouang-Si* (Paris, Challemel, 1895).

Fauvel, Albert-Auguste, *La Guerre sino-japonaise aujourd'hui et demain* (extrait du *Correspondant*, 10 décembre 1894) (Paris, De Saye et fils, 1894).

—— *L'Enseignement français en Orient et en Chine* (Paris, Revue politique et parlementaire, 1903).

—— *La Société étrangère en Chine* (extrait du *Samedi-Revue*) (Paris, Société de typographie, 1889).

Ferry Jules, *Lettres, 1846–1893* (Paris, Calmann-Lévy, 1914).

—— *Le Tonkin et la Mère-patrie* (Paris, V. Havard, 1890).

Fontpertuis, Adalbert Frout de, *Chine, Japon, Siam et Cambodge* (Paris, Bibliothèque générale de vulgarisation, 1882).

Foster, John W., *Diplomatic Memoirs* (Boston, Haughton, Mifflin and Co., 1909).

Franquet, Eugène, *De l'Importance du fleuve Rouge comme voie de pénétration en Chine, suivie d'une notice sur le cercle de Lao-Kay (2 avril 1896)* (Paris, Charles-Lavauzelle, 1897).

Gaffarel, Paul, *Les colonies françaises*, (Paris, G. Ballière, 1899 sixth edition).

—— *Les Explorations françaises depuis 1870* (Paris, Librairie générale de vulgarisation, 1882).

Garnier, Francis, *Voyage d'exploration en Indo-Chine* (Paris, Hachette, 1873, Vols. 1–2.)

Gautier, Hippolyte-Albert, *Les Français au Tonkin 1787–1883* (Paris, Challemel aîné, 1884).

Gérard, Auguste, *Ma Mission en Chine 1894–1897* (Paris, Plon Nourrit et cie, 1918).

—— *La Vie d'un diplomate sous la Troisième République* (Paris, Plon Nourrit et cie, 1928).

Gundry, R. S., *China and her neighbours: France in Indo-China, Russia and China, India and Thibet* (London, Chapman and Hall, 1893).

—— *China present and past. Foreign intercourse, progress and resources, the missionary question, etc.* (London, Chapman, 1895).

Jenkins, Roberts, *The Jesuits in China and the legation of cardinal de Tournon. An examination of conflicting evidence and an attempt at an impartial judgement* (London, D. Nutt, 1894).

Jernigan, T. R., *China in law and commerce* (London, Macmillan, 1905).

Kent, Percy H. B., *Railway Enterprise in China* (London, Edward Arnold, 1907).

Koulomzine, A. N. de, *Le Transsibérien* (Paris, Hachette, 1904).

Krausse, Alexis S., *China in decay: the story of a disappearing empire*, (London, Chapman, 1900, third edition).

Laboulaye, Edouard de, *Les Chemins de fer de Chine* (Paris, E. Larose, 1911).

Lanessan, Jean Louis de, *L'Expansion coloniale de la France* (Paris, F. Alcan, 1886).

—— *L'Indo-Chine française* (Paris, F. Alcan, 1889).

—— *Les Missions et leur protectorat* (Paris, F. Alcan, 1907).

Launay, Le P. Adrien, *Histoire des missions en Chine: mission du Kouang-Si* (Paris, V. Lecoffre, 1903).

—— *Histoire générale de la Société des Missions étrangères depuis sa formation jusqu'à nos jours* (Paris, Téqui, 1894, Vols, 1–3).

—— *Histoire des missions de Chine, mission du Kouy-tchéou* (Vannes, Lafolye frères, 1907–8, Vols. 1–3).

—— *Les Missionaires français au Tonkin* (Paris, J. Briguet, 1900).

Lefèvre-Pontalis, Pierre, *Voyages dans le Haut Laos et sur les frontières de Chine et de Birmanie* (Paris, E. Leroux, 1902).

Lefèvre, Raymond, *Les Chemins de fer de pénétration dans la Chine méridionale* (Thèse pour le doctorat du Faculté de droit de l'Université de Paris, Paris, 1902).

Legendre, Aimé François, *Au Yunnan et dans le massif du Khin-ho* (Paris, Plon Nourrit et cie, 1913).

—— *Le Far-west Chinois* (Paris, Plon Nourrit et cie, 1910).

Leroy-Beaulieu, Pierre, *La Rénovation de l'Asie: Sibérie, Chine, Japon* (Paris, Colin, 1900).

Letailleur, Eugène, *Contre l'Oligarchie financière en France* (Paris, La Revue, 1908).

Luzeux, le général Alexandre-Charles, *Notre Politique en Chine* (Paris, Charles-Lavauzelle, 1901).

Madrolle, Claudius, *Guide Madrolle. La ligne du Yunnan, Tonkin, Chine. Excursions et itinéraires* (Paris, Hachette, 1913).

Mahan, A. T., *The Problem of Asia and its effect upon international policies* (London, Sampson Low, Marston, 1900).

Maillard, Robert, *De l'Influence française en Chine aux points de vue historiques et économiques* (Paris, Chaix, 1900).

Matignon, J. J., *Dix Ans au pays du dragon* (Paris, A. Maloine, 1910).

Mayers, W. F. (ed.), *Treaties between the Empire of China and foreign powers together with regulations for the conduct of foreign trade.* (Shanghai, North China Herald, 1903, fifth edition).

Millet, R., *Notre Politique extérieure de 1898 à 1905* (Paris, F. Juven, 1905).

Mission laïque française, Comité lyonnais, *La France et le développement des idées modernes en Chine* (Lyon, Mission laïque française, 1911).

Moidrey, Joseph de, S. J., *La Hiérarchy catholique en Chine, en Corée et au Japon, 1307–1914* (Shanghai, Orphelinat de T'ou-sè-wè, 1914).

Monnier, Marcel, *La Tour d'Asie. l'Empire du Milieu* (Paris, Plon Nourrit et cie, 1899, Vol. 2).

Morse, Hosea Bellou, *Trade and administration of the Chinese Empire* (London, Longmans, Green and Co., 1908).

Norman, Charles Boswell, *Tonkin, or France in the Far East* (London, Allen, 1884).

Norman, Henry, *The Peoples and Politics of the Far East* (New York, Scribner, 1895).

Orléans, Prince Henri Philippe Marie d', *Around Tonkin and Siam*, translated by C. B. Pitman (London, Chapman and Hall, 1894).

Pelacot, M. de, *Expédition de Chine de 1900 jusqu'à l'arrivée du général Voyron* (Paris, Charles-Lavauzelle, 1903).

Pichon, Louis, *Un Voyage au Yunnan* (Paris, Plon Nourrit et cie, 1893).
Pinon, René, and Marcillac, Jean de, *La Chine qui s'ouvre* (Paris, Perrin, 1900).
Piolet, J.-B., and Vadot, Charles, *La Religion catholique en Chine* (Paris, Bloud, 1905).
Pouvourville, Albert de, *L'Affaire du Siam 1886–1896* (Paris, Schleicher, 1897).
——— *La Question d'Extrême-Orient* (Paris, Schleicher, 1900).
Rambaud, Alfred, *Les nouvelles Colonies de la République française* (Paris, A. Colin, 1889).
Rapport sur la création d'une société commerciale au Tonkin et dans la Chine méridionale (Lyon, A. H. Storck, 1897).
Reinach, J. (ed.), *Discours et plaidoyers politiques de M. Gambetta* (Paris, G. Charpertier, 1880–1885, Vols. 1–11).
Rocher, Emile, *La Province chinoise du Yün-nan* (Paris, E. Leroux, 1879–1880, Vols. 1–2).
Rousset, Léon, *A travers la Chine* (Paris, Hachette, 1878).
Sabatier, Capitaine A., *Etudes sur les établissements militaires créés en Chine par les étrangers (1900–1907)* (Paris, Berger Levrault, 1909).
Sargent, A. J., *Anglo-Chinese commerce and diplomacy* (Oxford, Clarendon Press, 1907).
Séauve, Capitaine, *Les Rélations de la France et du Siam* (Paris, Charles-Lavrauzelle, 1907).
Simon, J., *Note sur les chemins de fer chinois* (Paris, Dunod et Pinat, 1911).
Thalamas, A., *Les Colonies françaises et la géographie* (Paris, Imprimér-ies réunies, 1893).
Ular, Alexandre, *Un Empire Russo-chinois* (Paris, F. Juven, 1903).
Van Loo, Rodolphe, *La Belgique industrielle et la Chine commerciale* (Bruxelles, Imprimérie de l'Expansion belge, 1911).
Verignon, A., *Les Chemins de fer en Indo-Chine* (Paris, H. Jouve, 1904).
Vignon, Louis, *L'Exploitation de notre empire colonial* (Paris, Hachette, 1900).
Villenoisy, F. de, *La Guerre sino-japonais et ses conséquences pour l'Europe* (Paris, Charles-Lavauzelle, 1895).
Witte, Sergei Y., *The Memoirs of Count Witte*, translated by A. Yarmol-insky (London, Heinemann, 1921).
Wolferstan, Bertram, *The Catholic church in China from 1860 to 1907* (London, Sands, 1909).

Selected Contemporary Articles

Brenier, Henri, 'Rapport général de la mission lyonnaise d'exploration en Chine', *Questions diplomatiques et coloniales*, 1, 15 décembre 1897, 513–35, 585–613.

Coppet, Maurice de, 'L'Action économique des puissances en Chine', *Annales de l'Ecole libre des sciences politiques*, 15, 1900.

Fauvel, Albert Auguste, 'Les Chemins de fer en Chine, dernières concessions', *Questions diplomatiques et coloniales*, 1, 15 août 1898, 413–20, 457–63.

——— 'Le Transsinien et les chemins de fer chinois', *Revue politique et parlementaire*, Septembre 1899, 453–92.

François, A., 'De Canton à Yunnan-sen', *Revue de Paris*, 15 juillet 1900.

Garnier, Francis, 'Des nouvelles Routes de commerce avec la Chine', *Bulletin de la Société de Géographie*, Paris, janvier 1872.

Giret, E., 'La France et l'Angleterre au Yunnan', *Revue Indochinese*, 5, 12, 19 novembre 1900.

Hallett, Holt S., 'France and Russia in China', *Nineteenth Century*, 61, March, 1897, 487–502.

L. O., 'Notre Avenir commerciale au Tonkin', *Questions diplomatiques et coloniales*, 1 mai 1898, 35–42.

Lanessan, Jean Louis de, 'Ce qu'il faut faire en Indo-Chine pour notre expansion politique et économique', *Questions diplomatiques et coloniales*, 15 octobre 1898, 193–203.

——— 'Les Chemins de fer de l'Indo-Chine et de la Birmanie', *Questions diplomatiques et coloniales*, 15 décembre 1898, 449–57.

——— 'La France en Extrême-Orient', *Questions diplomatiques et coloniales*, 1 août 1898, 385–93.

Leroy-Beaulieu, Pierre, 'Les Chemins de fer chinois et l'ouverture du Celeste Empire', *Revue des deux mondes*, No. 155, 1 septembre 1899, 99–131.

Madrolle, Claudius, 'A propos du Chemin de fer français au Yunnan', *Questions diplomatiques et coloniales*, 15 juin 1899, 210–26.

Marcillac, Jean de, 'Les Chemins de fer en Chine', *Questions diplomatiques et coloniales*, 1, 15 juillet 1899, 265–74, 321–31.

Orléans, Henri d', 'L'Insurrection des Boxeurs', *Questions diplomatiques et coloniales*, 15 juillet 1900, 65–85.

Pila, Ulysse, 'La Mission lyonnaise d'exploration en Chine', *Questions diplomatiques et coloniales*, 1 septembre 1897, 130–43.

——— 'Nécessité de l'expansion française en Chine', *Questions diplomatique et coloniales*, 15 juin 1897, 449–57.

Pinon, René, and Marcillac, Jean de, 'Qui exploitera la Chine?'. *Revue des deux mondes*, No. 143, 15 septembre 1897, 331–66.

Plauchut, Edmond, 'Le Tonkin et les relations commerciales', *Revue des deux mondes*, No. 44, 1 mai 1874, 147–71.

Saglio, Charles, 'Les Intérêts français dans la Chine méridionale', *La Chine nouvelle*, No. 4, 15 novembre 1899, 552–553.

Schumacher, Hermann, 'Eisenbahnbau und Eisenbahnpläne in China', *Archiv für Eisenbahnwesen*, No. 22, 1899, 901–78.

Silvestre, J., 'La France à Kouang-Tchéou-Ouan', *Annales des Sciences Politiques*, 15 juillet 1902, 473–93.

Tony, Charles, 'Les Routes commerciales du Yunnan et la pénétration française', *La Chine nouvelle*, No. 2, 15 juin 1899, 253–80.

Zimmermann, Maurice, 'Lyon et la colonisation française', *Questions diplomatiques et coloniales*, 1 juillet 1900, 1–21.

Unpublished Theses

Dorland, A. A., 'A Preliminary study of the role of the French protectorate of Roman Catholic Missions in Sino-French diplomatic relations' (unpublished M. A. thesis, Cornell University, 1951).

Hartel, William Clark, 'The French colonial party 1895–1905' (unpublished Ph. D. thesis, Ohio State University, 1962).

Hsieh, Pei-chih, 'Diplomacy of the Sino-French war, 1883–1885' (unpublished Ph. D. thesis, University of Pennsylvania, 1968).

Kuo, S. P., 'Chinese reaction to foreign encroachment; with special reference to the first Sino-Japanese war and its immediate aftermath' (unpublished Ph. D. thesis, Columbia University, 1953).

Laffey, Ella S., 'Relations between Chinese provincial officials and the Black Flag army, 1883–1885' (unpublished Ph. D. thesis, Cornell University, 1971).

Laffey, John F., 'French imperialism and the Lyon mission to China' (unpublished Ph. D. thesis, Cornell University, 1966).

Lagana, Marc Laurent, 'Eugène Etienne and the strategy of capitalist expansion: colonialism and militarism in the Third Republic, 1881–1918' (unpublished Ph. D. thesis, University of Wisconsin, 1974).

Sims, R. L., 'French policy towards Japan 1854–1894' (unpublished Ph. D. thesis, School of Oriental and African Studies, London University).

More Recent Books and Pamphlets

Ageron, Charles-Robert, *L'Anticolonialisme en France de 1871 à 1914* (textes choisis et présentés par C.-R. Ageron) (Paris, Presses universitaires de France, 1973).

——— *France coloniale ou parti colonial* (Paris, Presses universitaires de France, 1978).

Alberti, J. B., *L'Indochine d'autrefois et d'aujourd'hui* (Paris, Société d'éditions géographiques, maritimes et coloniales, 1934).

Anderson, David L., *Imperialism and idealism: American diplomats in China, 1861–1898* (Bloomington, Indiana University Press, 1985).

Andrew, C. M., *Théophile Delcassé and the making of the Entente Cordiale 1898–1905* (London, Macmillan, 1968).

Auffray, Bernard, *Pierre de Margerie, 1861–1942, et la vie diplomatique de son temps* (Paris, Presses du Palais-Royal, 1976).

Baumgart, Winfried, *Imperialism: the idea and reality of British and French colonial expansion 1880–1914* (Oxford, Oxford University Press, 1982).

Baumont, Maurice, *L'Essor industriel et l'impérialisme colonial 1878–1914* (Paris, Presses universitaires de France, 1965).

Bennigson, Alexandre, *Russes et Chinois avant 1917* (Paris, Flammarion, 1974).

Bergeron, Louis, *Les Capitalistes en France* (Paris, Gallimard/Julliard, 1978).

Bonningue, Alfred, *La France à Kouang-Tchéou-Wan* (Paris, Berger-Levrault, 1931).

Bouvier, Jean, Furet, François, and Gillet, Marcel, *Le Mouvement du profit en France au XIXe siècle* (Paris, Mouton, 1965).

Bouvier, Jean, and Girault, René (eds.), *L'Impérialisme français d'avant 1914* (Paris, Ecole des Hautes Etudes en Sciences Sociales, 1976).

Brogan, Denis W., *The Development of modern France (1870–1938)* (London, Hamish Hamilton, 1967, second edition).

Brötel, Dieter, *Französischer Imperialismus in Vietnam* (Freiburg, Atlantis, 1971).

Brown, Roger Glenn, *Fashoda reconsidered: the impact of domestic politics on French policy in Africa* (Baltimore, John Hopkins Press, 1970).

Brunschwig, H., *French colonialism, 1871–1914. Myths and realities*, translated by W. G. Brown with an introduction by R. E. Robinson (London, Pall Mall Press, 1964).

Cady, J. F., *The Roots of French imperialism in Eastern Asia* (Ithaca, Cornell University Press, 1954).

Cameron, Rondo (ed.), *Essays in French economic history* (Homewood, Illinois, 1970).

Cameron, Rondo, *France and the economic development of Europe 1800–1914* (Princeton, Princeton University Press, 1961).

Carroll, Eber Malcolm, *French public opinion and foreign affairs, 1870–1914* (New York, Frank Cass and Co., 1931).

Carsalade du Pont, G. de, *La Marine française sur le haut Yang-tsé* (Paris, Académie de Marine, 1964).

Casey, J. E. D., *The Politics of French imperialism in the early Third Republic: the case of Bordeaux* (Columbia, University of Missouri Press, 1973).

Chan, Wellington K. K., *Merchants, Mandarins and modern enterprise in late Ch'ing China* (Cambridge, Harvard University Press, 1977).

Chandran, Jeshurun, *The Burma–Yunnan railway: Anglo-French rivalry in mainland Southeast Asia and South China, 1895–1902* (Athens, Ohio University Center for International Studies, 1971).

—— *The Contest for Siam, 1889–1902* (Kuala Lumpur, Penerbit Universiti Kebangsaan Malaysia, 1977).

Chao Zoo-Biang, *Etude sur le commerce extérieur de la Chine de 1864 à 1932* (Lyon, Imprimérie franco-suisse, 1935).

Chapman, G., *The Third Republic in France. The first phase, 1871–1894* (London, Macmillan, 1962).

Chastenet, Jacques, *Histoire de la Troisième République*, 2, *La République des républicains 1879–1893* (Paris, Hachette, 1954).

Chen Chang-bin, *La Presse française et les questions chinoises (1894–1901), étude sur la rivalité des puissances étrangères en Chine* (Paris, Thèse pour le doctorat de la Faculté des lettres de l'Université de Paris, 1941).

Cheng Tien-fong, *A History of Sino-Russian relations* (Washington, Public Affairs Press, 1957).

Chesneaux, Jean, *Contribution à l'histoire de la nation vietnamienne* (Paris, Editions sociales, 1955).

Chih, André, *L'Occident 'chrétien' vu par les Chinois vers la fin du XIXe siècle, 1870–1900* (Paris, Presses universitaires de France, 1962).

Clapham, J. H., *The Economic development of France and Germany, 1815–1914* (Cambridge, Cambridge University Press, 1955).

Clements, P. H., *The Boxer Rebellion, a political and diplomatic review* (New York, AMS Press, 1915).

Clough, S. B., *France: a history of national economics, 1789–1939* (New York, Charles Scribner and Sons, 1939).

Cohen, William B., *Rulers of empire: the French colonial service in Africa* (Palo Alto, Stanford University Press, 1971).

Colas, R. L. M., *Les Relations commerciales entre la France et l'Indo-Chine* (Paris, F. Lovition, 1933).

Le Comité des Forges, *La Sidurgie française, 1864–1914* (Paris, Comité des Forges, 1919).

Cooke, James J., *New French imperialism, 1880–1910, the Third Republic and colonial expansion* (Newton Abbot, David and Charles, 1973).

Coons, Arthur G., *The Foreign public debt of China* (Philadelphia, University of Pennsylvania Press, 1930).

Cordier, G. G., *Province de Yunnan* (Hanoï, Imprimérie Le-Van-Tran, 1928).

Cordier, Henri, *Mélanges d'histoire et de géographie orientales* (Paris, J. Maisonneuve, 1920, Vols. 1–2).

Dallin, David J., *The Rise of Russia in Asia* (New Haven, Yale University Press, 1949).

Delavignette, R., and Julien, C. A., *Les Constructeurs de la France d'outre-mer* (Paris, Corrêa, 1946).

Duchêne, Albert, *Histoire des finances coloniales de la France* (Paris, Payot, 1938).

—— *La Politique coloniale de la France* (Paris, Payot, 1928).

Duroselle, J.-B., *La France et les Français, t.1: La France de la belle époque (1900–1914)* (Paris, Richlieu, 1972).

Eastman, Lloyd E., *Throne and mandarins: China's search for a policy during the Sino-French controversy, 1880–1885* (Cambridge, Harvard University Press, 1967).

Edwards, E. W., *British Diplomacy and Finance in China, 1895–1914* (Oxford, Clarendon Press, 1987).

Eng, Robert Y., *Economic Imperialism in China: Silk Production and Exports, 1861–1932* (Berkeley, Institute of East Asian Studies, University of California, 1986).

Fairbank, John K., and Reischauer, Edwin O., *China, tradition and transformation* (Sydney, Allen and Unwin, 1979).

Fairbank, John K., and Teng, Ssu-yu, *China's response to the West: a documentary survey 1839–1923* (New York, Atheneum, 1973).

Faure, Félix, *Le Ministère Léon Bourgeois et la politique étrangère de Marcelin Berthelot au Quai d'Orsay, 2 novembre 1895–29 mars 1896* (Paris, A. Pedone, 1957).

Feis, Herbert, *Europe, the world's banker 1870–1914: an account of European financial investment and the connection of world finance with diplomacy before the war* (New Haven, Yale University Press, 1930).

Feuerwerker, Albert, *China's early industrialization: Sheng Hsüan-huai and mandarin enterprise* (New York, Atheneum, 1970).

—— *The Chinese economy, circa 1870–1911* (Ann Arbor, University of Michigan, 1969).

—— *Money and monetary policy in China, 1845–1895* (Cambridge, Harvard University Press, 1965).

Fieldhouse, David Kenneth, *The Colonial Empires: a comparative survey from the eighteenth century* (London, Weidenfeld and Nicolson, 1966).

—— *Economics and empire, 1830–1914* (London, Weidenfeld and Nicolson, 1973).

—— *The Theory of capitalist imperialism* (London, Longman, 1967).

Fredet, Jean, *Quand la Chine s'ouvrait ... Charles de Montigny, consul de France* (Paris, Sociéte de l'histoire des colonies françaises, 1953).

Frochisse, J. M., *La Belgique et la Chine: Relations diplomatiques et économiques* (Bruxelles, l'Edition universelle, 1937).

Ganiage, Jean, *L'Expansion coloniale de la France sous la troisième République (1871–1914)* (Paris, Payot, 1968).

Gifford, Prosser, and Louis, William Roger, (eds), *France and Britain in*

Africa: imperial rivalry and colonial rule (New Haven, Yale University Press, 1971).

Girault, Réné, *Diplomatie européenne et impérialisme* (Paris, Mosson, 1979).

—— *Emprunts russes et investissements français en Russie, 1887–1914* (Paris, A. Colin, 1974).

Heggoy, A. A., *The African politics of Gabriel Hanotaux 1894–1898* (Athens, Georgia, University of Georgia Press, 1972).

Hibbert, Christopher, *The Dragon wakes, China and the west, 1793–1911* (London, Longman, 1970).

Hobson, J. A., *Imperialism: a study* (London, Allen and Unwin, 1902).

Hou, C. M., *Foreign investment and economic development in China, 1840–1937* (Cambridge, Harvard University Press, 1965).

Howard, John Elred, *Parliament and foreign policy in France* (London, Cresset Press, 1948).

Hsiao Liang-lin, *China's foreign trade statistics 1864–1949* (Cambridge, Harvard University Press, 1974).

Hsü, Immanuel C. Y., *The Rise of modern China* (New York, Oxford University Press, 1975, second edition).

Hu Sheng, *Imperialism and Chinese Politics* (Beijing, Foreign Languages Press, 1981).

Huenemann, Ralph William, *The Dragon and the Iron Horse: the economics of railroads in China, 1876–1937* (Cambridge, Harvard University Press, 1984).

Hummel, Arthur W., *Eminent Chinese of the Ch'ing period (1644–1912)* (Washington, US Government Printing Office, 1943).

Iiams, T. M., jnr, *Dreyfus, diplomats and the dual alliance: Gabriel Hanotaux at the Quai d'Orsay 1894–1898* (Geneva, E. Droz, 1962).

Jolly, Jean, *Dictionnaire des parlementaires français: notices biographiques sur les ministres, deputés et sénateurs français 1889 à 1940* (Paris, Presses universitaires de France, 1960–1977, Vols. 1–8).

Joseph, Philip, *Foreign diplomacy in China, 1894–1900: a study in political and economic relations with China* (London, Allen and Unwin, 1928).

Julien, Charles André (ed.), *Les Politiques d'expansion impérialiste* (Paris, Presses universitaires de France, 1949).

Kanya-Forstner, Alexander Sydney, *The Conquest of the Western Sudan: a study in French military imperialism* (London, Cambridge University Press, 1969).

Keyder, Caglar and O'Brien, Patrick, *Economic growth in Britain and France 1780–1914. Two paths to the twentieth century* (London, Allen and Unwin, 1978).

Kia Yu-tong, *Essai d'un aperçu historique des relations politiques (industrielles, commerciales, religieuses) de la France et de la Chine, depuis le XVIIe siècle jusqu'à nos jours* (Paris, Thèse pour le doctorat de la Faculté de droit de l'Université de Paris, 1920).

Kiernan, V. G., *British diplomacy in China, 1880–1885* (Cambridge, Cambridge University Press, 1939).

——— *Marxism and imperialism* (London, Edward Arnold, 1974).

King, Frank, H. H., *Money and monetary policy in China 1845–1895* (Cambridge, Harvard University Press, 1965).

King, Frank H. H. and Clarke, Prescott, *A Research guide to China coast newspapers, 1822–1911* (Cambridge, Harvard University Press, 1965).

Koryakov, Vladimir Pavlovich, *Politika Frantsii v Kitae v Kontse XIXV* (On French Policy in China at the end of the nineteenth century) (Moskva, Izd-vo 'Nauka', 1985).

Kurgan-van Hentenryk, Ginette, *Jean Jadot: artisan de l'expansion belge en Chine* (Bruxelles, Académie royale des Sciences, 1975).

——— *Léopold II et les groupes financières belges en Chine. La politique royale et ses prolongements* (Bruxelles, Palais des Académies, 1972).

Langer, William Leonard, *The Diplomacy of Imperialism, 1890–1902* (New York, Alfred A. Knopf, 1957, second edition).

——— *The Franco-Russian alliance* (Cambridge, Harvard University Press, 1929).

Latourette, K. S., *A History of Christian missions in China* (London, Macmillan, 1929).

Le Fevour, Edward, *Western enterprise in late Ch'ing China: a selective survey of Jardine, Matheson and Company's operations, 1842–1895* (Cambridge, Harvard University Press, 1968).

Lee En-han, *China's Quest for Railway Autonomy, 1904–1911: a study of the Chinese railway-rights recovery movement* (Singapore, Singapore University Press, 1977).

Lenin, V. I., *Imperialism: the highest stage of capitalism* (Moscow, Progress Press, 1966, thirteenth edition).

Lensen, George Alexander, *Balance of Intrigue: international rivalry in Korea and Manchuria, 1884–1899* (Talahassee, University Presses of Florida, 1982).

Léon, Pierre, Lévy-Leboyer, Maurice, Armengaud, André, Broder, André, Bruhat, Jean, Daumard, Adeline, Labrousse, Ernest, Laurent, Robert, and Soboul, Robert, *Histoire économique et sociale de la France* t.3, *L'avènement de l'ère nouvelle (1789–années 1880)*, Vol. 2 (Paris, Presses universitaires de France, 1976).

Levy, Roger, Lacam, Guy, and Roth, Andrew, *French interests and policies in the Far East* (New York, Institute of Political Relations, 1941).

Lévy-Leboyer, Paul (ed.), *La Position internationale de la France, aspects économiques et financiers,· XIXe–XXe siècles* (Paris, Ecole des Hautes Etudes en Sciences Sociales, 1977).

Liao Kuang-sheng, *Anti-foreignism and Modernization in China, 1860–1980*, (Hong Kong, Chinese University Press, 1986, second edition).

Lomenie, E. Beau de, *Les Responsabilités des dynasties bourgeoises*, Vol. II, *De Mac-Mahon à Poincaré* (Paris, De Noël, 1963).

Louis, William Roger (ed.), *Imperialism: the Robinson and Gallagher controversy* (New York, New Viewpoints, 1976).

Lowe, Peter, *Britain in the Far East. A survey from 1819 to the present* (London, Longman, 1981).

Lung Chang, *La Chine à l'aube du XXe siècle, les relations diplomatiques de la Chine avec les puissances depuis la guerre sino-japonaise jusqu'à la guerre russo-japonaise* (Paris, Nouvelles éditions latines, 1962).

MacMurray, J. V. A. (ed.), *Treaties and agreements with and concerning China, 1894–1919* (New York, H. Fertig, 1921, Vols. 1–2).

Malozemoff, Andrew, *Russian Far Eastern policy, 1881–1904* (Berkeley, University of California Press, 1958).

Mancall, Marle, *China at the Centre: 300 years of foreign policy* (New York, Free Press, 1984).

Marr, David G., *Vietnamese anticolonialism 1885–1925* (Berkeley, University of California Press, 1971).

Masson, André, *Hanoi pendant la période heroïque, 1873–1888* (Paris, E. Guethner, 1929).

Maybon, C. B., *La Concession française d'autrefois* (Paris, Plon Nourrit et cie, 1924).

Maybon, C. B., and Fredet, J., *Histoire de la concession française de Changhaï* (Paris, Plon Nourrit et cie, 1929).

McAleavy, H., *Black Flags in Vietnam: the story of a Chinese intervention, the Tonkin war of 1884–1885* (London, Allen and Unwin, 1968).

McCordock, Robert Stanley, *British Far Eastern policy, 1894–1900* (New York, Octagon, 1931).

Meng, S., *The Tsungli Yamen: its organization and functions* (Cambridge, Harvard University Press, 1962).

Morse, Hosea Bellou, *The International relations of the Chinese Empire* (London, Longmans, Green and Co., 1910–1918, Vols, 1–3).

Murphy, Agnes, *The Ideology of French imperialism, 1871–1881* (Washington, Catholic University of America, 1948).

Osborne, Milton E., *The French presence in Cochinchina and Cambodia;*

rule and response (1859–1905) (Ithaca, Cornell University Press, 1969).

—— *River road to China, the Mekong River expedition 1866–1873* (London, Allen and Unwin, 1975).

Owen, E. R. J., and Sutcliffe, R. B. (eds.), *Studies in the theory of imperialism* (London, Longman, 1972).

Palmade, Guy P., *French capitalism in the nineteenth century*, translated by G. M. Holmes (Newton Abbot, David and Charles, 1961).

Parr, John Francis, *Théophile Delcassé and the practice of the Franco-Russian Alliance, 1898–1905*, (Moret-sur-loing, Thèse présentée à la Faculté des lettres de l'Université de Fribourg, 1952).

Pelcovits, Nathan Albert, *Old China hands and the Foreign Office* (New York, American Institute of Pacific Relations, 1948).

Pisani-Ferry, Fresnette, *Jules Ferry et le partage du monde* (Paris, Grasset, 1962).

Pouyanne, A. A., *Les Travaux publics de l'Indo-Chine* (Hanoï, Imprimérie d'Extrême-Orient, 1926).

Power, Thomas Francis, *Jules Ferry and the renaissance of French imperialism* (New York, King's Crown Press, 1944).

Priestley, Herbert Ingram, *France overseas: a study of modern imperialism* (New York, D. Appleton-Century Co., 1938).

Reclus, Maurice, *Emile de Girardin, le créateur de la presse moderne* (Paris, Hachette, 1934).

—— *Jules Ferry, 1832–1893* (Paris, Flammarion, 1947).

Renouvin, Pierre, *La Politique extérieure de Th. Delcassé, 1895–1905* (Paris, Tournier et Constans, 1954).

—— *La Question d'Extrême-Orient 1840–1940* (Paris, Hachette, 1946).

Ristelhueber, René, *Notre Conflict avec la Chine au sujet du Tonkin, 1884–1885* (Paris, A. Padone, 1955). (Extrait de la *Revue d'histoire diplomatique*, 3, juillet–septembre, 1954.)

Robequain, Charles, *L'Evolution économique de l'Indo-Chine française* (Paris, P. Hartmann, 1939).

Roberts, S. H., *History of French colonial policy 1870–1925* (London, Frank Cass, 1929).

Robinson, R. E., and Gallagher, J., *Africa and the Victorians: the official mind of imperialism* (London, Macmillan, 1961).

Romanov, B. A., *Russia in Manchuria, 1892–1906*, translated by S. W. Jones (Ann Arbor, Michigan, J. W. Edwards, 1952).

Ross, Angus, *New Zealand aspirations in the Pacific in the nineteenth century* (Oxford, Clarendon Press, 1964).

Sarraut, Albert, *La Mise en valeur des colonies françaises* (Paris, Payot, 1923).

Schreuder, Deryck Marshall, *The Scramble for Southern Africa, 1877–*

1895: the politics of partition reappraised (Cambridge, Cambridge University Press, 1980).

Schuman, F. L., *War and diplomacy in the French republic* (New York, Fertig, 1969).

Siegfried, André, *Mes Souvenirs de la IIIe République. Mon père et ses temps. Jules Siegfried, 1836–1922* (Paris, Editions du grand siècle, 1946).

Sun, E-tu Zen, *Chinese railways and British interersts, 1898–1911* (New York, King's Crown Press, 1954).

Sze Tsung-yu, *China and the most-favoured-nation clause* (T'aipei, Ch'eng Wen Press, 1971).

Taboulet, Georges, *La Geste française en Indo-Chine: histoire par les textes de la France en Indo-Chine des origines à 1914* (Paris, Librairie d'Amérique et d'orient, 1955–1956, Vols. 1 and 2).

Tamagna, Frank M., *Banking and finance in China* (New York, Institute for Pacific Relations, 1942).

Taylor, A. J. P., *The Struggle for mastery in Europe, 1848–1918* (London, Oxford University Press, 1957).

Tcheng Tse-sio, *Les Relations de Lyon avec la Chine: étude d'histoire et de géographie économique* (Paris, L. Rodstein, 1937).

Temperley, H., and Colville, A. (eds.), *Studies in Anglo-French history* (Cambridge, Cambridge University Press, 1935).

Thobie, Jacques, *La France impériale 1880–1914* (Paris, Mégrelis, 1982).

—— *Intérêts et impérialisme français dans l'Empire ottoman (1895–1914)* (Paris, Publications de la Sorbonne, 1977).

Thompson, Roger Clark, *Australian imperialism in the Pacific: the expansionist era, 1820–1920* (Melbourne, Melbourne University Press, 1980).

Tramond, J., and Reussner, A., *Eléments d'histoire maritime et coloniale contemporaine (1815–1914)* (Paris, Editions géographiques, maritimes, et coloniales, 1947).

Trolliet, P., Nguyen, A., and Rachline, M., *Noms propres de géographie, d'histoire et de littérature modernes de la Chine* (Paris, P. Geuthner, 1968).

Tung Lin (William L.), *China and the Foreign Powers: the impact of and reaction to unequal treaties* (Dobbs Ferry, Oceana Publications, 1970).

Von Laue, Theodore H., *Sergei Witte and the industrialization of Russia* (New York, Columbia University Press, 1963).

Wakeman, Frederic, *The Fall of Imperial China* (New York, Free Press, 1975).

Wang Wen-Yuan, *Les Relations entre l'Indo-Chine française et la Chine: étude de géographie économique* (Paris, P. Bousset, 1937).

Wei, Louis Tsing-sing, *Le Saint-Siège, la France et la Chine sous le pontificat de Léon XIII, le projet de l'Établissement d'une nonciature à Pékin et l'affaire du Pei-t'ang, 1880–1886* (Schöneck-Beckenried, Nouvelle Revue de Science missionnaire, 1966).

Woolf, Leonard Sidney, *Empire and commerce in Africa: a study in economic imperialism* (London, Fabian Society, 1919).

Wu Chao-Kwang, *The International aspect of the missionary movement in China* (New York, AMS Press, 1930).

Young, L. K., *British policy in China, 1895–1902* (Oxford, Clarendon Press, 1970).

More Recent Articles

Abrams, L., and Miller, D. J., 'Who were the French colonialists? A reassessment of the *Parti colonial*, 1890–1914', *The Historical Journal*, No. 19, part 3, September 1976, 685–726.

Bastid, Marianne, 'La Diplomatie française et la révolution chinoise de 1911', in J. Bouvier and R. Girault (eds.), *L'Impérialisme français d'avant 1914*, (Paris, Ecole des Hautes Etudes en Sciences Sociales, 1976), pp. 127–52.

Bouvier, Jean, 'Les Traits majeurs de l'impérialisme français avant 1914', in J. Bouvier and R. Girault (eds.), *L'Impérialisme français d'avant 1914*, (Paris, Ecole des Hautes Etudes en Sciences Sociales, 1976), pp. 305–33.

―――― 'The Banking mechanism in France in the late nineteenth century', in Rondo Cameron (ed.), *Essays in French economic history* (Homewood, Illinois, 1970), pp. 341–69.

Braisted, William R., 'The United States and the American China Development Company', *Far Eastern Quarterly*, No. 11, part 2, February 1952, 147–67.

Bruguière, Michel, 'Le Chemin de fer de Yunnan. Paul Doumer et la politique d'intervention française en Chine (1889–1902)', *Revue d'histoire diplomatique*, No. 77, 1963, 23–61, 129–62, 252–78.

Bury, J. P. T., 'Gambetta and overseas problems', *English Historical Review*, No. 82, 1967, 277–95.

Cady, J. F., 'The Beginnings of French imperialism in the Pacific Orient', *Journal of Modern History*, No. 14, 1942, 71–87.

Caron, François, 'French railroad investment, 1850–1914', in Rondo Cameron (ed.), *Essays in French economic history* (Homewood, Illinois, 1970), pp. 315–40.

Chang Sung-hwan, 'Russian designs on the Far East', in Taras Hanczak, *Russian imperialism from Ivan the Great to the Revolution* (New Brunswick, New Jersey, 1974). pp. 299–321.

'Etude statistique sur le développement de l'Indo-Chine de 1899 à 1923',

Bulletin économique de l'Indo-Chine, 28e année, 171 n.s., Hanoï, 1925.

Feuerwerker, Albert, 'Economic conditions in the late Ch'ing period', in Albert Feuerwerker (ed.), *Modern China* (Englewood Cliffs, New Jersey, 1964), pp. 105–25.

Gagnier, D., 'French loans to China 1895–1914: the alliance of international finance and diplomacy', *Australian Journal of Politics and History*, No. 18, part 2, August 1972, 229–49.

Hamilton, Keith, 'An Attempt to form an Anglo-French industrial entente', *Middle East Studies*, No. 11, 1975, 47–60.

Hargreaves, J. D., 'Entente manquée; Anglo-French relations 1895–1896', *The Cambridge Historical Journal*, No. 11, 1953, 65–92.

Kurgan-van Hentenryk, Ginette, 'Philippe Berthelot et les intérêts ferroviaires franco-belges en Chine (1912–1914)', *Revue d'histoire moderne et contemporaine*, No. 23, 1975, 269–92.

Lacam, Guy, 'The Economic relations of Indo-China with Southern China', in R. Levy, G. Lacam, and A. Roth, *French interests and policies in the Far East* (New York, Institute of Political Relations, 1941), pp. 85–111.

Laffey, Ella S., 'French adventurers and local bandits in Tonkin: the Garnier affair in its local context', *Journal of Southeast Asian Studies*, No. 6, part 1, 1975, 38–51.

Laffey, John F., 'Municipal imperialism in nineteenth century France', *Historical Reflections*, No. 1, June 1974, 81–114.

—— 'Les racines de l'impérialisme français en Extrême-Orient. A propos des thèses de J. F. Cady', *Revue d'histoire moderne et contemporaine*, No. 16, 1969, 282–99. Also in J. Bouvier and R. Girault (eds.), *L'Impérialisme français d'avant 1914* (Paris, Ecole des Hautes Etudes en Sciences Sociales, 1976), pp. 15–37.

—— 'Roots of French imperialism in the nineteenth century: the case of Lyon', *French Historical Studies*, No. 6, part 1, 1969, 78–92.

Landes, David S., 'Some thoughts on the nature of economic imperialism', *Journal of economic history*, No. 21, part 4, 1961, 496–512.

Leaman, B. R., 'The Influence of domestic policy on foreign affairs in France, 1898–1905', *Journal of modern history*, No. 14, 1942, 449–79.

McKay, D. V., 'Colonialism in the French geographical movement', *The Geographical Review*, No. 33, 1943, 214–32.

Metzgar, H. Michael, 'The Crisis of 1900 in Yunnan: Late Ch'ing Militancy in transition', *Journal of Asian Studies*, No. 35, part 2, February, 1976, 185–201.

Munholland, J. Kim, '"Collaboration strategy" and the French pacification of Tonkin, 1885–1897', *The Historical Journal*, No. 24, part 3, 1981, 629–50.

———— 'Admiral Jauréguiberry and the French scramble for Tonkin, 1879–1883', *French Historical Studies*, No. 11, part 1, 1979, 81–107.

Stengers, J., 'King Leopold's imperialism', in E. R. J. Owen and R. B. Sutcliffe (eds.), *Studies in the theory of imperialism* (London, Longman, 1972), pp. 248–76.

Stokes, Eric, 'Late nineteenth century colonial expansion and the attack on the theory of economic imperialism: a case of mistaken identity?', *Historical Journal*, No. 12, part 2, 1969, 285–301.

Walsh, W. B., 'The Yunnan myth', *Far Eastern Quarterly*, No. 2, part 3, May 1943, 272–85.

Index

ADAM, ACHILLE, 261–3
Africa, 16, 269
Agliardi, Monsignor, 3, 4
Algeria, 13
Alsace, 10
American China Development Company, 59, 76, 86–7, 89–94, 289n.41
American Sugar Refining Company, 87
Amsterdam, 113
Andigné, d', 257
Anethon, Baron d', 62, 73
Anglo-Chinese Convention, 4 February 1897, 150–1, 163, 248
Anglo-French Convention, 15 January 1896, 164, 181–2, 197, 199, 204, 209, 256, 258, 264–5, 268, 271; text, 301–2n.32
Anglo-French Syndicate, 228, 256–63
Anglo-Russian agreement 28 April 1899, 91, 219–20, 223, 288n.31
Angoulvant, 192, 209, 211–2
Antwerp, 76
Anzer, Father, 7
Armade, Commander d', 183
Army, 102, 104–5, 141; see also Military officers
Arnhold Karberg, 50
Association industrielle française, 37–41, 44
Auriac, Captain, 190, 203, 210, 214
Australia, 9, 96, 100

BAEYENS, 73
Baghdad railway, 272–3
Baker, Sir Benjamin, 46, 49
Balfour, Lord Arthur, 80–2
Banque de l'Indo-Chine, 108, 111, 123–6, 130; and Yunnan railway, 242
Banque de Paris et des Pays-Bas, and Beijing–Hankou railway, 62, 64, 70, 72, 77, 85, 89, 98; and indemnity loans, 112, 116; and Russo-Chinese Bank, 114; and Xi'an railway, 272; and Yunnan railway, 241; and Yunnan mining projects, 260, 262
Banque d'Outremer, 93

Banque française de l'Afrique du sud, 259–60
Banque Impériale Ottomane, 259
Banque Internationale de Paris, 65, 259
Banque Parisienne, 65, 73
Baoding, 75–6, 78, 102–4
Bardac, Noel, 259
Baring, 260
Barnes, Thurlow Weed, 86–7
Bash, Albert W., 59–60
Batignolles, Société de construction des, 70, 72, 100, 241, 247
Baüer, Chevalier Raphael de, 67, 89
Bauzon, Commander, 203, 209, 214
Beau, Paul, 244, 247, 265
Beauclerk, 150
Beihai, 144, 156, 158–9, 163, 178
Beihai–Nanning railway, 166–7
Beihe River, 56, 81
Beijing, 103–6
Beijing–Hankou railway, 55, 56–82 passim, 83–109 passim, 121, 126, 129, 134–5, 171, 225, 268, 271, 273
Belard, Marcel, 252–4, 257, 263, 270
Belgium and Belgian policy, 16, 48, 53, 60–82 passim, 83–109 passim, 268
Bellot, Captain, 203
Berger, Georges, 259
Bert, Paul, 147
Berthelot, Marcelin, 49, 152–3, 155, 181, 268, 302n.33
Berthelot, père, 8, 166
Berthémy Convention, 20 February 1865, 6
Bertie, Sir Francis, 261
Bertrand, Georges, 160, 167, 172
Besnard, Admiral, 38, 303n.56
Beylié, Colonel de, 236
Bezaure, Georges Servan de, Consul in Tianjin, 37, 40–2, 44–6; and indemnity loan, 124; and Lenin compared, 270, 274; and negotiations for Beijing–Hankou railway, 63–4, 66, 68, 75, 95, 109; and religious Protectorate, 7
Bhamo, 146